EMERGING CONCEPTS IN RIBOSOME STRUCTURE, BIOGENESIS, AND FUNCTION

EMERGING CONCEPTS IN RIBOSOME STRUCTURE, BIOGENESIS, AND FUNCTION

Edited by

VIJAY KUMAR

Department of Molecular and Cellular Medicine,
Institute of Liver and Biliary Sciences, New Delhi, India

ELSEVIER

ACADEMIC PRESS

An imprint of Elsevier

Academic Press is an imprint of Elsevier
125 London Wall, London EC2Y 5AS, United Kingdom
525 B Street, Suite 1650, San Diego, CA 92101, United States
50 Hampshire Street, 5th Floor, Cambridge, MA 02139, United States
The Boulevard, Langford Lane, Kidlington, Oxford OX5 1GB, United Kingdom

Library of Congress Cataloging-in-Publication Data
A catalog record for this book is available from the Library of Congress

British Library Cataloguing-in-Publication Data
A catalogue record for this book is available from the British Library

ISBN 978-0-12-816364-1

Publisher: Andre Gerhard Wolff
Editorial Project Manager: Samantha Allard
Production Project Manager: Sreejith Viswanathan
Cover Designer: Mark Rogers

Typeset by SPi Global, India

Contents

Contributors

Sukriti Baweja
Department of Molecular and Cellular Medicine, Institute of Liver and Biliary Sciences, New Delhi, India

Sunita Giri
Department of Molecular and Cellular Medicine, Institute of Liver and Biliary Sciences, New Delhi, India

Savneet Kaur
Department of Molecular and Cellular Medicine, Institute of Liver and Bilary Sciences, New Delhi, India

Jitendra Kumar
Department of Molecular and Cellular Medicine, Institute of Liver and Biliary Sciences, New Delhi, India

Vijay Kumar
Department of Molecular and Cellular Medicine, Institute of Liver and Biliary Sciences, New Delhi, India

Rekha Puria
School of Biotechnology, Gautam Buddha University, Greater Noida, India

Sumati Rohilla
School of Biotechnology, Gautam Buddha University, Greater Noida, India

Akhilesh Kumar Saini
Department of Molecular and Cellular Medicine, Institute of Liver and Biliary Sciences, New Delhi, India

M.G. Sanal
Department of Molecular and Cellular Medicine, Institute of Liver and Biliary Sciences, New Delhi, India

Sarita
Department of Molecular and Cellular Medicine, Institute of Liver and Biliary Sciences, New Delhi, India

Zaheenul Islam Siddiqui
Department of Microbiology, College of Medicine, Imam Abdulrahman Bin Faisal University, Dammam, Saudi Arabia

P. Debishree Subudhi
Department of Molecular and Cellular Medicine, Institute of Liver and Biliary Sciences, New Delhi, India

Purnima Tyagi
Department of Molecular and Cellular Medicine, Institute of Liver and Biliary Sciences, New Delhi, India

CHAPTER 1

Introduction to ribosome factory, origin, and evolution of translation

Vijay Kumar
Department of Molecular and Cellular Medicine, Institute of Liver and Biliary Sciences, New Delhi, India

What are ribosomes?

Ribosomes are small, rounded, dense particles of ribonucleoproteins found in prokaryotic and eukaryotic cells, in mitochondria, chloroplasts, and bacteria. Ribosomes were first observed around the mid-1950s by the cell biologist **George Emil Palade** under an electron microscope as dense particles or granules frequently associated with the endoplasmic reticulum of eukaryotic cells (Fig. 1.1A) (Palade, 1964). The term "ribosome" was proposed by the team of **Richard B. Roberts** wherein *ribo* is for ribonucleic acid and "*soma*" being a Latin word for body (McQuillen et al., 1959). The size of the ribosomes within cells varies and depends on the cell type and stage of the cell cycle. Those found in prokaryotes are generally smaller than those in eukaryotes. Ribosomes in mitochondria and chloroplasts are similar in size to those in bacteria. The average ribosome of *Escherichia coli* (a prokaryote) measures about 20nm in diameter. It exists both as free particles as well as particles attached to the membranes of the endoplasmic reticulum (ER) in eukaryotic cells (Palade and Siekevitz, 1956) to form "rough" ER (Fig. 1.1A). The association between ribosome and ER is well known to facilitate the processing, secondary modifications, and quality control of newly synthesized proteins.

Ribosomes are remarkably abundant in cells and their number may vary among different cell and tissue types. As many as 10 million ribosomes may be present in a single rapidly growing mammalian cell whereas an *E. coli* may carry ~20,000 ribosomes, constituting about 25% of the dry weight of the cell (Leslie, 2017). Ribosomes have only a temporary existence. After synthesizing a polypeptide, the two subunits separate and are either recycled or degraded.

Emerging Concepts in Ribosome Structure, Biogenesis, and Function
https://doi.org/10.1016/B978-0-12-816364-1.00002-0

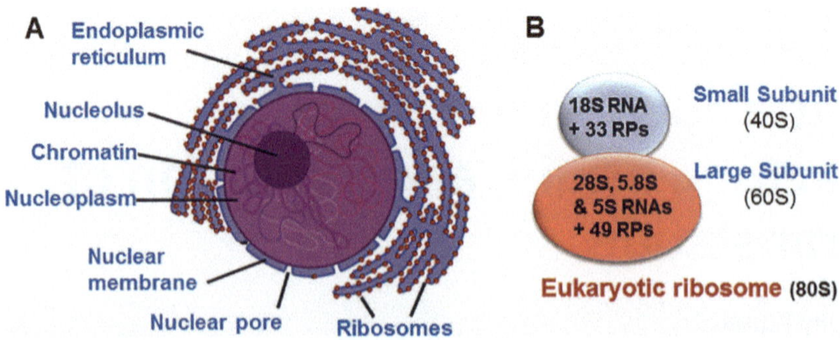

Fig. 1.1 Eukaryotic ribosome—Intracellular localization and subunit structure. (A) Schematic organization of the network on endoplasmic reticulum (ER) surrounding the nucleus. The ribosomes are present as dense particles (red dots) on the surface of rough ER. (B) Subunit structure and molecular components. S, sedimentation coefficient; RPs, ribosomal proteins.

Basic components and structure of the ribosome

The ribosome is composed of ribosomal proteins (RPs) and ribosomal RNAs (rRNAs) arranged in two subunits of unequal size—**small** and **large**—each of which has a characteristic shape (Fig. 1.1B). The large subunit is about twice the size of the small subunit. The two subunits remain separated by a narrow cleft but remain adhered to each other owing to high concentration (1mM) of Mg^{++}. The two ribosomal subunits fall apart when the concentration of Mg^{++} ion declines in the matrix.

The size of ribosomes, ribosomal subunits, and rRNAs are usually designated by their sedimentation coefficient expressed in the Svedberg or "S" units in a centrifugal field. The intact ribosome is called 70S in prokaryotes (e.g., blue green algae and bacteria) and 80S in eukaryotes (e.g., yeast, plant and mammalian cells) (Baßler and Hurt, 2019; Kaczanowska and Rydén-Aulin, 2007; Sáez-Vásquez and Delseny, 2019; Woolford Jr and Baserga, 2013). The small and large subunits of eukaryotes are designated 40S and 60S, respectively, whereas the prokaryotic ribosome contains a small 30S subunit and a large 50S subunit (Fig. 1.2). The fine structures of ribosomes are described in Chapter 2. The ribosomal subunits are assembled at the nucleolus, where newly transcribed and processed rRNAs associate with the RPs acquired into the nucleus from the cytoplasm (Boisvert et al., 2007). These two ribosomal subunits are then exported to the cytoplasm through a nuclear pore. Besides, the ribosomal subunits are recycled and reused through the cycles of translation. After completion of peptide synthesis, the ribosomes split into the individual subunits, thus completing the cycle.

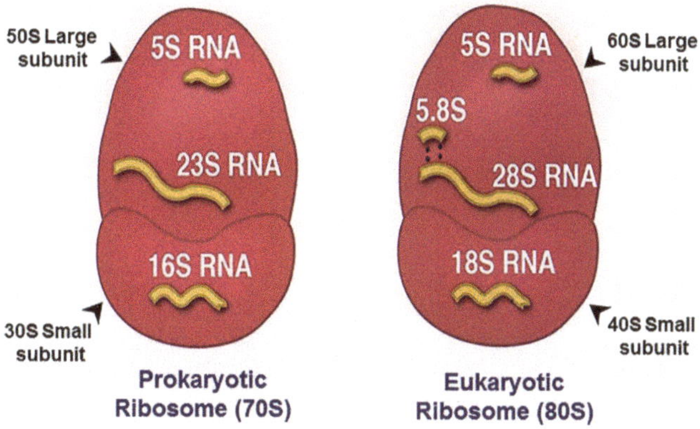

Fig. 1.2 Prokaryotic and eukaryotic ribosomes. Comparison of sedimentation coefficients (S) and molecular components (rRNAs).

Since new ribosome assembly is an energy-consuming process, its recycling is considered to be a vital biological process in the cell (Hellen, 2018). The current knowledge on ribosome assembly, degradation, and recycling is described in Chapter 6.

Although the prokaryotic and eukaryotic ribosomes have very similar structural organization, these differ in their protein and RNA composition and functions. Ribosomes are usually made up of three or four rRNA molecules and nearly 55–80 different RPs (Bommer and Stahl, 2005). Some are characteristic for all domains of life (universal ribosomal proteins), whereas others are specific for bacteria, archaea, or eukaryotes. In addition, ribosome heterogeneity could arise due to differences in the constituent RPs and rRNAs of ribosomal subunits. As described in Chapter 5, such diversity may allow ribosomes to adapt and adjust to the diverse environmental conditions and perform specialized functions (Briggs and Dinman, 2017; Ferretti and Karbstein, 2019). However, the precise role of most RPs in the assembly and functioning of the ribosome remains unclear.

Interestingly, a functional ribosome can be assembled in vitro under appropriate conditions with purified RP and rRNA constituents (Traub and Nomura, 1968). The ability of ribosomes to self-assemble in vitro has provided a strong experimental tool to investigate the roles of individual proteins and rRNAs. The composition of the ribosomal subunits is given in Fig. 1.3. The 30S small ribosomal subunit of *Escherichia coli* ribosomes carries one copy of the 16S rRNA and 21 proteins whereas the 50S large subunit consists of one copy each of the 23S and 5S rRNAs and 34 proteins. Each bacterial ribosome contains one copy of each of RPs except for one

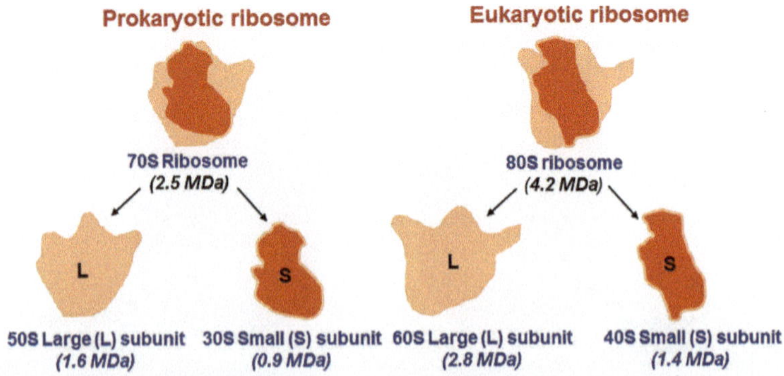

rRNA	23S RNA (120 nt) 5S RNA (2900 nt)	16S RNA (1540 nt)	28S RNA (4700 nt) 5.8S RNA (160 nt) 5S RNA (120 nt)	18S RNA (1900 nt)
No. of RPs	34	21	~49	~33
Function	Catalyzes the peptide bond formation	Decodes information on mRNA	Catalyzes the peptide bond formation	Decodes information on mRNA

Fig. 1.3 Components of prokaryotic and eukaryotic ribosomes and their functions. Both prokaryotic and eukaryotic ribosomes carry two subunits—large (L, *light brown*) and small (S, *brown*) subunits made up of ribonucleoproteins held together by noncovalent bonds. The small subunits of both prokaryotes and eukaryotes carry only one ribosomal ribonucleic acid (rRNA) molecule, whereas the large subunits of prokaryotes carry two rRNAs and that of eukaryotes have three rRNAs. The size of complete ribosomes, their subunits as well as all the associated rRNAs are shown as mega Daltons (MDa) and sedimentation coefficient (S). The number of associated ribosomal proteins (RPs) are also indicated. The two ribosomal subunits perform different roles in protein synthesis. Whereas the large subunit catalyzes the peptide bond formation, the small subunit participates in decoding the information on mRNA.

protein of the large subunit that is present in four copies. The eukaryotic ribosomes are larger and carry more proteins than bacterial ribosomes. The 40S small subunit of eukaryotic ribosomes is composed of single 18S rRNA and about 33 proteins, the 60S large subunit contains three rRNAs, viz., 28S, 5.8S, and 5S, and approximately 49 proteins (Fig. 1.3). The details of ribosomal RNA and protein genes can be found in Chapters 3 and 4.

Functions of ribosome

There are nearly 10 billion protein molecules in a mammalian cell and ribosomes produce most of them. Ribosomes are the sites at which information carried in the genetic code in the messenger RNAs (mRNAs) is converted into protein molecules and thus, serve as an elegant and efficient bio-machine for protein synthesis in the cell. Ribosomes perform multiple functions during translation, such as

(a) enhancing the accuracy of codon-anticodon pairing between the mRNA transcript and the aminoacyl-tRNAs,
(b) polymerizing the growing peptide chain (via peptidyl transferase),
(c) acting as energy transducers converting chemical energy into the mechanical energy during translocation of amino acids from tRNA carriers,
(d) protecting the growing peptide from attack by proteases possibly by forming a long protective tunnel, and
(e) assisting in the hydrolysis of the amino acid-tRNA bond during termination of the growing peptide chain.

Protein biosynthesis requires mRNA with an initiation site ("**Start**" codon), translation machinery (ribosome and aminoacyl-tRNA), and termination site ("**Stop**" codon) in mRNA (Fig. 1.4). If one is missing, the genetic information cannot be translated into functional proteins. During protein synthesis ribosomes appear as beads on a string of individual mRNA molecules; they are known as polyribosomes or polysomes (Afonina and Shirokov, 2018). Each ribosome has one binding site for mRNA and three binding sites for tRNA: A, P, and E sites (for aminoacyl-tRNA, peptidyl-tRNA, and exit, respectively) (Fig. 1.5). During protein synthesis, ribosomes move along an mRNA molecule, reading one codon at a time. The mRNAs determine the order of transfer RNA (tRNA) binding to nucleotide triplets (codons) and ultimately determine the amino acid sequence of a protein. The amino acids are brought to the ribosome by tRNA molecules that serve as the adapter molecules for each of the 20 amino acids during translation (reviewed by Green and Noller, 1997). The two subunits are known to play different roles in protein synthesis. Whereas the large subunit catalyzes

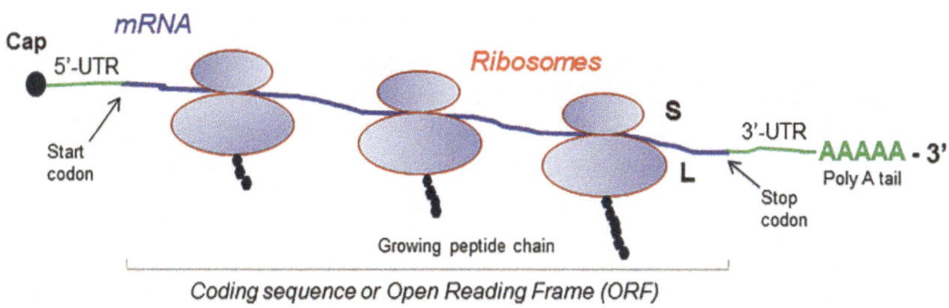

Fig. 1.4 Schematic representation of role of ribosomes in protein biosynthesis (translation).

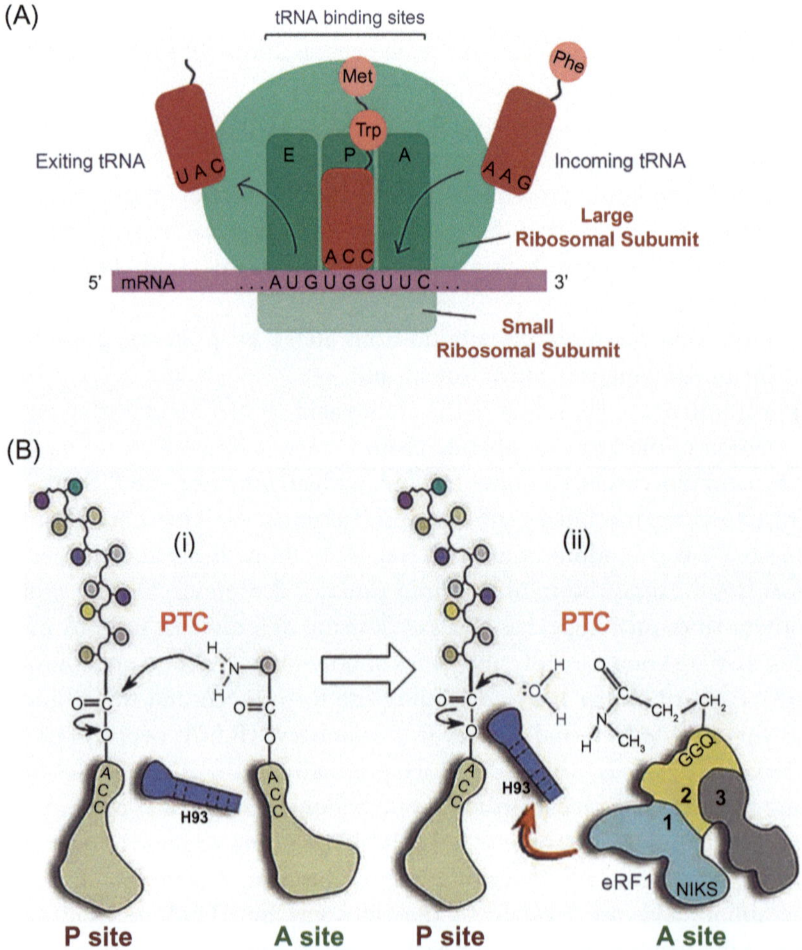

Fig. 1.5 Translation of mRNA's message into protein by the ribosomal machinery. (A) Each ribosome has one binding site for mRNA and three binding sites for tRNA, viz., A (aminoacyl-tRNA,), P (peptidyl-tRNA) and E (Exit) sites. The path of mRNA (purple) through the small ribosomal subunit. The A and P sites constrain tRNAs to form base pairs with adjacent codons on the mRNA molecule that maintains the correct reading frame on the mRNA. (B) The peptidyl transferase reaction catalyzed at the peptidyl transferase center (PTC). (i) The ribosomal PTC catalyzes the formation of a peptide bond. The α-amino group of the amino-acyl tRNA bound at the "A site" attacks the carbonyl carbon atom of the ester bond linking the peptidyl residue to the 3′ end of the tRNA bound at the "P site." The result is a new peptide bond and the transfer of the peptidyl residue to the amino acid of the tRNA bound at "A site." (ii) The PTC catalyzes hydrolysis of the peptidyl-tRNA bound in the "P site" by a catalytic water molecule. Translation termination depends on the inclusion of a water molecule in the PTC and on the action of eRF1 release factor, which can also modulate the PTC activity. Helix 93 of the PTC plays a critical role during hydrolysis of peptidyl-tRNA and its orientation following its binding to eRF1 *(red arrow)*. 1–3, Domains of eRF1 with NIKS and GGQ motifs.

peptide bond formation, the small subunit participates in decoding the information on mRNA and locking-on to a large subunit. Earlier, rRNAs in the ribosomal subunits were considered to play a structural scaffold for the assembly of RPs. Following the discovery of the catalytic functions associated with ribozymes, investigations on the catalytic activity of rRNA revealed that the large ribosomal subunit could catalyze the formation of peptide bonds even after ~95% of the RPs were removed. Further, treatment with RNase completely abolishes the peptide bond formation indicating at the participation of 23S rRNA in the peptidyl transferase function (Noller et al., 1992). Subsequently, these results were confirmed by demonstrating that the peptidyl transferase reaction could be catalyzed by synthetic fragments of 23S rRNA in the absence of any RP (reviewed by Beringer and Rodnina, 2007). For this reason, RPs are now considered to participate in proper folding of rRNAs and to enhance ribosome function by properly positioning the tRNAs. The ability of RNA molecules to catalyze peptide synthesis as well as self-replication could be the earliest biochemical reactions adopted during the evolution of cells. Normally, ribosomes can join up amino acids at a rate of 200 per minute. Therefore, small proteins can be made rapidly, whereas it takes few hours for the synthesis of larger proteins like titin of the human striated muscle (with more than 34,000 amino acids). Once the ribosome reaches a stop codon on the mRNA, translation stops, the ribosomal subunits separate and detach from the mRNA and the newly formed proteins migrate to other parts of the cell for diverse functions. The process of translation is described in Chapter 9.

It is speculated that certain subpopulations of ribosomes may have unique properties that modify the functions of the proteins they make (Filipovska and Rackham, 2013). Besides, some ribosomal proteins can also perform nonribosomal functions (Zhou et al., 2015). In addition, the post-translational modifications of RPs and rRNAs might result in "specialized ribosomes" that serve a special function in a specific cell or tissue type through an additional level of regulation. Such intrinsic regulation may allow selective translation of subsets of mRNAs having unique cis-regulatory elements, such as internal ribosomal entry sites (IRESs) and upstream open reading frames (uORFs) (Xue and Barna, 2012). Besides, the posttranscriptional modifications in rRNAs have rekindled considerable interest on their impact on gene expression programs and disease development (Babaian et al., 2020). These modifications are highly tissue specific and include the addition of functional groups such as a base or ribose methylation or substitutions (reviewed by Wang and He, 2014). The impact of mutations in RP genes and posttranscriptional modifications of rRNAs on ribosome function and disease development is described in Chapter 11.

Ribosome biogenesis and assembly

Ribosome biogenesis is the process of making ribosomes. As ribosomes are not self-replicating particles, the synthesis of its various components such as rRNAs and proteins are under genetic control. As described in Chapter 4, the expression of RPs is well synchronized at the levels of transcription and posttranscriptional modification, translation as well as posttranslational modification to produce accurate stoichiometric ratio. Thus, ribosome biogenesis is a highly dynamic process of the cell articulated in multiple, highly coordinated steps beginning in the nucleolus and ending up in the cytoplasm. It involves transcription of the rRNA and RP genes; their processing, modifications, association, and proper folding; and finally their assembly as mature ribosomal subunits. In prokaryotes, ribosome production occurs in the cytoplasm with the involvement of many ribosome gene operons whereas in eukaryotes, it occurs both in the cytoplasm as well as in the nucleolus. Ribosome biogenesis is an energy-dependent process orchestrated by over 200 other nonribosomal proteins and 75 snoRNAs are required in the synthesis and processing of the rRNAs as well as in the assembly of rRNAs with the ribosomal proteins. Adenosine triphosphate (ATP)-dependent RNA helicases, AAA-ATPases, guanosine triphosphatase (GTPases), and other kinases play a major role in this process (Baßler and Hurt, 2019).

In *Escherichia coli*, the synthesis of ribosomal proteins is controlled at the translational level. Some of the ribosomal proteins that bind directly to rRNA can also bind to similar structure in their own mRNA (Kaczanowska and Rydén-Aulin, 2007). In eukaryotes, the biogenesis of ribosomes is the result of the coordinated assembly of several molecular products that converge upon the nucleolus. The 18S, 5.8S, and 28S RNAs are synthesized as part of a much longer precursor molecule in the nucleolus, 5S RNA is synthesized on the chromosomes outside the nucleolus, and ribosomal proteins are synthesized in the cytoplasm. All these components migrate to the nucleolus, where they are assembled into ribosomal subunits and transported to the cytoplasm (Baßler and Hurt, 2019). The process of ribosome biogenesis in eukaryotes and prokaryotes is described in Chapters 7 and 8, respectively.

Ribosomal gene mutations and ribosomopathies

It is now well known that human cells can fail to make sufficient amounts of ribosomes, causing a number of diseases collectively known as ribosomopathies. The ribosome-associated diseases can be caused by mutations in genes encoding RPs, ribosomal biogenesis factors (RBFs) or one or more components of the rDNA transcription machinery. Mutations that decrease the levels of specific RPs or impair the rRNA modification machinery may affect the production of tissue-specific proteins leading to the induction of the disease phenotype. Alternatively, these mutations do not allow the cellular translational machinery to perform optimally and meet the demand of ribosomes in rapidly proliferating cells. Although ribosomopathies are relatively rare, these can manifest with multiple disease phenotypes. These could share some overlapping features such as inherited bone marrow failure syndromes as in case of Diamond-Blackfan anemia (DBA) and Shwachman Diamond syndrome and/or craniofacial defects in case of Treacher Collins syndrome and Bowen-Conradi syndrome. DBA-associated mutations lead to low levels of ribosome production in hematopoietic cells, which impair their commitment to the erythroid lineage (Khajuria et al., 2018). As elaborated in Chapter 11, the DBA-associated mutations have been identified in 11 different genes encoding RPs of both the small and large ribosomal subunits (Boria et al., 2010). Many of the gene defects observed in ribosomopathies are considered to be p53 dependent and therefore, can be prevented by interference in this pathway (Jones et al., 2008). Besides, dysregulated ribosome biogenesis has been implicated in the development and progression of most spontaneous cancers and a causal association between ribosomopathies and elevated cancer risk is now well established (Aspesi and Ellis, 2019; Derenzini et al., 2017). These studies have led to the notion that inhibition of ribosome biogenesis could be a potential therapeutic avenue for treatment of ribosomopathies and cancer.

Inhibition of ribosome biogenesis

With ribosomal biogenesis being tightly linked with cell proliferation and hence oncogenesis, several lines of chemical drugs that inhibit ribosome biogenesis serve as potential therapeutics for cancer intervention. Similarly, many anticancer drugs have been shown to exert their effect in part by inhibiting ribosome biogenesis (Bhat et al., 2015; Dmitriev et al., 2020). Nevertheless, considering the fundamental role of ribosome biogenesis in cell growth and proliferation, it provides an excellent target to impede cell growth and induce cell death by inhibiting ribosome biogenesis (Fig. 1.6).

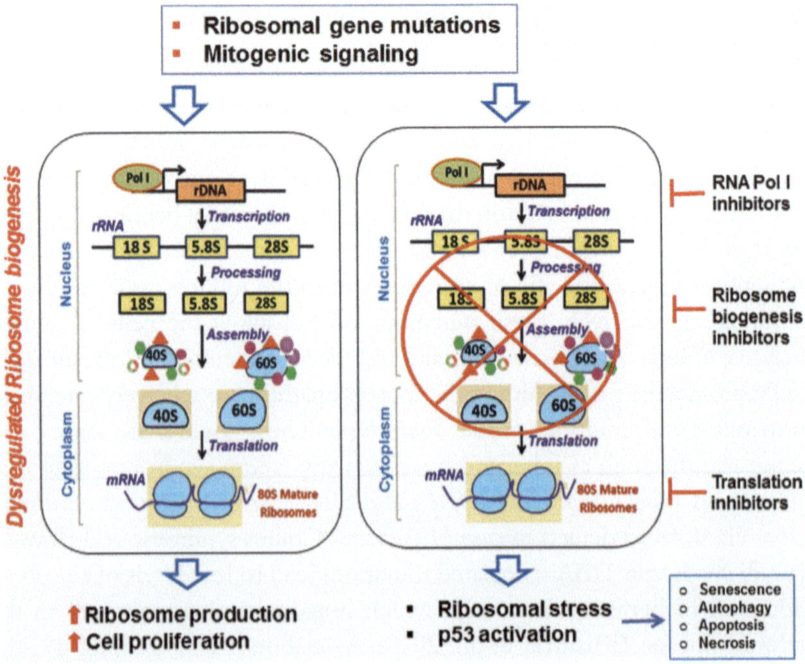

Fig. 1.6 Ribosome biogenesis as a promising target for treating ribosomopathies and cancer.

Most of the antimicrobial agents or antibiotics target ribosomes during translation. They prevalently inhibit either the peptidyl transferase center of the 50S large subunit of the ribosome or interfere with the aminoacyl-tRNA binding to the 30S small subunit of the ribosome (Poehlsgaard and Douthwaite, 2005). Since protein translation and ribosome biogenesis are interdependent, the effect of these antibiotics on the inhibition of ribosome biogenesis is not astounding. However, the enormous use of known antibiotics has strengthened the emergence of antibiotic-resistant microbes creating a never-ending search for new antibiotics. More information regarding the classical (e.g., chloramphenicol and neomycin) and new antibiotics can be found in Chapter 10.

In eukaryotes, each step of ribosome biogenesis is controlled by a different set of machinery in localized compartments. Therefore, impeding ribosome biogenesis in eukaryotes is directed at multiple levels such as transcription, assembly, and maturation. Among these, rRNA transcription is the primary target process because of two reasons—(a) rRNA transcription accounts for about 70% of the cell's transcription activity.

Importantly, the synthesis of rRNAs by RNA polymerase I is the first and the only rate-limiting step in ribosome production. The rate of rRNA synthesis thus plays a critical role in determining the rate of ribosome biogenesis; (b) RNA pol I and III transcribe rRNAs. Being structurally distinct from RNA polymerase II, it provides a unique opportunity to specifically inhibit rRNA transcription without any effect on mRNA synthesis. Besides, upstream binding factor (UBF), which is a major accessory protein of the Pol I initiation complex, is also a positive regulator of rRNA transcription. Hence, chemotherapeutic agents interfering with the expression and functions of UBF could be good candidates for the inhibition of ribosome biogenesis.

Identification of small molecule inhibitors such CX-5461 and CX-3543 has opened a whole new domain of selective chemotherapeutic interventions to block Pol I transcription (Drygin et al., 2009; Haddach et al., 2012). The observation that treatment of CX-5461 elicits a p53 response leading to apoptosis and autophagy only in cancer cells and not in normal cells is quite imperative. Such drugs have been found to be more effective in hematological malignancies despite the universal dependence of ribosome biogenesis in cancer cells. Diazaborine and ribozinoindoles (Rbins) are some other well studied and promising inhibitors of eukaryotic large subunit ribosome biogenesis by selectively targeting AAA-ATPase Drg1 or Mdn1 (Kawashima et al., 2016; Loibl et al., 2014). It has been over 5 decades since ribosomes were identified as the main cellular machinery for protein synthesis. However, the ribosome as a potential target for cancer therapy and treatment of other ribosomopathies is beginning to be realized now.

Summary and conclusions

Mature ribosomes are indispensable for protein production and hence critical for cellular growth and proliferation. The process of making ribosomes is called ribosome biogenesis, which is a highly orchestrated and conserved cellular process in prokaryotes and eukaryotes. Production of mature ribosomes requires coordinated action of three different processes—transcription and processing of rRNAs, synthesis of RPs, and finally successful assembly of the two ribosomal subunits. Dysregulation of ribosome biogenesis due to genetic mutations could cause a range of ribosomopathies and even may lead to cancer. Several lines of chemotherapeutic agents are now available, which can suppress ribosome biogenesis at transcription or assembly levels.

References

Afonina ZA, Shirokov VA. Three-dimensional organization of polyribosomes—a modern approach. Biochemistry (Mosc) 2018;83(Suppl. 1):S48–55.

Aspesi A, Ellis SR. Rare ribosomopathies: insights into mechanisms of cancer. Nat Rev Cancer 2019;19:228–38.

Babaian A, Rothe K, Girodat D, Minia I, Djondovic S, Milek M, Spencer Miko SE, Wieden HJ, Landthaler M, Morin GB, Mager DL. Loss of m1acp3Ψ ribosomal RNA modification is a major feature of cancer. Cell Rep 2020;31:107611.

Baßler J, Hurt E. Eukaryotic ribosome assembly. Annu Rev Biochem 2019;88:281–306.

Beringer M, Rodnina MV. The ribosomal peptidyl transferase. Mol Cell 2007;26:311–21.

Bhat M, Robichaud N, Hulea L, Sonenberg N, Pelletier J, Topisirovic I. Targeting the translation machinery in cancer. Nat Rev Drug Discov 2015;14:261–78.

Boisvert FM, van Koningsbruggen S, Navascués J, Lamond AI. The multifunctional nucleolus. Nat Rev Mol Cell Biol 2007;8:574–85.

Bommer UA, Stahl J. Ribosomal proteins in eukaryotes. In: Encyclopedia of life sciences. John Wiley & Sons, Ltd; 2005. p. 1–13.

Boria I, Garelli E, Gazda HT, Aspesi A, Quarello P, Pavesi E, Ferrante D, Meerpohl JJ, Kartal M, Da Costa L, Proust A, Leblanc T, Simansour M, Dahl N, Fröjmark AS, Pospisilova D, Cmejla R, Beggs AH, Sheen MR, Landowski M, Buros CM, Clinton CM, Dobson LJ, Vlachos A, Atsidaftos E, Lipton JM, Ellis SR, Ramenghi U, Dianzani I. The ribosomal basis of diamond-Blackfan Anemia: mutation and database update. Hum Mutat 2010;31:1269–79.

Briggs JW, Dinman JD. Subtractional heterogeneity: a crucial steptoward defining specialized ribosomes. Mol Cell 2017;67:3–4.

Derenzini M, Montanaro L, Trerè D. Ribosome biogenesis and cancer. Acta Histochem 2017;119:190–7.

Dmitriev SE, Vladimirov DO, Lashkevich KA. A quick guide to small-molecule inhibitors of eukaryotic protein synthesis. Biochemistry (Mosc) 2020;85:1389–421.

Drygin D, Siddiqui-Jain A, O'Brien S, Schwaebe M, Lin A, Bliesath J, Ho CB, Proffitt C, Trent K, Whitten JP, Lim JK, Von Hoff D, Anderes K, Rice WG. Anticancer activity of CX-3543: a direct inhibitor of rRNA biogenesis. Cancer Res 2009;69:7653–61.

Ferretti MB, Karbstein K. Does functional specialization of ribosomes really exist? RNA 2019;25:521–38.

Filipovska A, Rackham O. Specialization from synthesis: how ribosome diversity can customize protein function. FEBS Lett 2013;587:1189–97.

Green R, Noller HF. Ribosomes and translation. Annu Rev Biochem 1997;66:679–716.

Haddach M, Schwaebe MK, Michaux J, Nagasawa J, O'Brien SE, Whitten JP, Pierre F, Kerdoncuff P, Darjania L, Stansfield R, Drygin D, Anderes K, Proffitt C, Bliesath J, Siddiqui-Jain A, Omori M, Huser N, Rice WG, Ryckman DM. Discovery of CX-5461, the first direct and selective inhibitor of RNA polymerase I, for cancer therapeutics. ACS Med Chem Lett 2012;3:602–6.

Hellen CUT. Translation termination and ribosome recycling in eukaryotes. Cold Spring Harb Perspect Biol 2018;10:1–18.

Jones NC, Lynn ML, Gaudenz K, Sakai D, Aoto K, Rey JP, Glynn EF, Ellington L, Du C, Dixon J, Dixon MJ, Trainor PA. Prevention of the neurocristopathy Treacher Collins syndrome through inhibition of p53 function. Nat Med 2008;14:125–33.

Kaczanowska M, Rydén-Aulin M. Ribosome biogenesis and the translation process in *Escherichia coli*. Microbiol Mol Biol Rev 2007;71:477–94.

Kawashima SA, Chen Z, Aoi Y, Patgiri A, Kobayashi Y, Nurse P, Kapoor TM. Potent, reversible, and specific chemical inhibitors of eukaryotic ribosome biogenesis. Cell 2016;167:512–524.e14.

Khajuria RK, Munschauer M, Ulirsch JC, Fiorini C, Ludwig LS, McFarland SK, Abdulhay NJ, Specht H, Keshishian H, Mani DR, Jovanovic M, Ellis SR, Fulco CP, Engreitz JM, Schütz S, Lian J, Gripp KW, Weinberg OK, Pinkus GS, Gehrke L, Regev A, Lander ES, Gazda HT, Lee WY, Panse VG, Carr SA, Sankaran VG. Ribosome levels selectively regulate translation and lineage commitment in human hematopoiesis. Cell 2018;173:90–103.e19.

Leslie M. https://www.sciencemag.org/news/2017/06/there-are-millions-protein-factories-every-cell-surprise-they-re-not-all-same. 2017, https://doi.org/10.1126/science.aan6994.

Loibl M, Klein I, Prattes M, Schmidt C, Kappel L, Zisser G, Gungl A, Krieger E, Pertschy B, Bergler H. The drug diazaborine blocks ribosome biogenesis by inhibiting the AAA-ATPase Drg1. J Biol Chem 2014;289:3913–22.

McQuillen K, Roberts RB, Britten RJ. Synthesis of nascent protein by ribosomes in *Escherichia coli*. Proc Natl Acad Sci U S A 1959;45:1437–47.

Noller HF, Hoffarth V, Zimniak L. Unusual resistance of peptidyl transferase to protein extraction procedures. Science 1992;256:1416–9.

Palade GE. The organization of living matter. Proc Natl Acad Sci U S A 1964;52:613–34.

Palade GE, Siekevitz P. Liver microsomes; an integrated morphological and biochemical study. J Biophys Biochem Cytol 1956;2:171–200.

Poehlsgaard J, Douthwaite S. The bacterial ribosome as a target for antibiotics. Nat Rev Microbiol 2005;3:870–81.

Sáez-Vásquez J, Delseny M. Ribosome biogenesis in plants: from functional 45S ribosomal DNA organization to ribosome assembly factors. Plant Cell 2019;31:1945–67.

Traub P, Nomura M. Structure and function of *E. coli* ribosomes. V. Reconstitution of functionally active 30S ribosomal particles from RNA and proteins. Proc Natl Acad Sci U S A 1968;59:777–84.

Wang X, He C. Dynamic RNA modifications in posttranscriptional regulation. Mol Cell 2014;56:5–12.

Woolford Jr JL, Baserga SJ. Ribosome biogenesis in the yeast *Saccharomyces cerevisiae*. Genetics 2013;195:643–81.

Xue S, Barna M. Specialized ribosomes: a new frontier in gene regulation and organismal biology. Nat Rev Mol Cell Biol 2012;13:355–69.

Zhou X, Liao WJ, Liao JM, Liao P, Lu H. Ribosomal proteins: functions beyond the ribosome. J Mol Cell Biol 2015;7:92–104.

CHAPTER 2

Ribosome structure

Akhilesh Kumar Saini and Vijay Kumar
Department of Molecular and Cellular Medicine, Institute of Liver and Biliary Sciences, New Delhi, India

Introduction

Ribosomes are submicroscopic, dense, membraneless granular ribonucleoprotein organelles that serve as cell protein factories. In 1941, **Claude** first observed ribosome and called them microsomes and in 1958 **R.B. Roberts** coined the term ribosome. Each ribosome is a porous, hydrated structure with a flattened spherical shape and a diameter range of 15–25 nm. So far, two Nobel prizes have been awarded for ribosome and its structure. Of these the first Nobel prize was awarded to **George Emil Palade**, a cell biologist in 1974 (Palade, 1974) and the second was awarded in 2009 for chemistry to **Ada Yonath**, **Tom Steitz**, **Venkatraman Ramakrishnan** (Ramakrishnan, 2018). Ribosomes are composed of two different subunits, one is small and the other one is a large subunit, each with multiple proteins and RNA components that play specific functional roles during translation (Müller and Wittmann-Liebold, 1997). The eukaryotic 80S ribosome is comprised of two unequal subunits that contain four different rRNAs and around 80 ribosomal proteins. Eukaryotic 40S subunit contains 18S rRNA and 33 proteins and the 60S subunit contains 25S, 5S, 5.8rRNA, and 47 ribosomal proteins. Bacterial ribosomal 50S subunit contains 23S, 5SrRNA, and 33 proteins (Fig. 2.1). The large subunit of ribosome catalyzes peptide bond formation and the small subunit plays a pivotal role in mRNA binding and decoding. During translation (or protein synthesis), the ribosome also interacts with many other macromolecules, such as transfer RNAs (tRNAs) carrying specific amino acids, translation factors, and regulatory factors.

Although ribosome composition is conversed among various species, it varies to some extent. In bacterial ribosome the mitoribosome rRNA/protein ratio varies from 2:1 to 1:3 and in terms of molecular weight it varies from 2.3 MDa in prokaryote to 4.3 MDa in eukaryotes (Melnikov et al., 2012). X-ray and cryo-electron microscopy (EM) studies made possible the advancement of ribosomal studies and elucidation of ribosomal

Emerging Concepts in Ribosome Structure, Biogenesis, and Function
https://doi.org/10.1016/B978-0-12-816364-1.00007-X

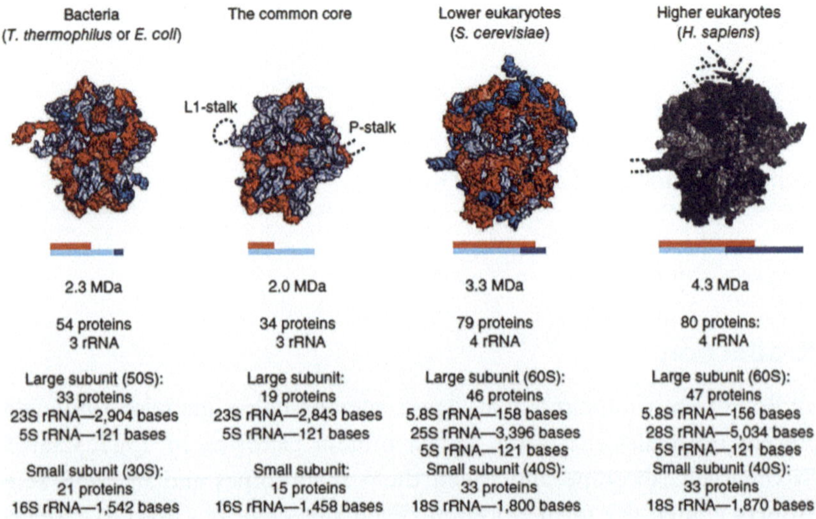

Fig. 2.1 Composition of bacterial and eukaryotic ribosomes and the common core. *(Reproduced with permission from Melnikov S, Ben-Shem A, Garreau de Loubresse N, Jenner L, Yusupova G, Yusupov M. One core, two shells: bacterial and eukaryotic ribosomes. Nat Struct Mol Biol 2012;19:560–7.)*

components such as rRNAs, proteins, and subunits. Bacterial ribosome and its subunits structure were first reconstituted by using negatively stained electron microscope images (Lake, 1976; Vasiliev, 1974). Conserved components of ribosomes among various species play important role in protein synthesis (Melnikov et al., 2018). Despite a detailed study of protein synthesis, the 3D structure of ribosome was not resolved.

Early low-resolution cryo-EM reconstructions of the bacterial 70S revealed distinct morphological features, such as the central protuberance, and L1 and L7/L12 stalks in the 50S subunit. First detailed three-dimensional images of the ribosomes were published independently in 1995 by two research groups (Frank et al., 1995; Stark et al., 1995). Subsequently, the cryo-EM maps of the ribosome at high, often near-atomic resolution, which provided new mechanistic insights into different stages of protein synthesis and ribosome recycling with the help of inhibitors (Jobe et al., 2019). Nevertheless, the ribosomal dynamics is observed through a systematic comparison of different models obtained by cryo-EM and X-ray studies. For example, an atomic model of the bacterial ribosome was fitted into the cryo-EM reconstructions of a pretranslocation (pre) and post-translocation (post) 70S ribosome to show the important role played by ribosomal proteins in facilitating the dynamics of translation (Gao et al.,

2003). Structural information from cryo-EM maps of eukaryotic ribosomes is now available for yeast, *Trypanosoma cruzi*, alga *Chlamydomonas reinhardtii*, and most recently for mammalian 80S ribosomes. Structural differences between rabbit 80S and *Escherichia coli* 70S ribosomes could be interpreted in terms of ribosomal RNA expansion segments in the 18S and 23S RNA (Jobe et al., 2019).

With the isolation of 50S and 30S subunits from *Bacillus stearothermophilus* and Haloarcula marismortui, respectively, ribosomal crystallography started. *Thermus thermophiles* 30S ribosome subunit (Wimberly et al., 2000) and the *H. marismortui 50S* subunit were one of the first determined crystal structure (Penczek et al., 1999) and these crystal structures were used for interpretation of *Thermus thermophilus* full ribosome X-ray electron density maps (Korostelev et al., 2006). Eukaryotic ribosomal crystal structure from yeast was determined at 4.2 and 3Å and it played a significant role in our understanding of protein synthesis mechanism and its regulation (Ben-Shem et al., 2011, 2010). Eukaryotic 40S and 60S ribosomal subunits were crystallized with their components at 3.8 and 3.5Å, respectively (Klinge et al., 2011; Rabl et al., 2011).

Several decades ago when structural studies on the ribosome were initiated in Dr. Joachim Frank's laboratory with 50–40Å cryo-EM maps, it was hard to imagine the achievement of near-atomic resolution (3Å) that is standard today. More recently, the Frank's lab has produced one of the highest resolution cryo-EM maps for a pathogenic ribosome (Liu et al., 2017). Such high-resolution ribosome maps could pave the way to the discovery of new drugs (Wong et al., 2017).

Structure of prokaryotic ribosome

The prokaryotic ribosome structure is an assembly of RNA chains and different proteins. The first of the high-resolution ribosome structures to be published was the structure of the 50S large (L) subunit from the archaean *Haloarcula marismortuii* at 2.4Å resolution (Ban et al., 2000). Subsequently, the structure of 50S subunit of a mesophilic bacterium *Deinococcus radioduarans* at 3.1Å resolution was published which showed a conformation quite similar to that originally reported for the *Haloarcula* 50S subunit (Harms et al., 2001). With central protuberance, L1, L7/L12 stalk 50 subunits make crown-like structure (Fig. 2.2). The L1 stalk is a domain of the E site of the large subunit that interacts with tRNA during translation (Fig. 2.2). Around the same time, two independent structures of the 30S small subunit from

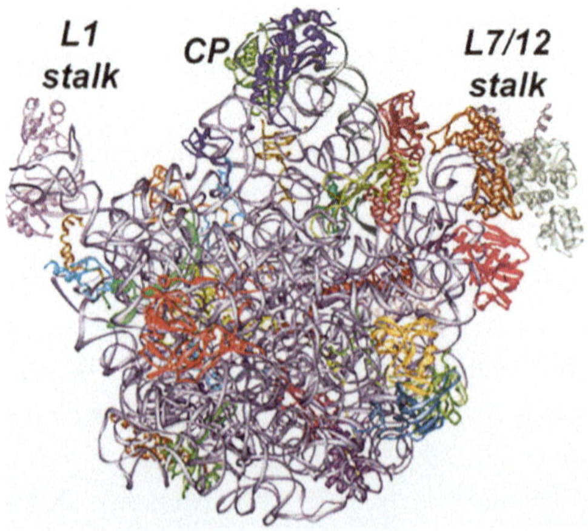

Fig. 2.2 Architecture of the 50S large subunit of a bacterial ribosome. *(Reproduced with permission from Harms J, Schluenzen F, Zarivach R, Bashan A, Gat S, Agmon I, Bartels H, Franceschi F, Yonath A. High resolution structure of the large ribosomal subunit from a mesophilic eubacterium. Cell 2001;107:679–88.)*

the bacterium *Thermus thermophilus*, one at 3.3Å (Schluenzen et al., 2000) and another at 3.05Å (Wimberly et al., 2000) were published.

Except for L7/L12 stalk, ribosomal components are present in a single copy. L7/L12 stalk is present in four or six copies. According to genetic data, prokaryotic and eukaryotic ribosomes have a common core structure which is composed of 34 conserved proteins and 4400 RNA bases (Fig. 2.1 and Table 2.1). The core contains the decoding site, PTC, and

Table 2.1 Properties of ribosomes from various sources.

Ribosome origin	Molecular weight	No. of associated proteins	Ribosomal subunits	rRNA subunits
Bacterial ribosome	2.3MDa	54	50S	23S (2904 bases), 5S (121bases)
			30S	16S (1542 bases)
Lower eukaryotes	3.3MDa	79	60S	25S (3396 bases), 5.8S (158 bases), 5S (121 bases)
			40S	18S (1800 bases)
Higher eukaryotes	4.3MDa	80	60S	28S (5034 bases), 5.8S (156 bases), 5S (121 bases)
			40S	18S (1870 bases)

tRNA-binding sites which are major functional centers of the ribosome. Domain-specific proteins, expansion segments of rRNAs, and extensions and insertions of conserved proteins present as specific moieties in each ribosome. The prokaryotic ribosome contains 20 bacteria-specific proteins, rRNA extensions such as helices h17, h68, and conserved protein extension such as S3, S4.

Protein and rRNA moieties envelop the core from the solvent side which is helpful in the study of their interaction with chaperons and transcription factors. There are some unusual structures related to mRNA such as turret and spire which are formed by rRNA expansion of the Trypanosoma cruzi 40S ribosomal subunits. Thx is a specific protein found in *Thermus thermophiles* ribosome which helps in the stabilization of its ribosome structure.

Ribosomal core subunits play a different role in protein synthesis. Small subunit has a role in decoding in which aminoacyl tRNA is selected for specific mRNA sequence. During translation, tRNA movement begins from A site to P site and finally from P to exit (E) site (Fig. 2.3). The A site binds to charged tRNA (aminoacyl tRNA), peptidyl (P) site binds to the growing polypeptide chain holding tRNA. At exit (E) site tRNA dissociates from the ribosome. The peptide channel extends through a large subunit body and peptidyl transferase center (PTC). PTC plays a major role in peptide bond formation (Fig. 2.4) and composed mainly of conserved rRNAs.

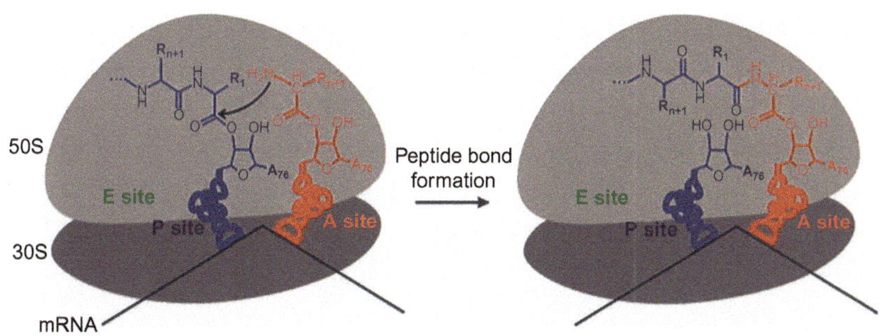

Fig. 2.3 Schematic representation of Peptide Bond Formation on the Ribosome. The α-amino group of aminoacyl-tRNA in the A site (red) attacks the carbonyl carbon of the peptidyl-tRNA in the P site (blue) to produce a new, one amino acid longer peptidyl-tRNA in the A site and a deacylated tRNA in the P site. The 50S large ribosomal subunit, where the PT center is located, is shown in light gray whereas the 30S small subunit in dark gray. A, P, and E sites of the ribosome are indicated. *(Reproduced with permission from Beringer M, Rodnina MV. The ribosomal peptidyl transferase. Mol Cell 2007;26:311–21.)*

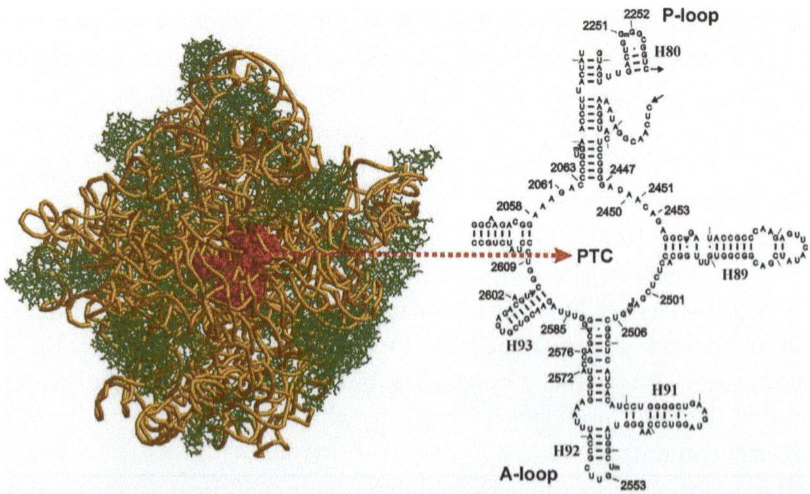

Fig. 2.4 **The peptidyl transferase center (PTC) in the large subunit of the ribosome.** *(Reproduced with permission from Polacek N, Mankin AS. The ribosomal peptidyl transferase center: structure, function, evolution, inhibition. Crit Rev Biochem Mol Biol 2005;40:285–311.)*

Major structural features of 30S subunits are known as head, body, beak, shoulder, and platform (Fig. 2.5). mRNA and tRNA binding sites location is ribosomal subunit interface. Head and shoulder make a tunnel for mRNA entry. mRNA exit side is located in between head and platform. The decoding center is composed of three domains from the head, penultimate stem, and shoulder.

There are seven bridges in the core of the ribosome which maintain the interaction between ribosomal subunits. Apart from conserved bridges specific prokaryotic and eukaryotic bridges have been discovered. In bacteria, one such unusual bridge is formed by protein L31 which modulate the swiveling movement of the small subunit head domain.

Structure of eukaryotic ribosome

Various specific proteins and rRNA expansion segments contribute to the eukaryotic ribosome structure diversity. The source of the first X-ray structures of the eukaryotic ribosome was *Tetrahymena thermophile* and, 40S.eIF1 complex (Rabl et al., 2011), yeast (Ben-Shem et al., 2011), and 60S.eIF1 complex (Klinge et al., 2011). The detailed structure of 80S ribosome of

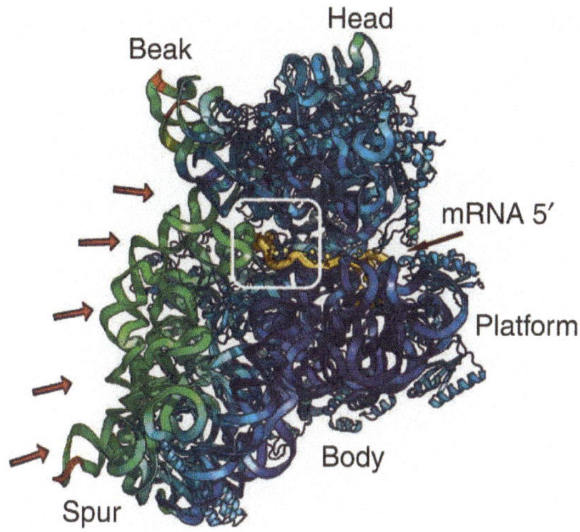

Fig. 2.5 Architecture of the 30S small subunit of the ribosome. *(Reproduced with permission from Jenner LB, Demeshkina N, Yusupova G, Yusupov M. Structural aspects of messenger RNA reading frame maintenance by the ribosome. Nat Struct Mol Biol 2010;17:555–60.)*

yeast has been reviewed by Yusupova and Yusupov (2014). The 80S yeast ribosome composed of 46 specific proteins and insertions and extensions in core proteins. Conserved rRNA chains have many extensions. The size of 80S yeast ribosome is within 0.5 MDa range which is due to ribosomal RNA expansion segments.

Major structural features of the 40S subunit are head, body, beak, and platform. tRNA- and mRNA-binding sites located in subunit interface (Fig. 2.6). The tRNA and mRNA binding sites are located on the subunit interface (Figs. 2.7 and 2.8). Head and shoulder makes a tunnel for mRNA entry. The location of mRNA exit is in between the head and the platform. Although there are many similarities in prokaryotic and eukaryotic ribosome structure there are many specific differences in small subunit ribosomal structure that explain the complex translation initiation in eukaryotes.

Novel features of the 80S Ribosome's structure

Many novel 80S ribosome features are long dynamic rRNA helices on both subunit solvent sides, protein clusters assembly around ribosomal RNA and interactions mediated by protein tails.

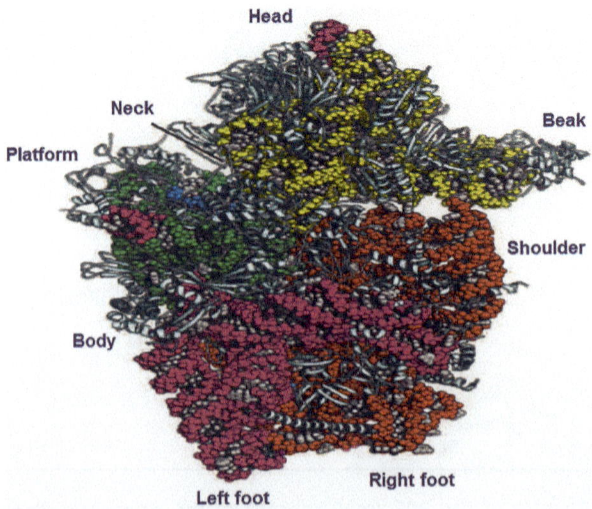

Fig. 2.6 Architecture of the small 40S subunit of the ribosome. The 18S rRNA as spheres and colored according to each domain (5′ domain, red; central domain, green; 3′ major domain, yellow; 3′ minor domain, blue; ESs, magenta), and the proteins as gray cartoons. *(Reproduced with permission from Rabl J, Leibundgut M, Ataide SF, Haag A, Ban N. Crystal structure of the eukaryotic 40S ribosomal subunit in complex with initiation factor 1. Science 2011;331:730–6.)*

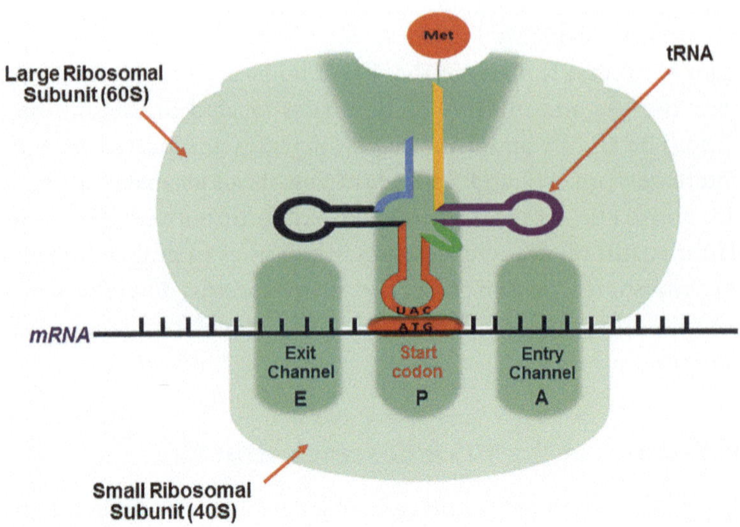

Fig. 2.7 Schematic representation of the A, P, and E tRNA binding sites in the eukaryotic ribosome.

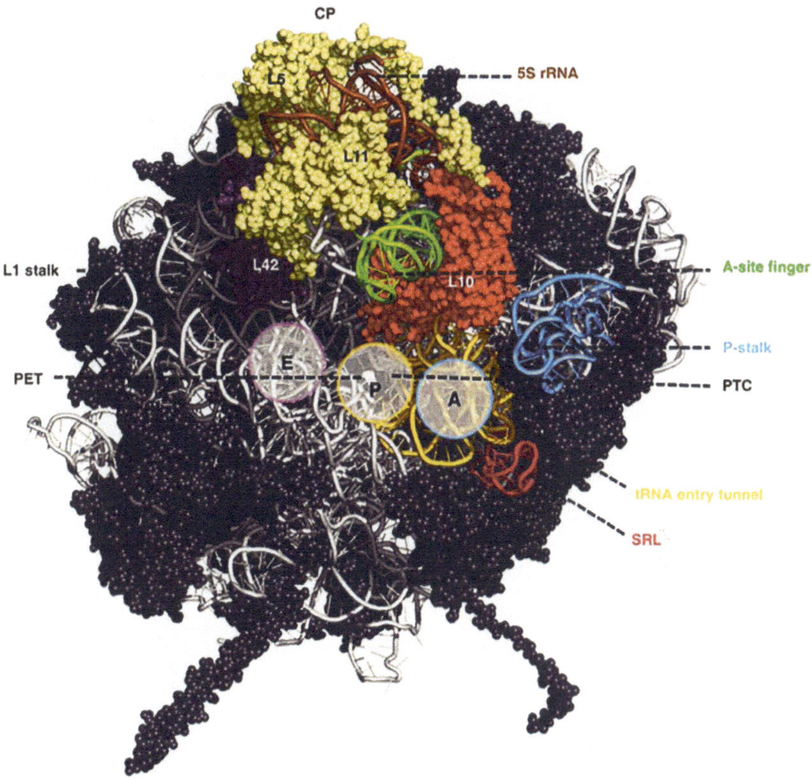

Fig. 2.8 Functional centers in the large ribosomal subunit. The P-stalk (blue), SRL (red), tRNA accommodation corridor beginning with helices 89 and 91 (yellow) containing the A- and P-sites, and ending with r-protein L42 (purple) at the end of the *E*-site, A-site finger (green), and CP containing 5S rRNA (brown) and r-proteins L5 and L11 (pale yellow) are shown. The r-protein L10 (orange-red) is also shown. *(Reproduced with permission from Konikkat S, Woolford JL Jr. Principles of 60S ribosomal subunit assembly emerging from recent studies in yeast. Biochem J 2017;474:195–214.)*

Eukaryote-specific protein tails

Many 80S ribosome proteins have very long tails and loops and not buried between rRNA as in the 70S ribosomes, but located on the surface. Advancement in the crystal structure of ribosome helped in the study of unusual protein folds in eukaryotic ribosome like extensions of L23 and S3. These unusual protein folds interact with other specific proteins such as L22 with ES39L, suggest their possible role in eukaryotic-specific component assembly during ribosome biogenesis.

Expansion segments

rRNA expansion elements are situated on the periphery of eukaryotic ribosome. Subunit interface area and Polypeptide exit tunnels have very less expansion segments and these areas are highly conserved. Cryo-electron microscopy structures of drosophila and human ribosome and crystal structure of yeast ribosome helped in the elucidation of the structure of expansion segments (Fujii et al., 2018). Although both human and *Drosophila* ribosomes carry a similar set of expansion segments as compared to yeast but have much longer in size as in yeast 110, 200, 160 and in humans 670, 170, 550 nucleotides long. Due to unstructured flexibility human rRNA expansion segments such as ES6S and ES30L could be partially interpreted.

Unfolded structure of human ribosome expansion segments made crystallization a challenging task. It has been shown that the expansion segment emerges on the solvent side of the platform where protein eL19 envelops it. Then ES6S extends in the shoulder direction and interacts with uS8 protein and its second arm extends toward the lower region of the small subunit. The second arm of this expansion segment extends toward the bottom of small subunits. Because ES6S is in contact entry and exit side of mRNA, it has a possible role in the translation.

Intersubunit bridges

Intersubunit bridges function as a communication mediator between ribosomal subunits during protein synthesis. Intersubunit rotation required for mRNA, tRNA translocation, and termination. There is a dynamic state in ribosomal subunits during translation which is evident when 80S ribosomes captured in a rotated state.

The various additional bridges on the 80S ribosome subunit interface increase the interaction surface which is a possible reason for the rotated state. One specific bridge is eB14 which is composed of a single alpha-helix eL41 protein in yeast. These bridges have two remarkable features. One is eL41-binding pocket in the small subunit is conserved in both eukaryotes and bacteria. Second, eL41 peptide has a stronger association with the 40S subunit than 60S subunit. Other specific features of 60S subunits are two protein helices extending from the left and right sides. These helices play role in creating bridges eB12 and eB13. These bridges are accessible from the solvent. eL24 is a key player in the reinitiation of the translation of polycistronic mRNAs.

The cryo-electron microscopy structures for *Trypanosoma cruzi* at 12Å (Gao et al., 2005) and for *Trypanosoma. brucei* at 5.5Å (Hashem et al., 2013) provided detailed architecture of these ribosomes structure. Ribosome structure elucidation in detail could be possible due to Cryo-electron microscopy reconstructions of 60S subunits from *T. cruzi* at 2.5Å (Liu et al., 2016, 2017), *Leishmania donovani* at 2.8Å (Shalev-Benami et al., 2016) and 80S ribosomes of L. *donovani* at 2.9Å (Zhang et al., 2016). It made it possible to decipher amino acid side chains, rRNA modifications, and solvent molecules.

28S rRNA fragmentation is a highly specific feature of the large ribosomal subunit of trypanosomatid which was discovered by nucleic acid electrophoresis (Campbell et al., 1987). Compared to other eukaryotic ribosomes, the cytosolic ribosome in trypanosomatids shows many specific protein and RNA components. Many ribosomal proteins have C and N terminal extensions, for example, protein eL19 contains a large C terminal extension. In the large subunit, various large rRNA kinetoplastid-specific expansion segments (ESs) are located in the periphery. The small subunit is more conserved in terms of rRNA as compared to the large subunit.

Specific features and structural dynamics of the *Plasmodium falciparum* 80S (Pf80S) ribosome were characterized by using cryo-EM (Wong et al., 2014, 2017; Sun et al., 2015). Initially, Wong et al. (2014) solved the Pf80S ribosome at 3.2Å resolution. As compared to the human ribosome the Pf80S ribosome has vast differences in the rRNA expansion segments and protein extensions. Fifteen ribosomal proteins are extended in *P. falciparum* as compared to yeast ribosomes and these extensions have highly specific interactions which are not observed in other ribosomes (Wong et al., 2014; Sun et al., 2015). Human parasite ribosome structure elucidation is helpful in parasite-specific features which lead to the development of new therapeutics against these parasites. Translation machinery in these parasites is not understood yet and in future high-resolution cryo-EM structures will be helpful in the characterization of parasite-specific elements and their interaction with a human. The transition from a pre- to a posttranslocation state induces several major rearrangements in the structure, mainly by ribosomal proteins eL24 and eL19 (which constitute some of the characteristic eukaryote-specific bridges (eB12 and eB13). The C-terminal part of eL24 relocates entirely and forms a new interaction site on the 40S subunit with ribosomal protein eS6 and 18S rRNA). In the case of eL19, a long protein helix protrudes from the 60S subunit to cross over to the 40S subunit. This is one of the rare 60S components (apart from the L1 region) that strongly changes conformation upon subunit rotation. The C-terminal helix of eL19 is kinked in the pretranslocation state (and extends by another seven residues), while it is straight in the posttranslocation state. This conformational transition results in the formation of salt bridges between the positively charged Arg172 and Arg176 side chains of protein eL19 and the phosphate moieties of 18S rRNA nucleotides G909 and G910. Compared to the yeast ribosome, the human ribosome contains an additional protein, eL28. The 40S subunit head shows a swivel movement away from the 60S, which leads to a partial closure of the latch region. The 40S subunit undergoes three characteristic types of conformational changes: (1) a universal rotation similar to 'ratcheting' in bacterial ribosomes; (2) a swivel movement of the 40S head and beak regions in the direction of the tRNA *E*-site; and (3) a rotation of the 40S body around a vertical axis perpendicular to the classical rotation axis termed 'subunit rolling. The subunit rolling leads to a slight opening of the ribosomal subunit interface on the side of the factor entry site where ribosomal GTPases bind (Khatter et al., 2015).

Structure of mitochondrial ribosome

Apart from their role in energy metabolism, mitochondria play an important role in different cellular processes, such as aging and apoptosis. Due to alpha-proteobacterial ancestry by endosymbiosis, mitochondria maintain a reduced genome and its corresponding protein synthesis machinery, the mitochondrial ribosomes also known as mitoribosomes. The human mitoribosome synthesizes 13 highly hydrophobic subunits of the oxidative phosphorylation complexes (Kaushal et al., 2014). Even though mitoribosomes are evolutionarily derived from bacterial ribosomes, they have strongly diverged from them in terms of composition, function, and structure. Mitoribosomes have acquired numerous mitochondrial-specific proteins and protein extensions. Mitoribosomes from various sources vary in their sedimentation coefficient, molecular weight, rRNAs, and proteins (Table 2.2). These structural changes have been accompanied by a strong functional specialization of the mitoribosome, which synthesizes predominantly membrane proteins, including hydrophobic components of highly important complexes of the mitochondrial respiratory chain.

The first cryo-EM reconstruction of the mammalian 55S mitoribosome was determined in a pioneering effort made by the laboratory of Rajendra K. Agrawal (Sharma et al., 2003; Kaushal et al., 2014, 2015) at a resolution of roughly 12–14Å. This landmark reconstruction revealed

Table 2.2 Properties of mitochondrial ribosomes from various sources.

Source of ribosomes	Molecular weight	Sedimentation coefficient	Subunits	rRNAs	Proteins
Bacterial	2.3 MDa	70S	30S and 50S	16S, 23S, and 5S	54
Eukaryotic cytosol	3.3–4.3 MDa	80S	40S and 60S	26S–28S, 5.8, 5S, 18S	82
Mammalian Mitochondria	2.7 MDa	55S	28S and 39S	12S and 16S	82
Yeast Mitochondria	3.3 MDa	74S	37S and 54S	155 and 21S	78

a highly divergent overall morphology of the 55S mitoribosome and its 28S small and 39S large subunits compared with the morphology of the bacterial ribosome. The development of direct electron detectors allowed the collection of much improved data using charge-coupled device cameras. This technique led to the determination of high-resolution architecture of the human and yeast mitoribosomal subunits in the laboratory of Venki Ramakrishnan (Amunts et al., 2014, 2015; Brown et al., 2014; Desai et al., 2017). The high-resolution structures of the mammalian 55S mitoribosome showed a strikingly different overall appearance of the mitoribosome from the ancestral bacterial ribosome. The rRNA of the mitoribosome is reduced to an innermost core, which is covered by a tightly interacting coat of mitoribosomal proteins that leaves very little rRNA exposed and projects far away from the ribosomal core (Fig. 2.9). Overall, 36 of the 82 proteins in the 55S mitoribosome are mitochondrial-specific, 22 in the large subunit and 14 in the small subunit. The interconnectivity of the protein–protein network has increased dramatically from the

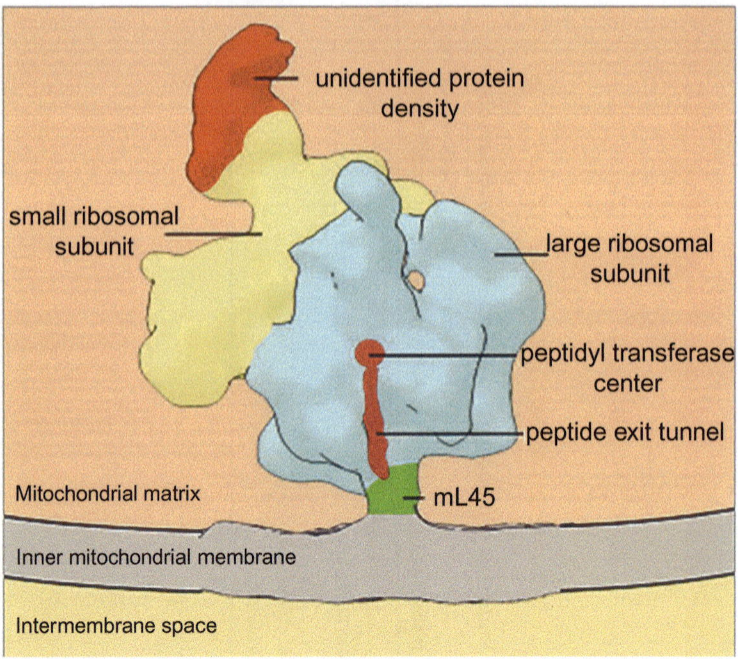

Fig. 2.9 The structure of human mitochondrial ribosome. *(Reproduced with permission from Englmeier R, Pfeffer S, Förster F. Structure of the human mitochondrial ribosome studied in situ by cryoelectron tomography. Structure 2017;25:1574–81.)*

bacterial to the mammalian mitoribosomes. Whereas bacterial ribosomal proteins have, on average, only 1.5 neighbor's, mitoribosomal proteins form contacts with an average of 4.5 neighbor's in the yeast 54S large subunit and 4.9 neighbor's in the mammalian 39S large subunit. The central protuberance (CP) of the 39S subunit is enlarged due to the acquisition of several mitochondrial-specific proteins, and two crescent-shaped pentatricopeptide repeat (PPR)-fold proteins dominate the overall appearance of the 28S small mitoribosomal subunit. The intersubunit bridges between the 39S and 28S subunits in the 55S mitoribosome are less extensive than in bacterial ribosomes, mostly due to the reduction of interacting rRNA segments. The reduced rRNA in mitoribosomes is limited to the innermost core of the ribosome, whereas numerous proteins cover the surface of the ribosome almost entirely and extend far away from the core. The reduction of peripheral contacts in the mitoribosome may lead to increased conformational flexibility of the mitoribosomal subunits, including tilting of the subunits relative to each other.

Summary and conclusions

The advancement in cryo-EM and X-ray crystallography have facilitated remarkable studies in the field of ribosomes structure by translation modeling at different atomic resolutions. It helped in the development of the electron density maps and provided mechanistic insights into ribosome functions and positions of functional groups. Much of the work done on the ribosomes since the 1960s has been directed to its structure. The crystal structures of the ribosome have already revolutionized the field of protein synthesis by providing a three-dimensional reference for interpreting existing biochemical and genetic data and designing new experiments, but nagging questions remain at each step. Since the 1950s, the structures of thousands of proteins have been solved by X-ray crystallography, but the work on the structure and function of the ribosome has allowed the scientific community to understand a fundamental component of translation with broad consequences in biology and medicine.

In future, advancements in the isolation, purification, and crystallization methods for the higher eukaryotic ribosomal subunits and ribosomal proteins will lead to structural studies of various ribosomal functional complexes with tRNA, mRNA, translation factors. This will enhance our understanding of ribosome structure and open a new horizon for the discovery of novel therapeutics against various infectious diseases.

References

Amunts A, Brown A, Bai XC, Llácer JL, Hussain T, Emsley P, Long F, Murshudov G, Scheres SHW, Ramakrishnan V. Structure of the yeast mitochondrial large ribosomal subunit. Science 2014;343:1485–9.

Amunts A, Brown A, Toots J, Scheres SHW, Ramakrishnan V. Ribosome. The structure of the human mitochondrial ribosome. Science 2015;348:95–8.

Ban N, Nissen P, Hansen J, Moore PB, Steitz TA. The complete atomic structure of the large ribosomal subunit at 2.4 A resolution. Science 2000;289:905–20.

Ben-Shem A, Garreau de Loubresse N, Melnikov S, Jenner L, Yusupova G, Yusupov M. The structure of the eukaryotic ribosome at 3.0 Å resolution. Science 2011;334:1524–9.

Ben-Shem A, Jenner L, Yusupova G, Yusupov M. Crystal structure of the eukaryotic ribosome. Science 2010;330:1203–9.

Brown A, Amunts A, Bai XC, Sugimoto Y, Edwards PC, Murshudov G, Scheres SHW, Ramakrishnan V. Structure of the large ribosomal subunit from human mitochondria. Science 2014;346:718–22.

Campbell DA, Kubo K, Clark CG, Boothroyd JC. Precise identification of cleavage sites involved in the unusual processing of trypanosome ribosomal RNA. J Mol Biol 1987;196:113–24.

Desai N, Brown A, Amunts A, Ramakrishnan V. The structure of the yeast mitochondrial ribosome. Science 2017;355:528–31.

Frank J, Verschoor A, Li Y, Zhu J, Lata RK, Radermacher M, Penczek P, Grassucci R, Agrawal RK, Srivastava S. A model of the translational apparatus based on a three-dimensional reconstruction of the *Escherichia coli* ribosome. Biochem Cell Biol 1995;73:757–65.

Gao H, Sengupta J, Valle M, Korostelev A, Eswar N, Stagg SM, Van Roey P, Agrawal RK, Harvey SC, Sali A, Chapman MS, Frank J. Study of the structural dynamics of the *E. coli* 70S ribosome using real-space refinement. Cell 2003;113:789–801.

Fujii K, Susanto TT, Saurabh S, Barna M. Decoding the function of expansion segments in ribosomes. Mol Cell 2018;72:1013–20.

Gao H, Ayub MJ, Levin MJ, Frank J. The structure of the 80S ribosome from Trypanosoma cruzi reveals unique rRNA components. Proc Natl Acad Sci U S A 2005;102:10206–11.

Harms J, Schluenzen F, Zarivach R, Bashan A, Gat S, Agmon I, Bartels H, Franceschi F, Yonath A. High resolution structure of the large ribosomal subunit from a mesophilic eubacterium. Cell 2001;107:679–88.

Hashem Y, des Georges A, Fu J, Buss SN, Jossinet F, Jobe A, Zhang Q, Liao HY, Grassucci RA, Bajaj C, Westhof E, Madison-Antenucci S, Frank J. High-resolution cryo-electron microscopy structure of the Trypanosoma brucei ribosome. Nature 2013;494:385–9.

Jobe A, Liu Z, Gutierrez-Vargas C, Frank J. New insights into ribosome structure and function. Cold Spring Harb Perspect Biol 2019;11:a032615.

Kaushal PS, Sharma MR, Booth TM, Haque EM, Tung CS, Sanbonmatsu KY, Spremulli LL, Agrawal RK. Cryo-EM structure of the small subunit of the mammalian mitochondrial ribosome. Proc Natl Acad Sci U S A 2014;111:7284–9.

Kaushal PS, Sharma MR, Agrawal RK. The 55S mammalian mitochondrial ribosome and its tRNA-exit region. Biochimie 2015;114:119–26.

Khatter H, Myasnikov AG, Natchiar SK, Klaholz BP. Structure of the human 80S ribosome. Nature 2015;520:640–5.

Klinge S, Voigts-Hoffmann F, Leibundgut M, Arpagaus S, Ban N. Crystal structure of the eukaryotic 60S ribosomal subunit in complex with initiation factor 6. Science 2011;334:941–8.

Korostelev A, Trakhanov S, Laurberg M, Noller HF. Crystal structure of a 70S ribosome-tRNA complex reveals functional interactions and rearrangements. Cell 2006;126:1065–77.

Lake JA. Ribosome structure determined by electron microscopy of *Escherichia coli* small subunits, large subunits and monomeric ribosomes. J Mol Biol 1976;105:131–9.

Liu Z, Gutierrez-Vargas C, Wei J, Grassucci RA, Ramesh M, Espina N, Sun M, Tutuncuoglu B, Madison-Antenucci S, Woolford Jr JL, Tong L, Frank J. Structure and assembly model for the Trypanosoma cruzi 60S ribosomal subunit. Proc Natl Acad Sci U S A 2016;113:12174–9.

Liu Z, Gutierrez-Vargas C, Wei J, Grassucci RA, Sun M, Espina N, Madison-Antenucci S, Tong L, Frank J. Determination of the ribosome structure to a resolution of 2.5 Å by single-particle cryo-EM. Protein Sci 2017;26:82–92.

Melnikov S, Ben-Shem A, Garreau de Loubresse N, Jenner L, Yusupova G, Yusupov M. One core, two shells: bacterial and eukaryotic ribosomes. Nat Struct Mol Biol 2012;19:560–7.

Melnikov S, Manakongtreecheep K, Söll D. Revising the structural diversity of ribosomal proteins across the three domains of life. Mol Biol Evol 2018;35:1588–98.

Müller EC, Wittmann-Liebold B. Phylogenetic relationship of organisms obtained by ribosomal protein comparison. Cell Mol Life Sci 1997;53:34–50.

Palade GE. Intracellular aspects of the process of protein secretion. In: Nobel Lecture; 1974.

Penczek P, Ban N, Grassucci RA, Agrawal RK, Frank J. Haloarcula marismortui 50S subunit-complementarity of electron microscopy and X-ray crystallographic information. J Struct Biol 1999;128:44–50.

Rabl J, Leibundgut M, Ataide SF, Haag A, Ban N. Crystal structure of the eukaryotic 40S ribosomal subunit in complex with initiation factor 1. Science 2011;331:730–6.

Ramakrishnan V. Gene machine: The race to decipher the secrets of the ribosome. New York: Basic Books, Hachette Book Group; 2018.

Schluenzen F, Tocilj A, Zarivach R, Harms J, Gluehmann M, Janell D, Bashan A, Bartels H, Agmon I, Franceschi F, Yonath A. Structure of functionally activated small ribosomal subunit at 3.3 angstroms resolution. Cell 2000;102:615–23.

Shalev-Benami M, Zhang Y, Matzov D, Halfon Y, Zackay A, Rozenberg H, Zimmerman E, Bashan A, Jaffe CL, Yonath A, Skiniotis G. 2.8-Å Cryo-EM structure of the large ribosomal subunit from the eukaryotic parasite Leishmania. Cell Rep 2016;16:288–94.

Sharma MR, Koc EC, Datta PP, Booth TM, Spremulli LL, Agrawal RK. Structure of the mammalian mitochondrial ribosome reveals an expanded functional role for its component proteins. Cell 2003;115:97–108.

Stark H, Mueller F, Orlova EV, Schatz M, Dube P, Erdemir T, Zemlin F, Brimacombe R, van Heel M. The 70S *Escherichia coli* ribosome at 23 Å resolution: fitting the ribosomal RNA. Structure 1995;3:815–21.

Sun M, Li W, Blomqvist K, Das S, Hashem Y, Dvorin JD, Frank J. Dynamical features of the *Plasmodium falciparum* ribosome during translation. Nucleic Acids Res 2015;43:10515–24.

Vasiliev VD. Morphology of the ribosomal 30S subparticle according to electron microscopic data. Acta Biol Med Ger 1974;33:779–93.

Wimberly BT, Brodersen DE, Clemons Jr WM, Morgan-Warren RJ, Carter AP, Vonrhein C, Hartsch T, Ramakrishnan V. Structure of the 30S ribosomal subunit. Nature 2000;407:327–39.

Wong W, Bai XC, Brown A, Fernandez IS, Hanssen E, Condron M, Tan YH, Baum J, Scheres SH. Cryo-EM structure of the plasmodium falciparum 80S ribosome bound to the anti-protozoan drug emetine. Elife 2014;3, e0308.

Wong W, Bai XC, Sleebs BE, Triglia T, Brown A, Thompson JK, Jackson KE, Hanssen E, Marapana DS, Fernandez IS, Ralph SA, Cowman AF, Scheres SHW, Baum J. Mefloquine targets the *Plasmodium falciparum* 80S ribosome to inhibit protein synthesis. Nat Microbiol 2017;2:17031.

Yusupova G, Yusupov M. High-resolution structure of the eukaryotic 80S ribosome. Annu Rev Biochem 2014;83:467–86.

Zhang X, Lai M, Chang W, Yu I, Ding K, Mrazek J, Ng HL, Yang OO, Maslov DA, Zhou ZH. Structures and stabilization of kinetoplastid-specific split rRNAs revealed by comparing leishmanial and human ribosomes. Nat Commun 2016;7:13223.

CHAPTER 3

rDNA gene structure, transcription, and its coregulation

Vijay Kumar
Department of Molecular and Cellular Medicine, Institute of Liver and Biliary Sciences, New Delhi, India

Introduction

Ribosome biogenesis is a highly regulated cellular process essential for growth and development. The proliferating eukaryotic cells require a continuous supply of new ribosomes for protein biosynthesis. Therefore, ribosome biogenesis normally parallels the growth rate of these cells. Production of a single human ribosome requires over 200 assembly factors, 80 RPs, and through the coordinated activity of all three RNA polymerases, viz., Pol I, II, and III and takes place in the nucleolus, nucleus, and cytoplasm of cells. The transcriptional activities of pol I and pol III dominate cellular transcription and accounts for ~80% of the total RNA synthesis in a growing cell. Among these, RNA Pol I is specifically devoted to transcription of only one large precursor 35–47S rRNA depending on the particular eukaryote and is based on their sedimentation coefficient during centrifugation. The 5S rRNA is transcribed by RNA pol III from multiple genes in the nucleus whereas the ribosomal protein genes that are present throughout the genome, are transcribed by RNA Pol II (Drygin et al., 2010; Grummt, 2003) (Fig. 3.1). The transcription of rRNA as well as RP genes and assembly of ribosomes show extremely complex regulation and is responsive to various cellular stimuli. The process stops following glucose depletion in the medium suggesting a direct link between the availability of nutrients and rRNA synthesis (Kusnadi et al., 2015). However, it is still unclear how the production of each constituent of the ribosome is regulated temporally and how pre-ribosomes are assembled posttranscriptionally. Note that dysregulation in Pol I transcription and ribosome biogenesis could be associated with different pathological consequences (Drygin et al., 2010; Hannan et al., 2013).

Emerging Concepts in Ribosome Structure, Biogenesis, and Function
https://doi.org/10.1016/B978-0-12-816364-1.00010-X

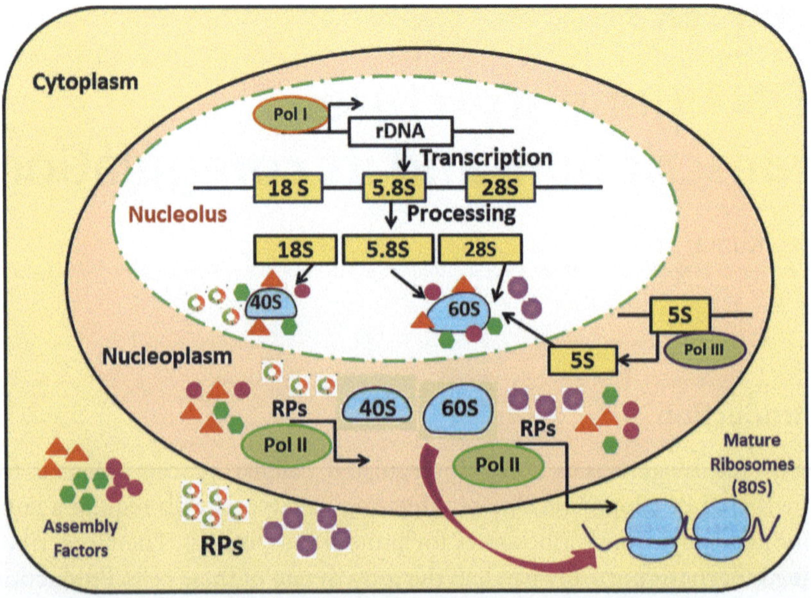

Fig. 3.1 Expression ribosomal RNAs and Ribosomal proteins, and their assembly into mature ribosomes. In eukaryotes, the ribosome biogenesis involves the coordinated activity of all three RNA polymerases, viz., Pol I, II, and III. Pol I synthesizes the precursor 47S ribosomal RNA (rRNA) in the nucleolus which is subsequently processed into mature 28S, 18S, and 5.8S rRNAs. Pol III synthesizes 5S rRNA whereas all ribosomal proteins (RPs) are synthesized by Pol II. The rRNAs and RPs assemble as ribonucleoproteins in the nucleolar compartment but the ribosomal subunits are formed in the nuclear compartment. The 80S mature ribosomes of Eukaryotes are made up of a small (40S) and large (60S) subunit and are formed in the cytoplasm after binding to mRNAs. The 40S small subunits carry one copy of 18S RNA and 33 RPs whereas the large subunits have one copy each of 5S, 28S, 5.8S rRNAs, and 46 RPs.

The synthesis of rRNA is usually regulated via changes in the rate of transcription initiation at active ribosomal DNA (rDNA) genes rather than through activation of silent transcription units (Stefanovsky and Moss, 2006). In fact in yeast, the number of active genes decreases from log to stationary phase, indicating that the proportion of active genes can be modulated between growth conditions. It has been observed that the promoter of active rRNA genes is free of CpG methylation and rich in acetylated histones whereas the silent genes show methylated CpG signatures and deacetylated histones (Lawrence et al., 2004). Apparently, the transcriptional repression is mediated via the recruitment of DNA methyltransferase and histone deacetylase activities on the promoter (Santoro and Grummt, 2005; Grummt and Längst, 2013). The silencing of rRNA gene is epigenetically regulated by a nucleolar remodeling complex (NoRC). However, the molecular mechanism governing the silencing and activation of individual rDNA transcription units during cell proliferation and differentiation also remains elusive.

Organization and regulation of ribosomal RNA genes

In bacteria, *Escherichia coli*, there are seven ribosomal RNA (rrn) operons located in noncontiguous sites around the genome and are transcribed in the same direction of genome replication (Fig. 3.2A). However, in eukaryotic cells, the ribosomal RNA genes or rDNA are organized in huge clusters of tandem repeats and are transcribed with high efficiency to cope with their metabolic needs and demand for ribosomes (Ide et al., 2010) (Fig. 3.2B and C). In the budding yeast *Saccharomyces cerevisiae*, there are ~150 repeats of ~9.1kb rDNA units that form a single cluster on chromosome XII (Kobayashi et al., 1998). In human cells, however, there are five rDNA gene clusters as tandem repeats, usually arranged head-to-tail and localized in the short arms of acrocentric chromosomes 13, 14, 15, 21, and 22. Each cluster has ~70 repeated units of ~43kb each (Sakai et al., 1995). The highly repetitive structure makes rDNA regions one of the most unstable regions in the genome often resulting in a reduction in their number via DNA recombination and causing cellular senescence. Interestingly, the reduced copy number is restored through a gene amplification strategy. For this reason, the length of rDNA region keeps changing through a cycle of contraction (reduction) and expansion (amplification) (Kobayashi, 2014). The sequence of the rRNA genes is highly conserved among eukaryotes. However, the intergenic spacer regions (IGS) is quite divergent in length and sequence but carry three unique regulatory regions including an origin of replication, a replication fork barrier site and a promoter that drives the expression of noncoding RNAs (Fig. 3.2B and C). One of the important roles of the replication fork barrier is to avoid collision between replication and transcription machineries (Takeuchi et al., 2003).

Ribosome biogenesis is initiated in the fibrillar centers of nucleoli through the active transcription of rDNA genes (Fig. 3.3A and B). There are several ways in which transcription of the rDNA can be measured. The Miller chromatin spreading technique is considered to be an elegant method to visualize transcription in chromatin under the electron microscope as spreads known as "Christmas tree." During active transcription,

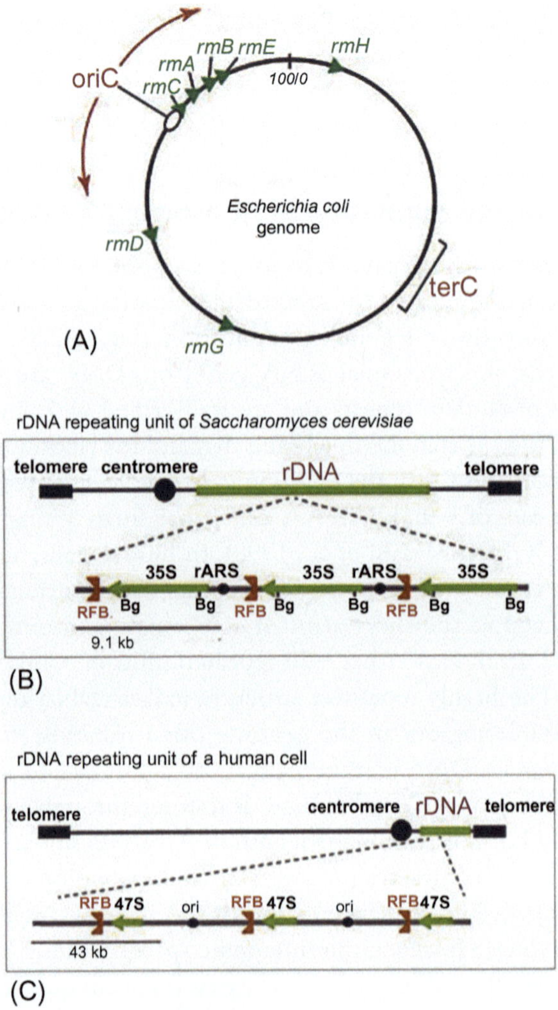

Fig. 3.2 Organization of ribosomal RNA genes in bacteria, yeast, and human cells. (A) In bacteria *Escherichia coli*, there are 7 rRNA operons (rrnA-rrnH) located primarily near the Ori region (oriC) of the genome. (B) In budding yeast (*Saccharomyces cerevisiae*), each rDNA transcriptional repeat unit contains the 35S rRNA precursor gene that is processed into three mature rRNA species (18S, 5.8S, and 25S rRNA). (C) In a human cell, the rDNA gene is expressed as a 47S precursor rRNA which is processed into three mature rRNA species (18S, 5.8S, and 28S rRNA). ARS, autonomously replicating sequences; Bg, Restriction enzyme *Bgl*II site; Ori, the origin of replication; RFB, replication fork barrier site; terC, Terminator C.

increasing lengths of rRNAs can be seen radiating from the rDNA genes in chromatin spreads of the repeated units (Fig. 3.3C). This technique was developed by Oscar Miller and his team at the Oak Ridge National Laboratory, Tennessee using amphibian oocytes (Miller Jr and Beatty, 1969; Meissner et al., 1991).

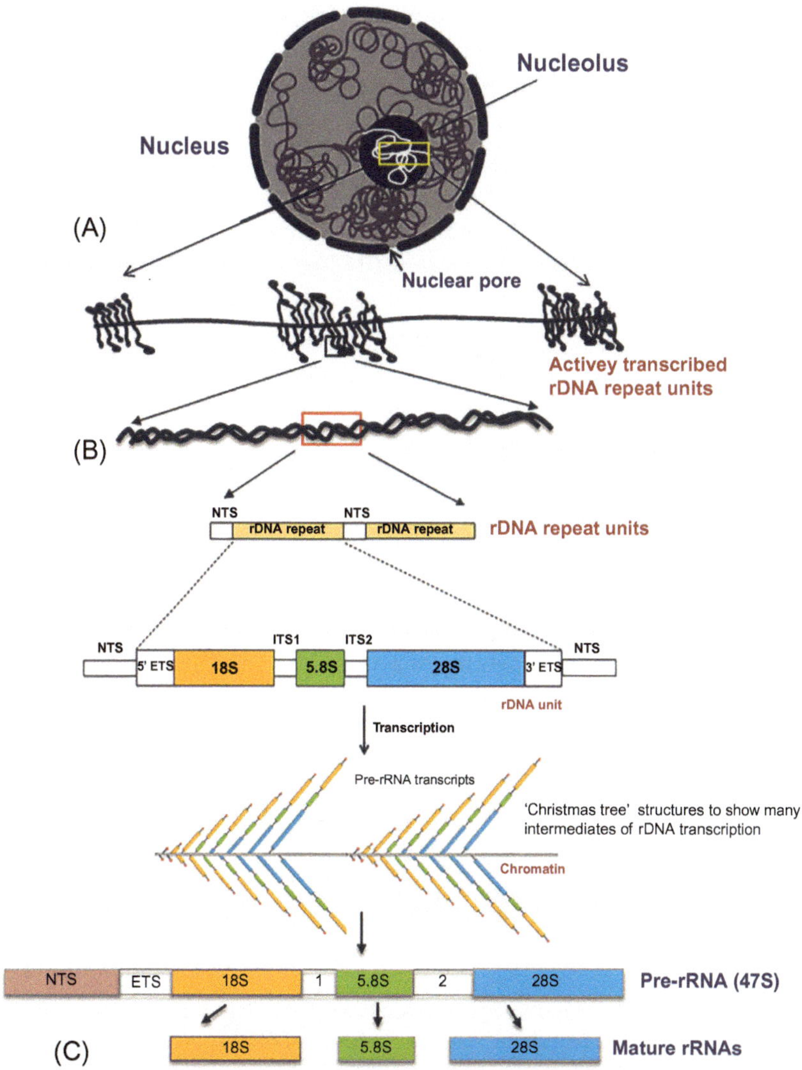

Fig. 3.3 Schematic representation of the ribosomal DNA (rDNA) transcriptional unit. (A) Nucleolar localization of the rDNA transcription units. (B) An actively transcribed segment of rDNA. (C) Expression of 47S pre-rRNA transcripts its processing into three mature rRNAs, viz., 18S, 5.8S, and 28S. NTS, nontranscribed spacer; 1 and 2, internal transcribed spacer 1 and 2; ETS, external transcribed spacer.

The promoter of ribosomal RNA genes spans about 50 base pairs upstream of the transcription start site (+1) and is localized in the intergenic spacer between the transcribed units (Fig. 3.3). The core promoter includes a conserved AT-rich ribosomal initiator (rInr) sequence surrounding +1 site. In addition, there could be an upstream promoter element

(UPE) nearly 150–200bp upstream of the start site. Though UPE is not an absolute requirement for transcription initiation, it significantly stimulates transcription driven by the core promoter (Moss and Stefanovsky, 1995; Reeder, 1989). The rRNA gene promoter can be primed for transcription by a single factor (Keener et al., 1998). The transcription is orchestrated by the synergistic action of RNA Pol I (a 14-subunit complex), the upstream binding factor (UBF) and the promoter selectivity factor SL1—a complex containing the TATA-binding protein (Bell et al., 1988; Paule and White, 2000). SL1 also recruits Pol I-specific initiation factor RNN3 (TIF-IA) to the promoter. Interestingly, yeasts with mutations in the TBP gene are defective not only for transcription by Pol II but also for transcription by Pol I and III. Thus, TBP is a common transcription factor required by all three eukaryotic RNA polymerases (Fig. 3.4).

Pol I-dependent transcription starts with the help of multiple initiation factors which involves phosphorylation of serine residues on UBF by G1 cyclin-dependent kinases (Grummt, 2003; Klein and Grummt, 1999). Other protein kinases including PI3K and mTOR also contribute to enhanced rDNA transcription (James and Zomerdijk, 2004; Russell and Zomerdijk, 2006). Sp1 transcription factor is also an important regulator of Pol I-dependent rDNA transcription. Sp1 engages the chromatin modifiers (Gadd45a-NER

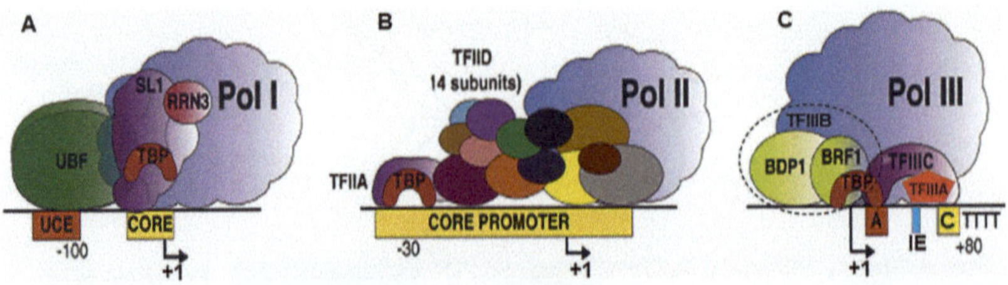

Fig. 3.4 Schematic representations of active RNA pol I-, pol II-, and pol III-transcription initiation complexes. (A) Pol I-dependent transcription of ribosomal RNA genes in higher eukaryotes involves an upstream control element (UCE) and a core promoter element (CORE) near the transcription start site (TSS or +1). Two transcription factors, upstream binding factor (UBF) and selectivity factor 1 (SL1) bind cooperatively to the rDNA promoter and recruit RNA pol I to form an initiation complex. SL1 contain TATA-binding protein (TBP) and four unique TBP-associated factors (TAFs). RRN3/TIF-IA interacts directly with SL1 and pol I. (B) Pol II transcription involves the core promoter region spanning −30 to +20bp relative to TSS (+1). TBP binds upstream of the TSS and is stabilized by TFIIA. The shared 14 subunits between TFIID and Spt-Ada-Gcn5-Acetyltransferase (SAGA) coactivator complex also bind to the core promoter. (C) Pol III-dependent promoter of a 5S rRNA gene that carries two "cis" elements—A and C blocks. The basal transcription factors TFIIIA and TFIIIC binds to these blocks and recruit TFIIIB composed of BRF1, BDP1, and TBP for transcriptional activation. IE, intermediate element.

complex) on the rDNA promoter in order to promote transcriptional activation through DNA demethylation (Nosrati et al., 2014). In addition, Pol I transcription can also be regulated by some oncogenes and tumor suppressor proteins. For example, c-Myc is reported to bind to the nucleolar "E box elements" of rRNA genes as well as associate with SL1/TIF-IB factors to induce rDNA transcription (Arabi et al., 2005; Grandori et al., 2005). On the other hand, tumor suppressors like pRb and p53 are reported to repress Pol I transcription (Quin et al., 2014). However, the involvement of other transcription factors or their "*cis*" elements in Pol I transcription is poorly understood. Nevertheless, it appears that there is a synchronized synthesis of both rRNAs and RPs during the G1 phase of the cell cycle (Nosrati et al., 2014).

Processing and maturation of pre-rRNA

In humans, the production of ribosomes begins with the transcription of a 47S precursor rRNA (pre-rRNA) by RNA Pol I. The processing of pre-rRNA can take place either cotranscriptionally or posttranscriptionally. However, in rapidly growing cells the majority of processing (~70%) is cotranscriptional. The expression of rRNA genes occurs in the G1 phase of the cell cycle, continues through S and G2 phases and stops during mitosis. The 47S precursor rRNA is processed into mature 28S, 18S, and 5.8S rRNAs before their assembly with the 5S rRNA into the mature ribosomes. The initial 47S pre-rRNA transcript maintains a secondary structure at the newly synthesized 5′-end that acts as a platform for the binding and association of an initial set of ribosomal proteins (RPs) forming the 90S ribonucleoprotein particle (RNP). The key role of the RPs is to stabilize rRNA folding and structure and thus, facilitate their processing. Two small subunit ribosomal proteins (RPS), viz., RPS7 and RPS24 first bind to the 5′ portion of the pre-rRNA which is destined to become 18S rRNA of the 40S small ribosomal subunit. These RPSs are required to initiate processing and cleavage of the pre-rRNA at the 5′-external transcribed spacer (ETS) (Fig. 3.3). Low levels of RPS7 or RPS24 result in the inhibition of rRNA processing and failure in the maturation of 5′-end of the 18S rRNA. The initial binding of RPs and processing at the 5′-ETS, combined with modifications of the rRNA by ribose methylases and polyuridylases facilitate the association of the next round of RPs. At the same time, the 3′-end of the 18S rRNA also undergoes processing and cleavage with the help of RPS17 and RPS19 within the internal transcribed sequence (ITS1) which results in the liberation of assembling 40S small subunit from the 60S large subunit. Depletion of RPS17 or RPS19 results in failure of the processing of the 3′-end of 18S rRNA. The maturation of 5.8S and 28S rRNAs with the ITS2 (between 5.8S and 28S sequences) follows a similar strategy. As 5.8S and 28S are destined to become part of the 60S large subunit, both rRNAs undergo several rounds of processing and folding along with the association of RPs. The above process continues until the last of the RPs for each subunit are finally incorporated in the cytoplasm. In addition to the RPs that participate in rRNA maturation or those that become part of the ribosome, a number of other proteins associate with the maturing ribosome in the nucleus to prevent their premature assembly (Fig. 3.5). The processing of 60S large ribosomal subunit is quite similar to the maturation of the 40S subunit. It starts with the incorporation of 5S rRNA with the help of large subunit

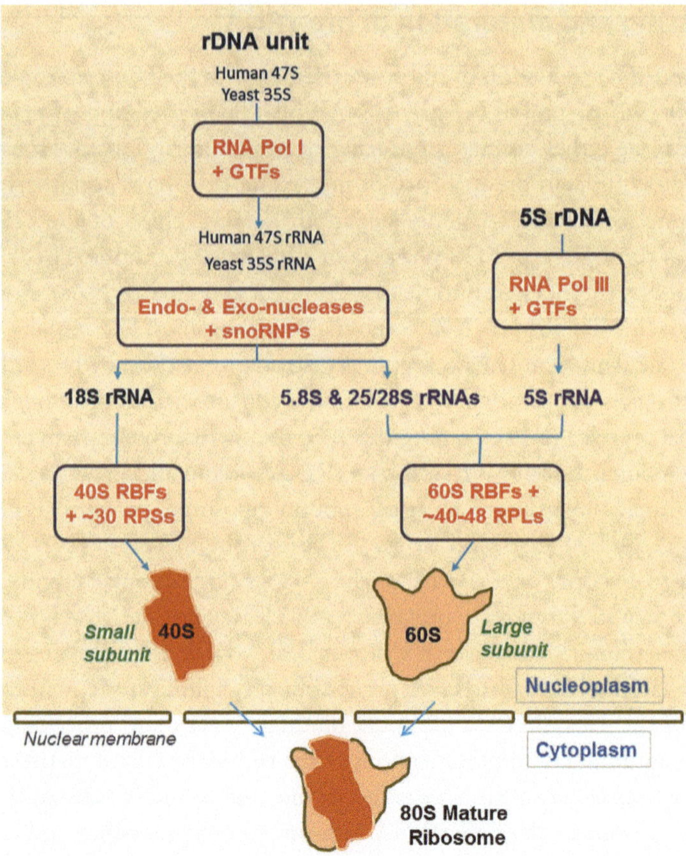

Fig. 3.5 Factors involved in the maturation of rRNAs and assembly of ribosomal subunits. GTFs, general transcription factors; RBFs, ribosome biogenesis factors; RPL, Large subunit ribosomal proteins; RPS, small subunit ribosomal proteins.

ribosomal proteins (RPLs). This step is facilitated by RPL5 and RPL11 resulting in the formation of 5S RNP. Interestingly, low levels of RPL5 or RPL11 may lead to reduced cyclin expression and thus, cell proliferation.

Coincident with transcription, rRNA undergoes multiple RNA-mediated RNA modifications with the help of different classes of factors including small nucleolar RNPs (snoRNPs), exonucleases, endonucleases, base methylases, etc. in order to fine-tune the structure and functions rRNA. For example, guide snoRNPs catalyze either pseudouridylation (by H/ACA snoRNPs) or 2′-O-ribose methylation (by box C/D snoRNPs) of pre-rRNAs. The U3 snoRNP and RNase MRP (mitochondrial RNA processing) help in posttranscriptional cleavage of pre-rRNA (Fig. 3.6). Multiple RNA helicases, including DEAD-box and DEAH-box helicases, also participate in ribosome assembly. In addition, a number of other proteins (chaperons) associate with the maturing ribosome in the nucleus to prevent their premature assembly. Once exported into the cytoplasm, these chaperons disassociate from the ribosomal subunits leading to the formation of mature ribosomal subunits. Finally, the association of mRNA and translation initiation factors (IFs) with ribosomes leads to the formation of the preinitiation complex for protein synthesis.

Organization and regulation of RNA pol III promoter

The 5S rRNA is transcribed by RNA pol III following its recruitment on the 5S rDNA promoter present as multiple gene copies in the genome. Confocal and electron microscopy studies have shown that pol III transcription can start simultaneously at nearly 2000 sites a HeLa cell (Pombo et al., 1999). The most distinguishing feature of the 5S rDNA promoter is that its regulatory sequence elements are present downstream of the start site (within the transcribed region). It is characterized by the presence of three internal elements for efficient transcription—A and C blocks and an intermediate element (Fig. 3.4C). Productive recruitment of TFIIIC to 5S rRNA promoter requires the presence of the transcription factor TFIIIA which is a member of the zinc finger family with nine tandem, zinc-dependent DNA-binding domains (Miller et al., 1985). The binding of TFIIIA to the core promoter is the first step in the assembly of the pol III complex. The C block is recognized by the N-terminal three fingers (fingers 1–3) of TFIIIA whereas the binding of A block is through fingers 7–9 (Clemens et al., 1992). TFIIIA binding induces a minor bend in the DNA which facilitates the interaction of TFIIIC in the pol III complex. The primary function of TFIIIC is to recruit TFIIIB—a

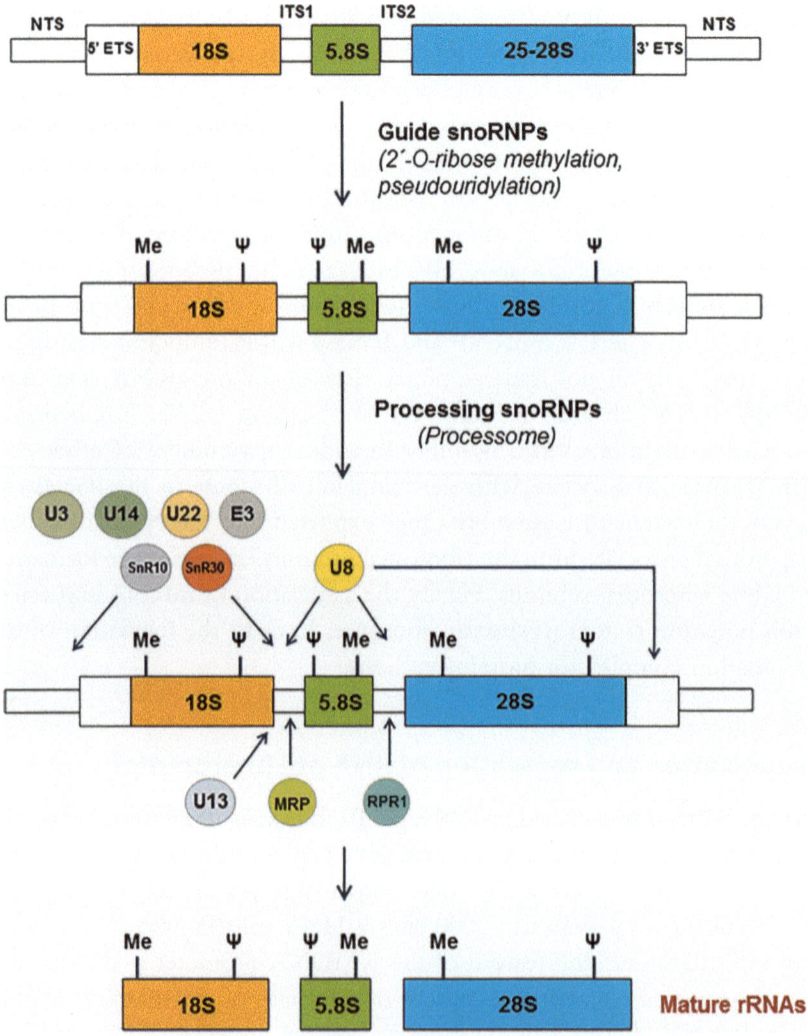

Fig. 3.6 Cotranscriptional processing/modifications of rRNA. The organization of an rDNA unit showing 18S, 5.8S, and 25-28S rRNA genes and their transcribed and non-transcribed segments shown in the top panel. rRNA undergoes multiple RNA-mediated RNA modifications (lower panels) to generate mature rRNAs with the help of different classes of factors including small nucleolar ribonuclear protein particles (snoRNPs), exo-nucleases, endonucleases, base methylases. ETS, external transcribed spacer; ITS, internal transcribed spacers 1 and 2; NTS, nontranscribed spacer.

complex of three polypeptides including BRF1, BDP1, and TBP (Fig. 3.4C). These complexes are sufficient to recruit RNA pol III on the promoter and position them over the transcription start site. The binding of TFIIIB to the promoter near the start site may be important in preventing the displacement of TFIIIC and TFIIIA from the DNA template as pol III transcribes through the internal promoter. Further, TFIIIB induces a major bend in the DNA at the transcriptional start site to facilitate the recruitment of Myc and HAT complex and induce transcription. On the other hand, binding of p53 or the pRB/p130 complex to the promoter can suppress RNA pol III-mediated transcription (Chesnokov et al., 1996; White et al., 1996). RNA pol III is also responsible for the transcription of tRNA genes needed later for translation initiation and elongation.

Summary and conclusions

The synchronized synthesis of rRNAs and ribosomal proteins is essential for ribosome biogenesis. It is not surprising that several physiological and pathological conditions are characterized by deregulated ribosome biogenesis. Ribosome production occurs in a cell cycle-dependent manner and coincides with the preparatory G1 phase that determines the growth and proliferation and thus the fate of the cell. In eukaryotes, it requires the concerted efforts of all three RNA polymerases. The Pol I-dependent transcription is dedicated to rDNA genes that are transcribed with high efficiency to meet the metabolic needs of growing cells. The 5S rRNA is transcribed by RNA pol III from multiple genes in the nucleus. The ribosomal protein genes are also present throughout the genome and are transcribed by RNA Pol II. Transcriptional regulation of these genes requires chromatin remodeling allowing the binding of activator and repressor molecules on the target gene promoters. The transcriptional activities of pol I and pol III exceed the total RNA synthesis in growing or proliferating cells. Although the transcription factors engaged by the two polymerases are different, functionally both are similar. Further, it is unclear if Pol II-associated transcription factors are shared during transcription by Pol I and Pol III. This assumes significance in light of the fact that the synchronized transcriptional machinery of all three RNA polymerases is essential for ribosome biogenesis.

References

Arabi A, Wu S, Ridderstråle K, Bierhoff H, Shiue C, Fatyol K, Fahlén S, Hydbring P, Söderberg O, Grummt I, Larsson LG, Wright AP. c-Myc associates with ribosomal DNA and activates RNA polymerase I transcription. Nat Cell Biol 2005;7:303–10.

Bell SP, Learned RM, Jantzen HM, Tjian R. Functional cooperativity between transcription factors UBF1 and SL1 mediates human ribosomal RNA synthesis. Science 1988;241:1192–7.

Chesnokov I, Chu WM, Botchan MR, Schmid CW. p53 inhibits RNA polymerase III-directed transcription in a promoter-dependent manner. Mol Cell Biol 1996;16:7084–8.

Clemens KR, Liao X, Wolf V. Definition of the binding sites of individual zinc fingers in the transcription factor IIIA-5S RNA gene complex. Proc Natl Acad Sci U S A 1992;89:10822–6.

Drygin D, Rice WG, Grummt I. The RNA polymerase I transcription machinery: an emerging target for the treatment of cancer. Annu Rev Pharmacol Toxicol 2010;50:131–56.

Grandori C, Gomez-Roman N, Felton-Edkins ZA, Ngouenet C, Galloway DA, Eisenman RN, White RJ. c-Myc binds to human ribosomal DNA and stimulates transcription of rRNA genes by RNA polymerase I. Nat Cell Biol 2005;7:311–8.

Grummt I. Life on a planet of its own: regulation of RNA polymerase I transcription in the nucleolus. Genes Dev 2003;17:1691–702.

Grummt I, Längst G. Epigenetic control of RNA polymerase I transcription in mammalian cells. Biochim Biophys Acta 2013;1829:393–404.

Hannan KM, Sanij E, Rothblum LI, Hannan RD, Pearson RB. Dysregulation of RNA polymerase I transcription during disease. Biochim Biophys Acta 2013;1829:342–60.

Ide S, Miyazaki T, Maki H, Kobayashi T. Abundance of ribosomal RNA gene copies maintains genome integrity. Science 2010;327:693–6.

James MJ, Zomerdijk JC. Phosphatidylinositol 3-kinase and mTOR signaling pathways regulate RNA polymerase I transcription in response to IGF-1 and nutrients. J Biol Chem 2004;279:8911–8.

Keener J, Josaitis CA, Dodd JA, Nomura M. Reconstitution of yeast RNA polymerase I transcription in vitro from purified components. TATA-binding protein is not required for basal transcription. J Biol Chem 1998;273:33795–802.

Klein J, Grummt I. Cell cycle-dependent regulation of RNA polymerase I transcription: the nucleolar transcription factor UBF is inactive in mitosis and early G1. Proc Natl Acad Sci U S A 1999;96:6096–7101.

Kobayashi T. Ribosomal RNA gene repeats, their stability and cellular senescence. Proc Jpn Acad Ser B Phys Biol Sci 2014;90:119–29.

Kobayashi T, Heck DJ, Nomura M, Horiuchi T. Expansion and contraction of ribosomal DNA repeats in *Saccharomyces cerevisiae*: requirement of replication fork blocking (Fob1) protein and the role of RNA polymerase I. Genes Dev 1998;12:3821–30.

Kusnadi EP, Hannan KM, Hicks RJ, Hannan RD, Pearson RB, Kang J. Regulation of rDNA transcription in response to growth factors, nutrients and energy. Gene 2015;556:27–34.

Lawrence RJ, Earley K, Pontes O, Silva M, Chen ZJ, Neves N, Viegas W, Pikaard CS. A concerted DNA methylation/histone methylation switch regulates rRNA gene dosage control and nucleolar dominance. Mol Cell 2004;13:599–609.

Meissner B, Hofmann A, Steinbeisser H, Spring H, Miller Jr OL, Trendelenburg MF. Faithful in vivo transcription termination of *Xenopus laevis* rDNA. Correlation of electron microscopic spread preparations with S1 transcript analysis. Chromosoma 1991;101:222–30.

Miller J, McLachlan AD, Klug A. Repetitive zinc-binding domains in the protein transcription factor IIIA from *Xenopus oocytes*. EMBO J 1985;4:1609–14.

Miller Jr and Beatty, 1969Miller Jr OL, Beatty BR. Visualization of nucleolar genes. Science 1969;164:955–7.

Moss T, Stefanovsky VY. Promotion and regulation of ribosomal transcription in eukaryotes by RNA polymerase I. Prog Nucleic Acid Res Mol Biol 1995;50:25–66.

Nosrati N, Kapoor NR, Kumar V. Combinatorial action of transcription factors orchestrates cell cycle-dependent expression of the ribosomal protein genes and ribosome biogenesis. FEBS J 2014;281:2339–52.

Paule MR, White RJ. Survey and summary: transcription by RNA polymerases I and III. Nucleic Acids Res 2000;28:1283–98.

Pombo A, Jackson DA, Hollinshead M, Wang Z, Roeder RG, Cook PR. Regional specialization in human nuclei: visualization of discrete sites of transcription by RNA polymerase III. EMBO J 1999;18:2241–53.

Quin JE, Devlin JR, Cameron D, Hannan KM, Pearson RB, Hannan RD. Targeting the nucleolus for cancer intervention. Biochim Biophys Acta 2014;1842:802–16.
Reeder RH. Regulatory elements of the generic ribosomal gene. Curr Opin Cell Biol 1989;1:466–74.
Russell J, Zomerdijk JC. The RNA polymerase I transcription machinery. Biochem Soc Symp 2006;7:203–16.
Sakai K, Ohta T, Minoshima S, Kudoh J, Wang Y, de Jong PJ, Shimizu N. Human ribosomal RNA gene cluster: identification of the proximal end containing a novel tandem repeat sequence. Genomics 1995;26:521–6.
Santoro R, Grummt I. Epigenetic mechanism of rRNA gene silencing: temporal order of NoRC-mediated histone modification, chromatin remodeling, and DNA methylation. Mol Cell Biol 2005;25:2539–46.
Stefanovsky V, Moss T. Regulation of rRNA synthesis in human and mouse cells is not determined by changes in active gene count. Cell Cycle 2006;5:735–9.
Takeuchi Y, Horiuchi T, Kobayashi T. Transcription-dependent recombination and the role of fork collision in yeast rDNA. Genes Dev 2003;17:1497–506.
White RJ, Trouche D, Martin K, Jackson SP, Kouzarides T. Repression of RNA polymerase III transcription by the retinoblastoma protein. Nature 1996;382:88–90.

CHAPTER 4

Ribosome proteins—Their balanced production

Jitendra Kumar and Vijay Kumar
Department of Molecular and Cellular Medicine, Institute of Liver and Biliary Sciences, New Delhi, India

Introduction

The eukaryotes carry 138 ribosomal protein genes (RPGs) which encode ~88 distinct ribosomal proteins (RPs). The RPGs exist in the form of tandem repeats spread throughout the genome of an organism. These genes also encode nearly 80 different proteins required for the biogenesis of mitochondrial ribosome (De Silva et al., 2015). The molecular mass of eukaryotic (80S), prokaryotic (70S), and mitochondrial (55S) ribosomes are 4.3, 3.0, and 2.3 MDa (Mega Dalton), respectively (Patel et al., 2001; Yusupova and Yusupov, 2014). Approximately 19 RPGs exist as single copies whereas more than 59 genes are in paralogous gene pairs. These paralogous gene pairs evolved 100 million years ago during evolution (Wolfe and Shields, 1997). During the biogenesis of ribosomes, RPGs transcription is well coordinated with the transcription of rRNAs (Reja et al., 2015). The structural analysis of the rRNA transcription complex revealed that all RPGs are driven by the presence of a common set of regulatory factors (Perina et al., 2011). The balanced production of RPs and rRNAs is one of the most significant molecular processes of the cell (Perry, 2007; Gupta and Warner, 2014; Albert et al., 2016). It depends on the promoter strength of RPGs, the copy number of genes, nonredundancy in paralogs, and protein levels (Chain et al., 2014; Reja et al., 2015). Interestingly, RPGs are haploinsufficient in the eukaryotes (Gilbert, 2011). Any alteration in the ratio of RPs and rRNAs or heterozygosity of RPGs can cause serious genetic disorders, diseases, developmental defects, and even in utero death of embryos. Ribosomopathies, a well-known disorder related to the ribosome, is mainly caused by a mutation in RPGs (Marygold et al., 2007). In yeast, the majority (~73%) of ribosomal genes carry introns. These intronic regions are really huge in comparison to nonribosomal genes (~5%) (Juneau et al., 2006; Pleiss et al., 2007; Parenteau et al., 2008; Teng et al., 2013). Moreover, a small

Emerging Concepts in Ribosome Structure, Biogenesis, and Function
https://doi.org/10.1016/B978-0-12-816364-1.00003-2

amount of alternative splicing may occur in the processing of preribosomal proteins mRNAs whereas 90% of RPGs are spliced. Approximately 34% of the total ribosomes of the cell are engaged in the translation of RP transcripts (Parenteau et al., 2008).

Some pioneering studies in yeast, drosophila, and mammalian cells have shown that transcription of ribosome components is regulated in a coordinated manner. Under favorable growth conditions, yeast cells activate a kinase called TORC1 (target of rapamycin 1) which drives the transcription of RPGs and rRNA by recruiting RNA polymerase II (RNAPII) and RNA polymerase I (RNAPI) on respective promoters along with accessory factors. In growing cells, RNAPI accounts for nearly 60% of the total RNA polymerase activity for the initiation of RPGs transcription (Albert et al., 2016). Studies on the promoters of RPGs have revealed that their promoter regions carry upstream activator sequences (UAS) for the activation of genes, and two Rap1/single Abf1 binding sites followed by one or maybe two T-rich elements for the regulation of transcription (Fig. 4.1). Rap1 exhibits dual behavior by activating or repressing transcription of RP genes which is responsible for the rapid and global change in RP levels (Fermi et al., 2016).

The transcriptomics data suggest that there a fivefold difference between the transcripts of most demanded and least demanded proteins. Not surprisingly, the equimolar production of proteins is balanced through the half-life of its mRNAs as well as translation and protein stability (Perry, 2007). Therefore, if unwanted proteins are translated then it is degraded rapidly owing to their highly basic nature and high affinity for DNA (Lott et al., 2013; Fedyukina et al., 2014). A feedback inhibition mechanism seems to regulate the balanced production of RPs. When RPs are synthesized in excess, they are unable to assemble in ribosomes and instead bind to 5′ splice

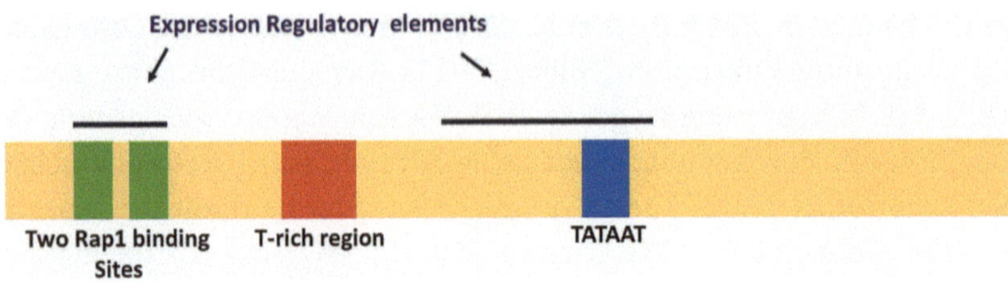

Fig. 4.1 Overview of ribosomal protein genes promoter region and transcription binding sites—RPGs have two Rap1 (repressor activator protein 1) binding site (in green) and its downstream these have one T (Thymidine)—rich region (in *red*) and so on TATAAT sites (in *blue*).

site of the nascent transcripts and thus, prevent the splicing and translation of these transcripts (Warner and McIntosh, 2009; Zhou et al., 2015; Petibon et al., 2016). When in excess, RP prevents the recruitment of splicing factors on the nascent transcripts and thus, blocks the maturity of spliceosomes leading to the degradation of self mRNAs (De and Brogna, 2010; Wang et al., 2015; Dvinge et al., 2016).

Structure of ribosomes

The eukaryotic ribosome is large than its bacterial counterparts, its both subunits contain more amount of proteins and rRNAs in comparison to bacterial ribosomes (Wilson and Doudna Cate, 2012). Cryo-electron microscopy (cryo-EM) studies of wheat germs and yeast 80S ribosomes at 4.15Å resolution reveal that some protein is homologs in a structure in comparison to bacteria (Armache et al., 2010b). High-resolution X-rays and Cryo-EM data have revealed the locations of RPs in ribosomes and thus, are useful in establishing the structure and function of ribosomes as well as RPs (Chandramouli et al., 2008; Armache et al., 2010a; Khusainov et al., 2016). These studies have revealed the arrangements of RPs and rRNAs in ribosomes. The size and complexity of ribosomes have increased during evolution. Nevertheless, the core of rRNAs is wrapped by the globular RPs that are conserved through evolution with a few specific changes in RPs and rRNAs in 80S ribosomes (Taylor et al., 2009).

Ribosomal proteins

The ribosome is a complex cellular machine of ribonucleoproteins that are necessary for the translation of mRNAs transcript to proteins. The 80S eukaryotic ribosome is made of two different subunits—large 60S and small 40S subunits whereas the 70S ribosome in bacteria is composed of large 50S and small 30S ribosomal subunits. The large 60S subunit of ribosomes is made up of three different rRNAs (28S, 5S, and 5.8S) with ~53 RPs, whereas the small 40S subunits are composed of single rRNA (18S) with ~34 RPs and RACK1 proteins (receptor for activated C kinase 1). Comparative analysis of the primary structure of RPs has revealed that there are 31 families of RPs present in the living organisms that have ribosomes suggesting that RPs are well conserved during evolution. Interestingly, 15 of 35 RPs of the small ribosomal subunit are conserved in both prokaryotes and eukaryotes with just 7 RPs being unique to prokaryotes, 17 RPs being specific to archaea and eukaryotes (Taylor et al., 2009). In the large subunit, only 18 of 53 proteins are shared to both prokaryotes and eukaryotes whereas 27 are exclusive to eukaryotes and archaea, and 17 proteins are unique to prokaryotes (Tables 4.1–4.3) (Prinz et al., 2000; Fox, 2010; Ghosh and Komar, 2015) till date. Information about RPs is incomplete without their fine structure and organization in the ribosomal subunits. The 70S ribosomes of bacteria have nearly 70 RPs whose length may vary from 50 to 150 amino acids whereas the 80S ribosome of eukaryotes contains nearly 88 RPs, which could be 35–421 amino acids in length and rich in arginine and lysine residues (Karlin et al., 2005). As a reason, RPs are highly positively charged (pI value = 11.05) that is required for their association with negatively charged rRNA with minimum electrostatic repulsion at physiological pH (Su et al., 2009; Fedyukina et al., 2014). One common feature of all RPs is the presence of clusters of basic amino acids in the form of 2–8 short amino acid sequence repeats, i.e., 15 dipeptides "RG" repeats are present in uS5 protein (Korobeinikova et al., 2012).

Nomenclature of RPs

Earlier, scientific groups working on different ribosomal proteins used names as per their convenience. In 1971, a group of American and European scientists proposed a standard naming system for ribosomal proteins from *Escherichia coli* based on experimentally identified and amino acid sequences (Wittmann et al., 1971). According to this naming system, the RPs of small (S) subunit and large (L) subunit of ribosomes have names starting with SX or LY, respectively, where X and Y are the integer number (such as S1, S2 or L1, L3). By the mid-1980s, this naming system becomes unsuitable after the atomic resolution crystal structures of ribosomal subunits and complete ribosomes were reported from different eubacteria species. Several homologs of RPs were identified by amino acid sequences that carried the same three-dimensional structures and binding partners at the same location. Such proteins were called functionally equivalent RPs. In 2016, the ribosome community published a new scheme for the nomenclature of ribosomal proteins in all domains of life without any conflict. This system was also extended for the nomenclature of mitochondrial and chloroplast ribosomal proteins. According to this scheme, as RPs are universally present in all domains of life, their name will have "u" prefix and bacterial numbering, e.g., uS4. If RPs are derived from just one domain of life then the nomenclature will have first letter prefix, for example, b, s, m, c, and e for RPs

Table 4.1 List of 40S subunit ribosomal proteins.

Sl. no.	Old name	New name[a]	Taxonomical range[b]	Full name	Size MW (Da)	No. of residues	Location (chromosome)	References
1	RPS2	uS5	B, A, E	Ribosomal protein S2	31,211	293	16p13.3	Suzuki et al. (1991a)
2	RPS3	uS3	B, A, E	Ribosomal protein S3	26,643	243	11q13.4	Chan et al. (1990a)
3	RPS3A	eS1	A, E	Ribosomal protein S3A	29,794	263	4q31.3	Chan et al. (1996)
4	RPS4X	–	A, E	Ribosomal protein S4 X-linked	29,596	263	Xq13.1	Watanabe et al. (1991)
5	RPS4Y1	eS4	A, E	Ribosomal protein S4 Y-linked 1	29,456	263	Yp11.2	Devi et al. (1989a)
6	RPS4Y2	eS41	E	Ribosomal protein S4 Y-linked 2	29,295	263	Yq11.223	Andres et al. (2008)
7	RPS5	uS7	B, A, E	Ribosomal protein S5	22,863	204	19q13.4	Kuwano et al. (1992)
8	RPS6	eS6	A, E	Ribosomal protein S6	28,683	249	9p22.1	Chan and Wool (1988)
9	RPS7	eS7	E	Ribosomal protein S7	22,113	194	2p25.3	Suzuki et al. (1990a)
10	RPS8	eS8	A, E	Ribosomal protein S8	23,928	207	1p34.1	Chan et al. (1987b)
11	RPS9	uS4	B, A, E	Ribosomal protein S9	22,360	193	19q13.42	Chan et al. (1993a)
12	RPS10	eS10	E	Ribosomal protein S10	18,917	165	6p21.31	Gluck et al. (1989)
13	RPS11	uS17	B, A, E	Ribosomal protein S11	18,247	157	19q13.3	Tanaka et al. (1985)
14	RPS12	eS12	E	Ribosomal protein S12	14,515	132	6q23.2	Lin et al. (1987)
15	RPS13	uS15	B, A, E	Ribosomal protein S13	17,080	150	11p15.1	Suzuki et al. (1990b)
16	RPS14	uS11	B, A, E	Ribosomal protein S14	16,248	151	5q33.1	Paz et al. (1989a)
17	RPS15	uS19	B, A, E	Ribosomal protein S15	17,040	145	19p13.3	Kitagawa et al. (1991)
18	RPS15A	uS8	B, A, E	Ribosomal protein s15a	14,698	129	16p12.3	Chan et al. (1994b)
19	RPS16	uS9	B, A, E	Ribosomal protein S16	16,304	145	19q13.1	Chan et al. (1990b)
20	RPS17	eS17	A, E	Ribosomal protein S17	15,368	134	15q25.2	Nakanishi et al. (1985)

Continued

Table 4.1 List of 40S subunit ribosomal proteins—cont'd

Sl. no.	Old name	New name[a]	Taxonomical range[b]	Full name	Size MW (Da)	No. of residues	Location (chromosome)	References
21	RPS18	uS13	B, A, E	Ribosomal protein S18	17,707	152	6p21.32	Chan et al. (1991a)
22	RPS19	eS19	A, E	Ribosomal protein S19	15,944	144	19q13.2	Suzuki et al. (1990c)
23	RPS20	uS10	B, A, E	Ribosomal protein S20	13,364	119	8q12.1	Chan and Wool (1990c)
24	RPS21	eS21	E	Ribosomal protein S21	91,21	83	20q13.33	Itoh et al. (1985)
25	RPS23	uS12	B, A, E	Ribosomal protein S23	15,666	142	5q14.2	Kitaoka et al. (1994)
26	RPS24	eS24	A, E	Ribosomal protein S24	15,413	133	10q22.3	Chan et al. (1990d)
27	RPS25	eS25	A, E	Ribosomal protein S25	13,733	125	11q23.3	Chan and Wool (1992a)
28	RPS26	eS26	E	Ribosomal protein S26	13,007	115	12q13.2	Kuwano et al. (1985)
29	RPS27	eS27	A, E	Ribosomal protein S27	9339	83	1q21.3	Chan et al. (1993e)
30	RPS27A	eS31	A, E	Ribosomal protein s27a	9397	80	2p16.1	Monia et al. (1989)
31	RPS27L	–	E	Ribosomal protein S27 like	9477		15q22.2	Xiong et al. (2011)
32	RPS28	eS28	A, E	Ribosomal protein S28	7836	69	19p13.2	Chan et al. (1991b)
33	RPS29	uS14	B, A, E	Ribosomal protein S29	6541	55	14q21.3	Chan et al. (1993e)
34	FAU	eS30	A, E	FAU, ubiquitin like and ribosomal protein S30 fusion	6643	59	–	Kas et al. (1992)
35	RACK1	RACK	E	Receptor of activated protein C kinase 1	35,077	317	4q34–35	Sengupta et al. (2004)

[a] *b*, bacterial; *e*, eukaryotic; *u*, universal.
[b] *B*, bacteria; *A*, archaea; *E*, eukaryotes.

Table 4.2 Ribosomal proteins of 60S subunit.

Sl.-no.	Old name	New name[a]	Taxonomical range[b]	Full name	Size MW (Da)	No. of residues	Location (chromosome)	References
1	RPLP0	uL10	B, A, E	Ribosomal protein lateral stalk subunit P0	34,178	316	12q24.23	Wool et al. (1991)
2	RPLP1	uL1	B, A, E	Ribosomal protein lateral stalk subunit P1	14,900	114	15q23	Wool et al. (1991)
3	RPLP2	uL2	B, A, E	Ribosomal protein lateral stalk subunit P2	11,684	115	11p15.5	Wool et al. (1991)
4	RPL3	uL3	B, A, E	Ribosomal protein L3	46,106	403	22q13.1	Kuwano and Wool (1992)
5	RPL3L	uL31	B, A, E	Ribosomal protein L3 like	46,296	407	16p13.3	Ou et al. (1987)
6	RPL4	uL4	B, A, E	Ribosomal protein L4	47,280	421	15q22.31	Bagni et al. (1993)
7	RPL5	uL18	B, A, E	Ribosomal protein L5	34,298	296	1p22.1	Tamura et al. (1987)
8	RPL6	eL6	E	Ribosomal protein L6	32,728	288	12q24.13	Shen et al. (1998)
9	RPL7	uL30	B, A, E	Ribosomal protein L7	30,310	260	8q21.11	Meyuhas and Klein (1990)
10	RPL7A	eL8	E	Ribosomal protein L7a	29,863	265	9q34.2	Nakamura et al. (1989)

Continued

Table 4.2 Ribosomal proteins of 60S subunit—cont'd

Sl.-no.	Old name	New name[a]	Taxonomical range[b]	Full name	Size MW (Da)	No. of residues	Location (chromosome)	References
11	RPL7L1	uL301	E	Ribosomal protein L7 like 1	28,661	246	6p21.1	Hemmerich et al. (1993)
12	RPL8	uL2	B, A, E	Ribosomal protein L8	28,007	257	8q24.3	Chan and Wool (1992b)
13	RPL9	uL6	B, A, E	Ribosomal protein L9	21,879	192	4p14	Suzuki et al. (1990e)
14	RPL10	uL16	B, A, E	Ribosomal protein L10	24,604	214	Xq28	Zanni et al. (2015)
15	RPL10A	–	E	Ribosomal protein L10a	24,831	217	6p21.31	Odintsova et al. (2003)
16	RPL10L	–	E	Ribosomal protein L10 like	24,519	214	14q21.2	Balasubramanian et al. (2009)
17	RPL11	uL5	B, A, E	Ribosomal protein L11	20,239	178	1p36.11	Chan et al. (1992)
18	RPL12	uL11	B, A, E	Ribosomal protein L12	17,834	165	9q33.3	Suzuki et al. (1990f)
19	RPL13	eL13	A, E	Ribosomal protein L13	24,261	211	6q24.3; 17p11.2	Olvera and Wool (1994)
20	RPL13A	uL13	B, A, E	Ribosomal protein L13a	23,330	202	19q13.33	Chan et al. (1994a)
21	RPL14	eL14	A, E	Ribosomal protein L14	23,432	215	3p22.1	Tanaka et al. (1998)
22	RPL15	eL15	A, E	Ribosomal protein L15	24,000	203	3p24.2	Kenmochi et al. (1998)

23	RPL17	uL22	B, A, E	Ribosomal protein L17	21,383	184	18q21.1	Suzuki and Wool (1991)
24	RPL18	eL18	A, E	Ribosomal protein L18	21,634	188	19q13.33	Devi et al. (1988)
25	RPL18A	eL20	E	Ribosomal protein L18a	20,716	176	19p13.11	Aoyama et al. (1989a)
26	RPL19	eL19	A, E	Ribosomal protein L19	26,971	196	17q12	Chan et al. (1987a)
27	RPL21	eL21	A, E	Ribosomal protein L21	18,322	159	13q12.2	Devi et al. (1989b)
28	RPL22	eL22	E	Ribosomal protein L22	14,779	128	1p36.31	Wool et al. (1995) and Chan and Wool (1995)
29	RPL23	uL14	B, A, E	Ribosomal protein L23	14,586	140	17q12	Chan et al. (1991c)
30	RPL23A	uL23	B, A, E	Ribosomal protein L23a	17,684	156	17q11.2	Suzuki and Wool (1993)
31	RPL24	eL24	A, E	Ribosomal protein L24	17,767	157	3q12.3	Chan et al. (1994c)
32	RPL26	uL24	B, A, E	Ribosomal protein L26	17,266	145	17p13.1	Paz et al. (1989b)
33	RPL26L1	uL241	B, A, E	Ribosomal protein L26 like 1	17,256	145	5q35.1	Guo et al. (2020)
34	RPL27	eL27	E	Ribosomal protein L27	15,666	135	17q21	Tanaka et al. (1988)

Continued

Table 4.2 Ribosomal proteins of 60S subunit—cont'd

Sl.-no.	Old name	New name[a]	Taxonomical range[b]	Full name	Size MW (Da)	No. of residues	Location (chromosome)	References
35	RPL27A	uL15	B, A, E	Ribosomal protein l27a	16,476	147	11p15	Wool et al. (1990)
36	RPL28	eL28	E	Ribosomal protein L28	15,707	136	19q13.4	Wool et al. (1990)
37	RPL29	eL29	E	Ribosomal protein L29	17,183	155	3p21.2	Chan et al. (1993b) and DeLabre et al. (2002)
38	RPL30	eL30	A, E	Ribosomal protein L30	12,652	114	8q22.2	Nakanishi et al. (1985)
39	RPL31	eL31	A, E	Ribosomal protein L31	14,331	124	2q11.2	Tanaka et al. (1987)
40	RPL32	eL32	A, E	Ribosomal protein L32	15,730	35	3p25.2	Rajchel et al. (1988)
41	RPL34	eL34	A, E	Ribosomal protein L34	13,498	135	4q25	Aoyama et al. (1989b)
42	RPL35	uL29	B, A, E	Ribosomal protein L35	14,412	122	9q33.3	Suzuki et al. (1990d)
43	RPL35A	eL33	A, E	Ribosomal protein l35a	12,422	109	3q29	Tanaka et al. (1986)
44	RPL36	eL36	E	Ribosomal protein L36	12,128	104	19p13.2	Chan et al. (1993c)
45	RPL36A	eL46	E	Ribosomal protein L36a	12,441	106	Xq22.1	Gallagher et al. (1988)

46	RPL36AL	–	E	Ribosomal protein L36a like	12,469	106	14q21.3	Uechi et al. (2002)
47	RPL37	eL37	A, E	Ribosomal protein L37	10,939	96	5p13.1	Chan et al. (1993d)
48	RPL37A	eL43	A, E	Ribosomal protein L37a	10,143	91	2q35	Tanaka et al. (1989)
49	RPL38	eL38	A, E	Ribosomal protein L38	8081	69	17q25.1	Kuwano et al. (1991)
50	RPL39	eL39	A, E	Ribosomal protein L39	6271	50	Xq24	Otsuka et al. (1996)
51	RPL39L	eL391	E	Ribosomal protein L39 like	14,728	128	3q27.3	Wong et al. (2014)
52	UBA52	–	E	Ubiquitin A-52 residue ribosomal protein fusion product 1	6177	52	19p13.1-p12	Kitoh et al. (2007)
53	RPL41	eL41	A, E	Ribosomal protein L41	3454	25	12q13	Stoffler et al. (1974)

[a] *B*, bacterial; *E*, eukaryotic; *u*, universal.
[b] *B*, bacteria; *A*, archaea; *E*, eukaryotes.

Table 4.3 List of 55S mitochondrial ribosomal proteins.

Sl. no.	New[a] abbrv.	Old abbrv.	Full name's	MW (kDa)	No. of residues	Location (chromosomes)	References
Mitochondrial ribosomal proteins(MRPs)of the 39S							
1	uL1	MRPL1	Mitochondrial ribosomal protein L1	32	325	4q21.1	Kenmochi et al. (2001)
2	uL2	MRPL2	Mitochondrial ribosomal protein L2	28.2	305	6p21.1	Fearon and Mason (1988), Kitakawa et al. (1997), and Otaka and Kobata (1978)
3	uL3	MRPL3	Mitochondrial ribosomal protein L3	34.4	348	3q22.1	Graack et al. (1988) and Kitakawa et al. (1997)
4	uL4	MRPL4	Mitochondrial ribosomal protein L4	33	311	19p13.2	Graack et al. (1988, 1995)
5	bL12	MRP7	Mitochondrial ribosomal protein L7	34–194			
6	bL9	MRPL9	Mitochondrial ribosomal protein L9	24.8	367	1q21.3	Graack et al. (1988, 1992)
7	uL10m	MRPL10	Mitochondrial ribosomal protein L10	26.4	261	17q21.32	Grohmann et al. (1991) and Kitakawa et al. (1997)
8	uL11m	MRPL11	Mitochondrial ribosomal protein L11	16.3	174	11q13.2	Grohmann et al. (1991) and Kitakawa et al. (1997)
9	bL12m	MRPL12	Mitochondrial ribosomal protein L12	16.5	198	17q25.3	Graack et al. (1999)
10	uL13m	MRPL13	Mitochondrial ribosomal protein L13	19	178	8q24.12	Grohmann et al. (1991, 1994)
11	uL14m	MRPL14	Mitochondrial ribosomal protein L14	12.6	145	6p21.1	Grohmann et al. (1991) and Kitakawa et al. (1997)

12	uL15m	MRPL15	Mitochondrial ribosomal protein L15	31.3	296	8q11.23	Grohmann et al. (1991) and Kitakawa et al. (1997)
13	uL16m	MRPL16	Mitochondrial ribosomal protein L16	25.3	251	11q12.1	Harrer et al. (1993) and Kitakawa et al. (1997)
14	uL17m	MRPL17	Mitochondrial ribosomal protein L17	19.4	175	11p15.5-p15.4	Graack et al. (1999)
15	uL18m	MRPL18	Mitochondrial ribosomal protein L18	18	180	6q25.3	Graack et al. (1999)
16	bL19m	MRPL19	Mitochondrial ribosomal protein L19	29	292	2p12	Grohmann et al. (1991) and Kitakawa et al. (1997)
17	bL20m	MRPL20	Mitochondrial ribosomal protein L20	16.4	149	1p36.33	Garcia-Cantalejo et al. (1994) and Kitakawa et al. (1990)
18	bL21m	MRPL21	Mitochondrial ribosomal protein L21	18.9	205	11q13.3	Graack et al. (1999)
19	uL22m	MRPL22	Mitochondrial ribosomal protein L22	19.4	206	5q33.2	Goldschmidt-Reisin et al. (1998)
20	uL23m	MRPL23	Mitochondrial ribosomal protein L23	14.1	153	11p15.5	Tsang et al. (1995)
21	uL24m	MRPL24	Mitochondrial ribosomal protein L24	23.9	216	1q23.1	Goldschmidt-Reisin et al. (1998)
22	bL27	MRPL27	Mitochondrial ribosomal protein L27	13	137	17q21.33	Graack et al. (1991), Grohmann et al. (1991), and Holmstrom et al. (1994)
23	bL28	MRPL28	Mitochondrial ribosomal protein L28	27	256	16p13.3	Goldschmidt-Reisin et al. (1998)
24	uL30m	MRPL30	Mitochondrial ribosomal protein L30	14.7	161	2q11.2	O'Brien et al. (2000)

Continued

Table 4.3 List of 55S mitochondrial ribosomal proteins—cont'd

Sl. no.	New[a] abbrv.	Old abbrv.	Full name's	MW (kDa)	No. of residues	Location (chromosomes)	References
25	bL32m	MRPL32	Mitochondrial ribosomal protein L32	12.8	188	7p14.1	Goldschmidt-Reisin et al. (1998)
26	bL33m	MRPL33	Mitochondrial ribosomal protein L33	6.7	65	2p23.2	Goldschmidt-Reisin et al. (1998)
27	bL34m	MRPL34	Mitochondrial ribosomal protein L34	5.4	92	19p13.1	Graack et al. (1999)
28	bL35	MRPL35	Mitochondrial ribosomal protein L35	21.5	188	2p11.2	Grohmann et al. (1991) and Kitakawa et al. (1997)
30	bL36m	MRPL36	Mitochondrial ribosomal protein L36	4.9	103	5p15.33	Goldschmidt-Reisin et al. (1998)
32	mL38	MRPL38	Mitochondrial ribosomal protein L38	41.2	380	17q25.1	Tsang et al. (1995)
33	uL40	MRPL40	Mitochondrial ribosomal protein L40	19.3	206	22q11.21	Grohmann et al. (1991) and Kitakawa et al. (1997)
35	mL41	MRPL41	Mitochondrial ribosomal protein L41	14	137	9q34.3	Goldschmidt-Reisin et al. (1998)
36	mL43	MRPL43	Mitochondrial ribosomal protein L43	20.7	215	10q24.31	Goldschmidt-Reisin et al. (1998)
37	mL44	MRPL44	Mitochondrial ribosomal protein L44	34.4	332	2q36.1	Grohmann et al. (1991) and Matsushita et al. (1989)
40	mL46	MRPL46	Mitochondrial ribosomal protein L46	33.7	306	17q12	Kitakawa et al. (1997)

41	mL48	MRPL48	Mitochondrial ribosomal protein L48	20.6	212	11q13.4	Koc et al. (2001a)
42	mL49	MRPL49	Mitochondrial ribosomal protein L49	16.4	166	11q13.1	Vandenbol et al. (1994)
43	mL50	MRPL50	Mitochondrial ribosomal protein L50	17.8	158	9q31.1	Goldschmidt-Reisin et al. (1998)
43	mL43	MRPL51	Mitochondrial ribosomal protein L51	11.7	128	12p13.31	Goldschmidt-Reisin et al. (1998)
44	mL52	MRPL52	Mitochondrial ribosomal protein L52	11.3	123	14q11.2	Goldschmidt-Reisin et al. (1998)
45	mL53	MRPL53	Mitochondrial ribosomal protein L53	12.1	112	2p13.1	Goldschmidt-Reisin et al. (1998)
46	mL54	MRPL54	Mitochondrial ribosomal protein L54	14.3	138	19p13.3	Goldschmidt-Reisin et al. (1998)
47	bL31m	MRPL55	Mitochondrial ribosomal protein L55	11.6	128	1q42.13	Goldschmidt-Reisin et al. (1998)
48	MRPL56	MRPL56	Mitochondrial ribosomal protein L56	–	–	not identified	Toivonen et al. (1999)
49	mL63	MRPL57	Mitochondrial ribosomal protein L57	11.8	102	13q12.11	Suzuki et al. (2001)
50	mL62	MRPL58	Mitochondrial ribosomal protein L58	23.3	205		Lao et al. (2016)
51	mL64	MRPL59	Mitochondrial ribosomal protein L59	25.4	222		Vahedi et al. (2015)

Continued

Table 4.3 List of 55S mitochondrial ribosomal proteins—cont'd

Sl. no.	New[a] abbrv.	Old abbrv.	Full name's	MW (kDa)	No. of residues	Location (chromosomes)	References
Mitochondrial ribosomal proteins (MRPs)of the 28S							
1	bS1m	MRPS28	Mitochondrial ribosomal protein S28	20.8	187	8q21.13	Amunts et al. (2015)
2	uS2m	MRPS2	Mitochondrial ribosomal protein S2	33.2	296	9q34.3	Myers et al. (1987)
3	uS3m	MRPS24	Mitochondrial ribosomal protein S24	18.9	167	–	Amunts et al. (2015)
4	uS5m	MRPS5	Mitochondrial ribosomal protein S5	48.0	430	2q11.1	Gopisetty and Thangarajan (2016)
5	bS6m	MRPS6	Mitochondrial ribosomal protein S6	14.1	125	21q22.11	Amunts et al. (2015)
6	uS7m	MRPS7	Mitochondrial ribosomal protein S7	24.1	242	17q25.1	Cavdar Koc et al. (1999)
7	uS9m	MRPS9	Mitochondrial ribosomal protein S9	45.8	396	2q12.1	Kotter and Entian (1995)
8	uS10m	MRPS10	Mitochondrial ribosomal protein S10	23.0	201	6p21.1	Kwasniak et al. (2013)
9	uS11m	MRPS11	Mitochondrial ribosomal protein S11	20.6	194	15q25.3	Kenmochi et al. (2001)
10	uS12m	MRPS12	Mitochondrial ribosomal protein S12	12.1	138	19q13.2	Shah et al. (1997)
11	uS13m	MRPS13	Mitochondrial ribosomal protein S13	8.9	79	–	Partaledis and Mason (1988)

12	uS14m	MRPS14	Mitochondrial ribosomal protein S14	15.1	128	1q25.1	Koc et al. (2000)
13	uS15m	MRPS15	Mitochondrial ribosomal protein S15	23.5	257	1p34.3	Amunts et al. (2015)
14	bS16m	MRPS16	Mitochondrial ribosomal protein S16	11.5	137	10q22.2	Emdadul Haque et al. (2008) and Miller et al. (2004)
15	uS17m	MRPS17	Mitochondrial ribosomal protein S17	12.3	30	7p11.2	Haffter and Fox (1992)
16	bS18c	MRPS18A	Mitochondrial ribosomal protein S18A	29.4	258	6p21.1	Kitakawa et al. (1997)
17	bS18b	MRPS18B	Mitochondrial ribosomal protein S18B	15.8	142	6p21.33	Darekar et al. (2015)
18	bS21	MRPS21	Mitochondrial ribosomal protein S21	10.7	87	1q21.2	Suzuki et al. (2001)
19	mS22	MRPS22	Mitochondrial ribosomal protein S22	41.2	360	3q23	Amunts et al. (2015)
20	mS23	MRPS23	Mitochondrial ribosomal protein S23	21.8	190	17q22	Amunts et al. (2015)
21	mS24	MRPS24	Mitochondrial ribosomal protein S24	15.3	167	7p13	Koc et al. (2000)
22	mS25	MRPS25	Mitochondrial ribosomal protein S25	20.1	173	3p25.1	Koc et al. (2000)
23	mS26	MRPS26	Mitochondrial ribosomal protein S26	21.3	205	20p13	Goldschmidt-Reisin et al. (1998)
24	mS27	MRPS27	Mitochondrial ribosomal protein S27	47.6	414	5q13.2	Koc et al. (2001b)

Continued

Table 4.3 List of 55S mitochondrial ribosomal proteins—cont'd

Sl. no.	New[a] abbrv.	Old abbrv.	Full name's	MW (kDa)	No. of residues	Location (chromosomes)	References
25	mS29	DAP3	Death associated protein 3	43.4	398	1q22	Jacques et al. (2009)
26	mS31	MRPS31	Mitochondrial ribosomal protein S31	38.1	395	13q14.11	Jourdain et al. (2013)
27	mS33	MRPS33	Mitochondrial ribosomal protein S33	12.5	106	7q34	Zhang and Gerstein (2003)
28	mS34	MRPS34	Mitochondrial ribosomal protein S34	25.5	218	16p13.3	O'Brien et al. (2000)
29	mS35	MRPS35	Mitochondrial ribosomal protein S34	36.8	323	12p11.22	Kenmochi et al. (2001)
30	mS37	MRPS37	Mitochondrial ribosomal protein S37	13.5	118	Not identified	Amunts et al. (2015)
31	mS38	MRPS38	Mitochondrial ribosomal protein S38	22.4	199	Not identified	Amunts et al. (2015)
32	mS39	MRPS39	Mitochondrial ribosomal protein S39	74.4	689	Not identified	Amunts et al. (2015)

[a] *b*, bacterial; *e*, eukaryotic; *u*, universal; *m*, mitochondria; *c*, chloroplast.

from bacteria, mitochondrial, chloroplast, and eukaryotes, respectively, and subunit identity will be just like the previous nomenclature system (Mager et al., 1997; Desmond et al., 2011; Ban et al., 2014).

The function of ribosomal proteins

Prior to 2000, the RP structure, function, and interaction with rRNAs were focused on individual proteins and investigated independently of each other. The protein structure and interactions were mapped using various biophysical techniques including protein footprinting, mutational analysis, and cross-linking with rRNAs (Stelzl et al., 2000). Subsequently, the structure of complete 80S ribosomes was worked out on electron cryomicroscopy and angular reconstitution (Stark, 2002; Wilson and Doudna Cate, 2012). RPs are associated with rRNAs through strong noncovalent interactions and after isolation, it is extremely difficult to handle ribosomal proteins in in vitro. Now we know that the tails of RPs cause hindrance in the crystallization and these tails are involved in protein-protein interactions and association among RPs (Jin et al., 2011).

During the ribosome biogenesis, rDNA is transcribed by RNAPI as two separate transcripts—one is for 5S, another for 18S, 5.8S, and 25S. The RPGs and accessory protein genes (transacting factors) are transcribed from the genome via RNAPII, translated into the cytoplasm, and assembled in the nucleolus (Palade, 1955; Anger et al., 2013). Mitochondrial ribosomes (55S) are different from other ribosomes and it is made up of the 39S large subunit and 28S small subunit. Interestingly, several RPs of the large and small subunits of mitochondrial ribosomes are encoded by nuclear genes rather than by the mitochondrial genome. After their translation in the cytoplasm, RPs are translocated into the mitochondria for the ribosome assembly (Liu and Spremulli, 2000; O'Brien, 2003). The majority of the universal RPGs are found in an operon in the bacteria and archaea genome (Karlin et al., 2005; Bratlie et al., 2010). Moreover, in *Escherichia coli*, 30 out of 32 RPGs are arranged in five large operons: sps, S10, Alpha, L11, and str. The operon system is quite simple and effective in the regulation of their synthesis (Karlin et al., 2005). RPs are considered to be essential proteins during evolution because if conserved proteins are removed or deleted from the ribosomes will have a deleterious effect or may lead to death, while some deletions do not cause any side effect (Lambertsson, 1998; Saeboe-Larssen et al., 1998).

The 40S small subunit ribosomal proteins

The small subunit of ribosomes has a major role in protein translation because it has the decoding center of messenger RNAs and is responsible for the fidelity of the translation process. The eukaryotic 40S ribosomal subunits are more complex than prokaryotes and carry an additional 15 RPs. Moreover, nearly half of the RPs of small subunits are well conserved and have eukaryotic-specific extension tails. These tails are a long extended region at both C- and N-terminus of RPs and are essential for their structure and function by increasing the interaction surface. For example, three RPs are positional analogs of bacterial small subunit RPs (e.g., eS1, eS4, and eS30 are replaced by S6 and S8, S1, and S4, respectively). Likewise, eS10 and eS12 are structural analogs of archaeal large subunit protein eL7A. The mutational studies have exposed, eS12 is liable for the correct decoding of tRNAs at the A site of ribosomes during mRNA translation. On the other hand, eS31 is a fusion protein expressed along with ubiquitin. When the ubiquitin partner is removed, eS31 covers the A-site (decoding center). However, failure in the removal of the ubiquitin partner leads to inhibition of the translation initiation process and thus, is lethal. Several RPs (such as eS27 and eS29) have zinc finger domains such as C-X2-CX, -C-X, and -C- for interaction with DNA and RNAs after binding to zinc ions. Some RPs (e.g., eS3, eS9, and eS30) are engaged at the site of mRNA binding and have a high content of basic amino acids. For instance, eS3 (bacterial homolog S3) present at the mRNA tunnel of the 40S subunit, has conserved amino acid and position (Arg^{119}) from bacteria to humans. Some RPs (e.g., eS6, eS4, and RACK1) are considered to be involved in the regulation and signaling of the small subunit of ribosome assembly, translation regulation (Nilsson et al., 2004). The linker sequence (residues 131–185) present in the mid-region of eS6 is rich in basic amino acids required for binding to the phosphate backbone of 18S rRNA. eS6 is phosphorylated by mTOR kinase at conserved five serine residues present in the C-terminal region of mammalian protein whereas, in yeast, two serines are phosphorylated by eS6 kinase (Dihazi et al., 2001). Phosphorylation of eS6 has a direct influence on protein translation by changing the interaction of eS6 to initiation factors, mRNAs, and tRNAs.

RACK1 is considered an interlink between the signaling and translation of mRNAs. RACK1 is associated with a 40S subunit in highly replicative cells whereas, in stationary phase cells, it is found in the cellular fraction. It interacts with not only the phosphate backbone of 18S rRNA but also with other RPs (eS16, eS17, and eS3) through salt bridges. RACK1 gets phosphorylated at the position of Tyr[52] by c-Abl (Abelson murine leukemia) tyrosine kinase for its regulatory function (Kiely et al., 2009).

RPS27A (eS31) is a unique protein encoded by the UBA80 fusion gene and is analogous to RPL40 of the large subunit. The UBCEP80 transcript is naturally translated as a Ub-RPS27a precursor protein which is rapidly processed as N-terminal monoubiquitin and C-terminal extension protein (CEP) (Redman and Rechsteiner, 1989; Komander et al., 2009). RPS27A is involved in cell cycle regulation and shows high expression in solid tumors (Wang et al., 2014).

The 60S large subunit ribosomal proteins

The large subunit of ribosomes is an attraction point of research because it is involved in peptide bond formation and has a nascent peptide exit tunnel. Crystallographic studies of complete 70S, including small 30S and large 50S subunit, and 40S subunit cocrystallized with eukaryotic translation initiation factor1 (eIF-I) are available (Yusupov et al., 2001; Tung and Sanbonmatsu, 2004; Wilson et al., 2005; Rabl et al., 2011). Further, scientists have cocrystallized the large 60S subunit of the ribosome together with translation initiation factor 6 (Klinge et al., 2011). These studies have helped understand the stoichiometry of all RPs and their location in the ribosomal subunit, the molecular mechanism of protein translation at a different stage, as well as the mode of action of various antibiotics that target the translational machinery of cells (Schmeing et al., 2009).

The large 60S ribosomal subunit contains 42 proteins including 16 conserved from bacteria, 20 from Achaea, and 6 are specific to eukaryotes. RPs of 60S subunit contain specific eukaryotic extension which is essential to stabilize the protein-rRNA network. Some RPs are involved in the formation of the conserved core region of the subunit including active sites and exit channel. These proteins could be specific (eL22, eL29, eL36, and eL28) or shared with archaea (eL13, eL34, and eL38) (Lecompte et al., 2002; Armache et al., 2010b). One unique feature of the large 60S subunit, which never gets noticed in prokaryotic and archaeal RPs, is the large extension of RPs and bridging to numerous rRNAs segments (Klinge et al., 2011). These interactions are occurring because of highly positive charge extensions of RPs to negatively charge rRNAs segment whereas some interaction by β-sheet of RPs (Fedyukina et al., 2014).

The e21 creating a secondary structure to link both eL7 and eL20 (RPL18A) as well as eL29 and rRNAs thus the sandwich structure is formed. The eL20 is necessary for the stabilization of a large 60S subunit and plays an important role in holding three different RPs (eL6, uL13, and eL14) by src-homology-3 (SH-3) domain. These three RPs are specific and form the cluster which is unique to prokaryotic ribosomes. Two RPs (eL13 and eL36) act as a bridge between two clusters of rRNAs by forming their longer extensions. A defect in eL36, uL18, uL5, and eL33 has been linked to congenital red blood anemia and Diamond-Blackfan aplasia (Ellis and Gleizes, 2011) because the processing of 5S rRNAs is regulated by uL18 and uL5 at the time of ribosome assembly. Out of 6 eukaryotic specific RPs, eL27 has an SH-3 domain for interaction with rRNA and two RPs, viz., eL8, and eL30 on the surface of large 60S subunit via their extensions.

Peptidyl transferase center and exit tunnel are extremely conserved regions of the 60S subunit of ribosomes in all domains of life. In bacteria, peptidyl transferase center has 23S rRNA whereas 26S rRNA in eukaryotes has the peptidyl transferase activity. Moreover, it carries all segments that are necessary for substrate binding, catalysis, and release products. The eL29 spreads from the surface to the core of the 60S subunit and at the active site, adenosine (A_{2820}) is essential for the stabilization of 26S rRNAs. Some other RPs (uL4, uL16, and eL21) are also present at the active site and interconnected with the surface of the 60S subunit by their eukaryotic-specific large extensions. If the exit tunnel is altered, it affects protein synthesis. The exit tunnel is surrounded by a flat area and does not have any eukaryotic specific bulky features added in this region and nascent proteins interact with eL22 which clench the nascent polypeptide at the exit tunnel along with rRNAs during proteins translation (Armache et al., 2010a, b). The eL27 is surrounded by eL38 and eL22 at the exit tunnel and these are involved in the selective recruitment of HOX gene mRNA on 60S subunit in mice (Kondrashov et al., 2011). UBA52 (RPL40) is a unique RPs in a large 60S subunit because it is ribosomal-ubiquitin fusion proteins. It expresses along with ubiquitin and has a zinc finger domain. UBA52 has N-terminal monoubiquitin and C-terminal ribosomal protein before the processing of polypeptide, after processing, C-terminal ribosomal protein associated with 60S subunit, and N-terminal region release in the cytosol. It presents nearby to the sarcin-ricin loop and the elongation factor binding site in the 60S subunit. If unprocessed UBA52 is associated with 60S subunit which affects the binding of eukaryotic elongation factor to the ribosomal subunit. It is documented that the N-terminal ubiquitin region is necessary for the maturation of the ribosome, and it helps to block the immature entering of the 60S subunit in the translation cycle (Archibald et al., 2003; Lacombe et al., 2009).

Eukaryotic mitochondrial ribosome or mitoribosomes

Mitochondria are cell organelles of eukaryotic cells that produce chemical energy in the form of adenosine triphosphate (ATP) (Kurland and Andersson, 2000). Mammalian mitochondrial ribosomes (mitoribosomes) are unique as compared to those in other domains of life because they synthesize 13 essential proteins of oxidative phosphorylation complexes. The study of the structure of mitochondria under the 3.5Å resolution via single-particle electron cryogenic microscopy, found 80 extensively interconnected proteins, in this 36 of which are specific to mitochondria, and three ribosomal RNA molecules, including 16s rRNA coordinated with 50 out of 52 mt-RPs (Amunts et al., 2015). Most of the mitochondrial RPs (mt-RPs) are share homology with the bacterial ribosome proteins and are organized in two subunits one is a large 39S subunit (mt-LSU) and another is a small 28S subunit (mt-SSU) which together constitute the 55S mature mitoribosomes. The mitoribosomes have a small content of rRNAs in comparison to bacterial ribosomes (Friedman and Nunnari, 2014). As cytoplasmic RPs, several mt-RPs contain long extension includes tails and loops in the globular domain of the 39S subunit and are highly basic for binding to rRNAs. The 55S mature mitoribosome has a large number of protein-protein interactions in comparison to the eukaryotic large 60S subunit but lesser than 70S bacterial ribosomes (Kaushal et al., 2015).

All 13 mitoribosome-specific RPs interact with rRNAs except mL43, 5 mt-RPs contain nucleic acid binding (mL49, mL44, mL58, mL60, and mL57), 2 nucleotides binding (mL38 and mL48), and 2 lipids binding (mL38 and mL50) properties (Wool et al., 1995). In mitoribosomes, 5S rRNA is absent and two mt-RPs (L18 and L25) are replaced by mitoribosome-specific nucleic acid-binding proteins (mL49, mL44, and mL57), and these mt-RPs binds to RNA to make RNA-proteins clusters for interaction with the extension of five homologous mt-RPs (uL5, uL16, bL27, bL31, and bL33) and the maintenance of the structure of mitoribosomes. At the exit site (*E*-site), there are two specific mt-RPs for holding tRNA and rRNAs, while bL33 holds tRNAs and stabilizes the tRNAs-E-site complex, mL59 and bL33 together stabilize the paths of rRNA at the E-site. The exit tunnels are well-conserved in ribosomes of all domains of life but in mitoribosomes, the exit tunnels are blocked by uL23. Further, the exit tunnel of mitoribosomes is wider than that of the bacterial ribosome as it allows the cotranslation of oxidative phosphorylation proteins (Amunts et al., 2014; Brown et al., 2014; Greber et al., 2014; Kaushal et al., 2015).

Extra ribosomal functions of ribosomal proteins

RPs are major housekeeping genes in all domains of life. Of late, RPs are identified with other functions without any interference of the ribosome, during stress conditions leads to the accumulation of RPs due to increased ribosome-free RPs. The expression of these proteins appears to be tightly regulated in the cell because any interference in ribosome assembly is associated with lethality. RPs stress may be induced under three different conditions: (i) by chemicals or radiation which increase rRNA production or decrease RPs degradation; (ii) nutrient deprivation, hypoxia, and glucose starvation; and (iii) gene deregulation due to malfunctioning of genes. The ribosome-free RPs interact with other cellular proteins and perform extra-ribosomal functions (Lu et al., 2015; Zhou et al., 2015; Buskirk and Green, 2017).

The ribosomal machinery components such as RPs, rRNA, and assembly factors are upregulated in most cancers. Moreover, the cancer cells are uncontrolled growing cells, therefore they require a large number of mature ribosomes for the synthesis of desired proteins. The defective ribosome assembly due to the altered expression of its components could be a reason for the development of cancers (Bundred et al., 2018; Molavi et al., 2019; Stepinski, 2018).

Balanced production of RPs

Protein synthesis is the central aspect of cell growth, proliferation, and the desired function of the cell, as well as the ribosome is the most abundant subcellular component of the cell. Single yeast cell contains nearly 2,00,000 ribosomes (Woolford and Baserga, 2013). Increasing demand for proteins is directly corelated with the requirements of ribosomes, whereas the unnecessary ribosome biogenesis is energetically a penalty for the cell. For this reason, ribosome biogenesis is maintained under a highly controlled process for the production of RPs and rRNAs. Cells increase their metabolic demand through the enhancement of ribosome biogenesis (Slomnicki et al.,

2016). Besides, the assembly of the functional ribosome is subject to the availability of all components of the ribosome in the cell.

The synthesis of RPs is regulated at transcription, translational, and posttranslational levels in both unicellular and multicellular organisms to minimize their aberrant function in the cell. In yeast, RPs are predominantly regulated at the level of transcription because the half-life of RP mRNA is much shorter than other gene transcripts (Perry, 2007; Narla and Ebert, 2010; Wang et al., 2015). In the exponential phase, yeast cells actively transcribe RP mRNAs from one or two copies of RPGs to compensate for the imbalance caused due to turnover of RPs (Woolford and Baserga, 2013). While in higher eukaryotes, only a single copy of RPG is expressed and the RP synthesis is primarily regulated at translation level by a ubiquitous sequence called 5′ terminal oligopyrimidine (TOP) in the 5′ region of RP mRNAs (Fonseca et al., 2018). A narrow feedback regulation mechanism operates at the level of processing and turnover of RP mRNA in yeast and some vertebrates (Sonenberg and Hinnebusch, 2009). RPs that are not assembled in ribosomes after achieving the complete stoichiometry are efficiently degraded (Perry, 2007). Some RPs behave as regulatory and stress sensors by activating p53-mediated responses such as cell cycle arrest, premature senescence, and apoptosis, as a result, exhibit anticancer activity. On the contrary, RPS13 and RPL13 are two well-known biomarkers of cancer (Guo et al., 2011; Dolezal et al., 2018). Interestingly, RPS3a (Zhang and Lu, 2009) and RPS27a (Fatima et al., 2012) levels are elevated in the hepatitis B virus-induced hepatocellular carcinoma (Fig. 4.2).

At the transcription level

Nearly 10% of all protein-coding genes in yeast account for RPs and transacting factor-encoding genes. RPGs are transcribed under the regulatory control of RNAPII that collectively respond to nutrient availability and environmental stimuli. The ribosome biogenesis genes are encompassed not only RPGs and transacting factors, but also some other genes that are functionally connected with ribosome biogenesis and function, such as genes encoding translation factors, aminoacyl-tRNA synthetases, tRNA-modifying factors, the enzymes involved in nucleotide metabolism, and the subunits of RNAP I and III (Jorgensen et al., 2002; Jorgensen et al., 2004). Non-RPGs are known as ribosome biogenesis regulon (RB-regulon) and the expression profile of RPs and RB-regulons are analogous to each other because both respond similarly under different

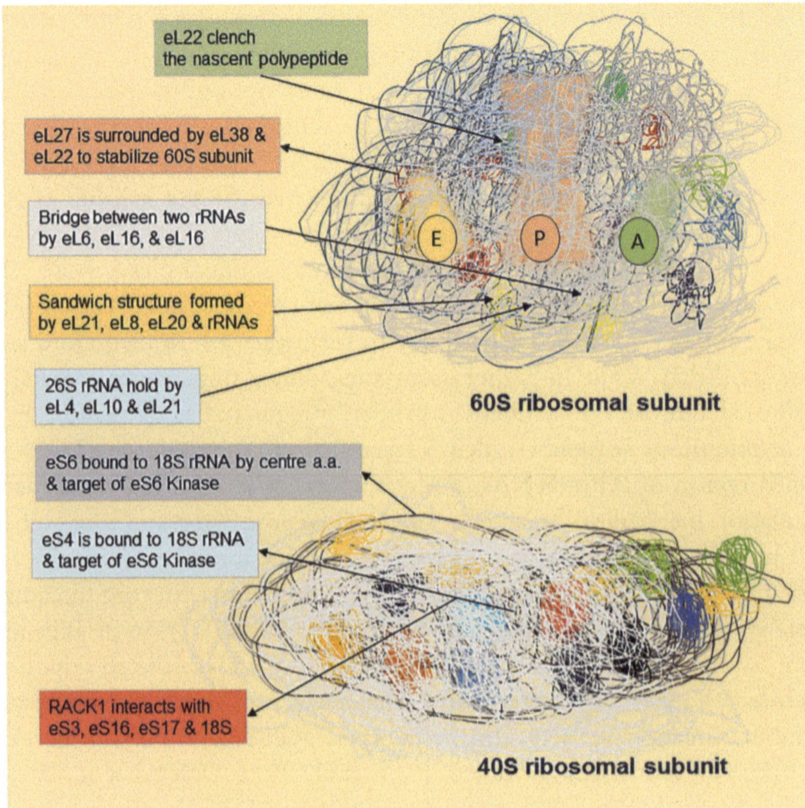

Fig. 4.2 Location of some Important RPs.

nutrients and environmental conditions. RPs and RB-regulons are under the control of TORC1 and Protein kinase A (PKA) that senses nutrient availability and environmental stresses. TORC1 is involved in indirect regulation of 35S pre-rRNAs by suppressing the degradation of Rrn3 (an RNAP I initiation factor) and by phosphorylating the Maf1, which directly interacts with RNAP III for the stimulate expression of 5S rRNA and related transcripts (Mayer and Grummt, 2006). The T-rich elements are present in the RPG promoters are responsible for the repression of RPG transcription in response to various signals (Juven-Gershon et al., 2006; Grove, 2018). Transcription of RPGs are tightly controlled by ribosomal RNA-processing elements (PRPE) that bind to transcription repressors such as Dot6, Tod6, and Stb3 (Liko et al., 2007; Lippman and Broach, 2009). The promoter of RPGs also have binding elements for transcriptional activators Abf1, Reb1 and Rap1, and Tbf1 that are essential for optimal expression of RPGs. Thus, both negative and positive regulation is responsible for the tight regulation of RPG expression (Fig. 4.3) (van Riggelen et al., 2010; Bosio et al., 2017). As discussed in previous sections, the level of RPs and rRNAs is in the equimolar ratio imbalance of this ratio is responsible for the nucleolar stress response and cell growth arrest at G1-phase. In yeast, this balance is essentially maintained by the coordinated transcription of rRNAs and RPGs in response to the mTOR pathway (Russo and Russo, 2017).

Cotranscription is a very common phenomenon in the case of rRNA transcription and ribosome biogenesis accounts for 70% of the global cellular transcriptional activity. The yeast genome contains 150–200 copies of tandem repeats of rDNA that are activated or repressed depending on the environmental condition of the cell (Dammann et al., 1993). The RPG regulation seems to be carried out by chromatin remodeling in two different ways—(i) by epigenetic regulation via histone deacetylase Rpd3 and FACT complex; (ii) by an actual number of RNA pol-I bound to rDNA to initiate transcription and carefully elongate rRNA (Johnson et al., 2013). In higher eukaryotes, the assembly of ribosomes starts cotranscriptionally. This was first noticed and called the knobs on Miller chromatin (Mougey et al., 1993). However, in yeast, posttranscriptional modifications of rRNA can occur either before or after completion of 35S pre-rRNA transcription distinct from higher eukaryotes. In rRNA modification, most of the transcripts are cotranscriptionally cleaved at site A2 in the growing condition of cells. The cleavage of sites A2 is important because it separates the large and small subunit synthesis pathways, which independently mature to 60S and 40S ribosomes, respectively. Cleavage of the transcript at the A2 site produces first 90S preribosome particle and start the transcription of 60S preribosome particle at 5′-region of rDNA before the termination of transcription (Gallagher et al., 2004). However, under unfavorable growth condition, the cotranscriptional cleavage diminishes leading to depletion or inactivation of 60S specific factors. Recent studies suggest that Rrp5—a factor which binds a conserved region near the 3′ end of A2 site in ITS1 (internal transcribed spacer 1) is responsible for the correct timing of cotranscriptional pre-rRNAs modifications. Therefore, the depletion of 60S subunit specific factors leads to improper folding at the A2 site of rRNA causing delay in cleavage and thus, permits the formation of 35S pre-rRNA. Interestingly, late steps in the formation of 60S subunit require transacting factors which are typically associated with late pre-60S particles (de la Cruz et al., 2018).

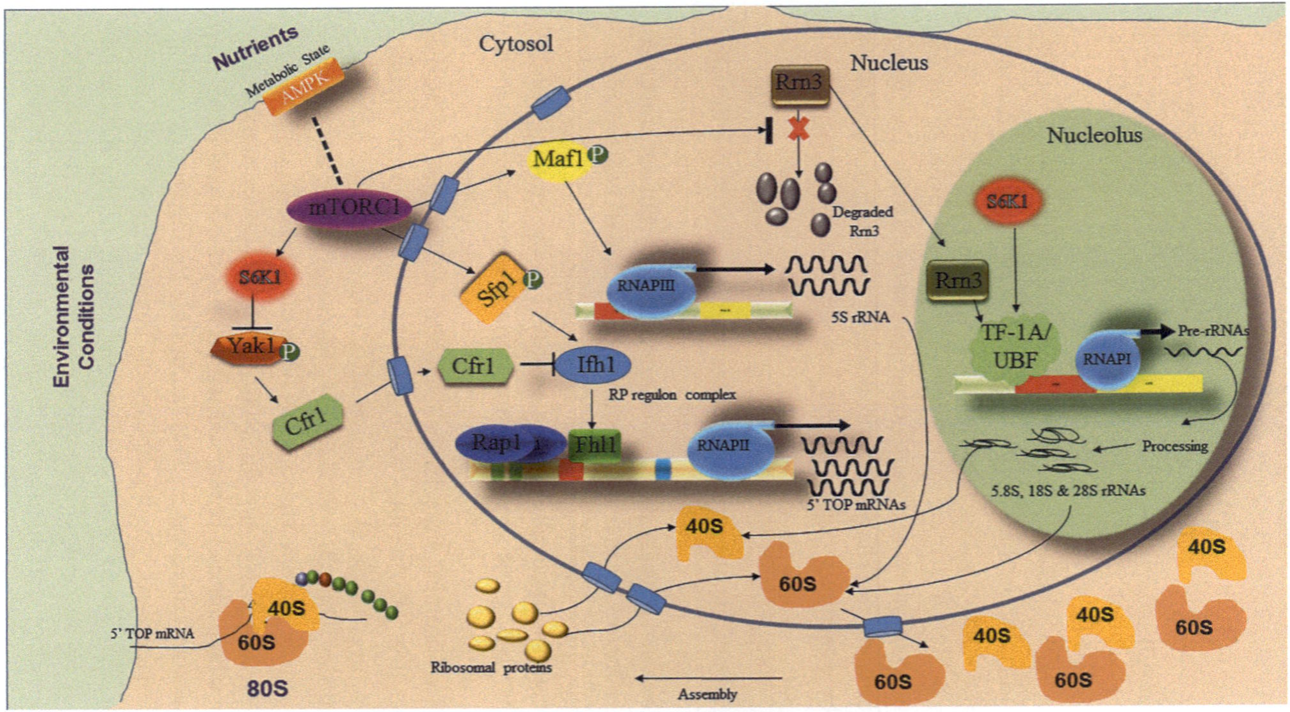

Fig. 4.3 Signaling cascade involved in the balanced production of ribosomal proteins in response to various environmental condition—suitable condition sensed by the cell and activate AMPK/mTORC1 Pathway, mTORC1 phosphorylates multiple target such as MAF1, SFP1, and S6K1. These proteins with the help of downstream signaling regulate Ribosomal protein genes transcription and translation as well as rRNAs transcription. *AMPK*, 5′ AMP-activated protein kinase; *mTORC1*, mammalian target of rapamycin complex 1; *S6K1*, ribosomal protein S6 kinase beta-1; *SFP1*, split finger protein; *YAK1*, yet another kinase1; *CFR1*, cell fusion protein1; *IFH1*, interacts with Fork Head1; *MAF1*, MAF1 negative regulator of RNA polymerase III; *RRN3*, RNA polymerase I-specific transcription initiation factor RRN3; *FHL1*, four and A half LIM domains protein 1; *5′ TOP mRNA*, 5′ terminal oligopyrimidine (5′ TOP) motif messenger RNA; *UBF*, upstream binding factor; *TF1A*, transcription factor 1A.

At the translational level

The translation of RPs is highly delicate to the availability of nutrient and cell energetics because it is controlled by the mTOR pathways. The primary regulation of RPs is at translation levels in prokaryotes. However, in eukaryotes, it is the mass transcription of RP mRNAs (50- to 100-fold more than house-keeping genes) (Lahr et al., 2017). The mTORC1 is a principal factor to control the translation through the phosphorylation of LARP1 and then phosphorylated LARP1 (pLARP1) interact with RAPTOR (regulatory-associated protein of mTOR) but not with RICTOR (rapamycin-insensitive companion of mTOR). After association with RAPTOR, pLARP1 can specifically bind to N7-methyl guanosine triphosphate (m7Gppp) cap moiety and the adjacent 5′terminal oligopyrimidine (5′TOP) sequence RP transcripts resulting in the blocking of translation initiation complex and thus, translation. However, under favorable conditions, mTORC1 does not phosphorylate LARP1 and the subsequent steps are omitted leading to active translation (Lippman and Broach, 2009; Philippe et al., 2018).

At the posttranslational level

Certain RPs are fusion proteins and their roles are not fully understood. It is believed that they have a role in the regulation of the assembly of the ribosome. These RPs are eS27A (Redman and Rechsteiner, 1989), eL40 (Grou et al., 2015), and eS30 (Michiels et al., 1993) encoded by RPS27A (UBA80), UBA52, and FAU genes, respectively. Furthermore, these genes also encode ubiquitin proteins, after translation. These polypeptides are processed by proteases to produce two separate proteins, at the N-terminal monomer of ubiquitin and C-terminal RPs except for FAU gene, it has ubiquitin-like protein (fubi) at C-terminal (Rossman et al., 2003). Ubiquitin is the most crucial protein in the higher domain of life for the degradation of ubiquitin-targeted proteins.

Summary and conclusions

Ribosomal proteins are actual crucial proteins for the cells, and their balanced production is supreme essential for survival and efficient purposes. RPs contributions to all activity of cells via directly or indirectly, because without RPs cells unable to execute protein synthesis and their activities as well as the determination of cells. Regulation of ribosomal protein production, association with accessory factors and rRNAs is achieved at every step of ribosomes biogenesis (i.e., gene, transcription, translation, protein folding, recycling, and degradation via ubiquitination, in addition to the association with designated rRNA and processing of ribosomal subunits) in healthy cells. Failure of a single step of regulation may lead to different types of ribosomopathies. The disorders may include all genetic abnormalities and some acquired defects due to the impaired ribosome biogenesis and functioning. It may show some specific clinical phenotype including life-threatening developmental defects along with cancers. The identification of additional proteins and regulation steps as well as diseases related to the overproduction of RPs or rRNAs. Therefore, we will also combat the therapeutics strategies for ribosomopathies and related deadly diseases.

References

Albert B, Knight B, Merwin J, Martin V, Ottoz D, Gloor Y, et al. A molecular titration system coordinates ribosomal protein gene transcription with ribosomal RNA synthesis. Mol Cell 2016;64:720–33.

Amunts A, Brown A, Bai XC, Llacer JL, Hussain T, Emsley P, et al. Structure of the yeast mitochondrial large ribosomal subunit. Science 2014;343:1485–9.

Amunts A, Brown A, Toots J, Scheres SHW, Ramakrishnan V. Ribosome The structure of the human mitochondrial ribosome. Science 2015;348:95–8.

Andres O, Kellermann T, Lopez-Giraldez F, Rozas J, Domingo-Roura X, Bosch M. RPS4Y gene family evolution in primates. BMC Evol Biol 2008;8:142.

Anger AM, Armache JP, Berninghausen O, Habeck M, Subklewe M, Wilson DN, et al. Structures of the human and Drosophila 80S ribosome. Nature 2013;497:80–5.

Aoyama Y, Chan YL, Meyuhas O, Wool IG. The primary structure of rat ribosomal protein L18a. FEBS Lett 1989a;242–6.

Aoyama Y, Chan YL, Wool IG. The primary structure of rat ribosomal protein L34. FEBS Lett 1989b;119–22.

Archibald JM, Teh EM, Keeling PJ. Novel ubiquitin fusion proteins: ribosomal protein P1 and actin. J Mol Biol 2003;328:771–8.

Armache JP, Jarasch A, Anger AM, Villa E, Becker T, Bhushan S, et al. Cryo-EM structure and rRNA model of a translating eukaryotic 80S ribosome at 5.5-A resolution. Proc Natl Acad Sci USA 2010a;107:19748–53.

Armache JP, Jarasch A, Anger AM, Villa E, Becker T, Bhushan S, et al. Localization of eukaryote-specific ribosomal proteins in a 5.5-A cryo-EM map of the 80S eukaryotic ribosome. Proc Natl Acad Sci USA 2010b;107:19754–9.

Bagni C, Mariottini P, Annesi F, Amaldi F. Human ribosomal protein L4: cloning and sequencing of the cDNA and primary structure of the protein. Biochim Biophys Acta 1993;1216:475–8.

Balasubramanian S, Zheng D, Liu YJ, Fang G, Frankish A, Carriero N, et al. Comparative analysis of processed ribosomal protein pseudogenes in four mammalian genomes. Genome Biol 2009;10:R2.

Ban N, Beckmann R, Cate JH, Dinman JD, Dragon F, Ellis SR, et al. A new system for naming ribosomal proteins. Curr Opin Struct Biol 2014;24:165–9.

Bosio MC, Fermi B, Spagnoli G, Levati E, Rubbi L, Ferrari R, et al. Abf1 and other general regulatory factors control ribosome biogenesis gene expression in budding yeast. Nucleic Acids Res 2017;45:4493–506.

Bratlie MS, Johansen J, Drablos F. Relationship between operon preference and functional properties of persistent genes in bacterial genomes. BMC Genomics 2010;11:71.

Brown A, Amunts A, Bai XC, Sugimoto Y, Edwards PC, Murshudov G, et al. Structure of the large ribosomal subunit from human mitochondria. Science 2014;346:718–22.

Bundred JR, Hendrix E, Coleman ML. The emerging roles of ribosomal histidyl hydroxylases in cell biology, physiology and disease. Cell Mol Life Sci 2018;75:4093–105.

Buskirk AR, Green R. Ribosome pausing, arrest and rescue in bacteria and eukaryotes. Philos Trans R Soc Lond B Biol Sci 2017;372:20160183.

Cavdar Koc E, Blackburn K, Burkhart W, Spremulli LL. Identification of a mammalian mitochondrial homolog of ribosomal protein S7. Biochem Biophys Res Commun 1999;266:141–6.

Chain FJ, Feulner PG, Panchal M, Eizaguirre C, Samonte IE, Kalbe M, et al. Extensive copy-number variation of young genes across stickleback populations. PLoS Genet 2014;10:e1004830.

Chan YL, Lin A, McNally J, Peleg D, Meyuhas O, Wool IG. The primary structure of rat ribosomal protein L19. A determination from the sequence of nucleotides in a cDNA and from the sequence of amino acids in the protein. J Biol Chem 1987a;262:1111–5.

Chan YL, Lin A, Paz V, Wool IG. The primary structure of rat ribosomal protein S8. Nucleic Acids Res 1987b;15:9451–9.

Chan YL, Wool IG. The primary structure of rat ribosomal protein S6. J Biol Chem 1988;263:2891–6.

Chan YL, Devi KR, Olvera J, Wool IG. The primary structure of rat ribosomal protein S3. Arch Biochem Biophys 1990a;283:546–50.

Chan YL, Paz V, Olvera J, Wool IG. The primary structure of rat ribosomal protein S16. FEBS Lett 1990b;263:85–8.

Chan YL, Wool IG. The primary structure of rat ribosomal protein S20. Biochim Biophys Acta 1990c;1049:93–5.

Chan YL, Paz V, Olvera J, Wool IG. The primary structure of rat ribosomal protein S24. FEBS Lett 1990d;262:253–5.

Chan YL, Paz V, Wool IG. The primary structure of rat ribosomal protein S18. Biochem Biophys Res Commun 1991a;178:1212–8.

Chan YL, Olvera J, Wool IG. The primary structure of rat ribosomal protein S28. Biochem Biophys Res Commun 1991b;15:314–8.

Chan YL, Paz V, Wool IG. The primary structure of rat ribosomal protein L23. Biochem Biophys Res Commun 1991c;178:1153–9.

Chan YL, Wool IG. The primary structure of rat ribosomal protein S25. Biochem Biophys Res Commun 1992a;186:1688–93.

Chan YL, Wool IG. The primary structure of rat ribosomal protein L8. Biochem Biophys Res Commun 1992b;185:539–47.

Chan YL, Olvera J, Paz V, Wool IG. The primary structure of rat ribosomal protein L11. Biochem Biophys Res Commun 1992c;185:356–62.

Chan YL, Paz V, Olvera J, Wool IG. The primary structure of rat ribosomal protein S9. Biochem Biophys Res Commun 1993a;182:106–12.

Chan YL, Olvera J, Paz V, Wool IG. The primary structure of rat ribosomal protein L29. Biochem Biophys Res Commun 1993b;192(192):583–9.

Chan YL, Paz V, Olvera J, Wool IG. The primary structure of rat ribosomal protein L36. Biochem Biophys Res Commun 1993c;192:849–53.

Chan YL, Paz V, Olvera J, Wool IG. The primary structure of L37—a rat ribosomal protein with a zinc finger-like motif. Biochem Biophys Res Commun 1993d;192:590–6.

Chan YL, Suzuki K, Olvera J, Wool IG. Zinc finger-like motifs in rat ribosomal proteins S27 and S29. Nucleic Acids Res 1993e;21:649–55.

Chan YL, Olvera J, Gluck A, Wool IG. A leucine zipper-like motif and a basic region-leucine zipper-like element in rat ribosomal protein L13a. Identification of the tum- transplantation antigen P198. J Biol Chem 1994a;269:5589–94.

Chan YL, Olvera J, Paz V, Wool IG. The primary structure of rat ribosomal protein S15a. Biochem Biophys Res Commun 1994b;200:1498–504.

Chan YL, Olvera J, Wool IG. The primary structure of rat ribosomal protein L24. Biochem Biophys Res Commun 1994c;202:1176–80.

Chan YL, Wool IG. The primary structure of rat ribosomal protein L22. Biochim Biophys Acta 1995;1260:113–5.

Chan YL, Olvera J, Paz V, Wool IG. The primary structures of rat ribosomal proteins S3a (the V-Fos transformation effector) and of S3b. Biochem Biophys Res Commun 1996;228:141–7.

Chandramouli P, Topf M, Menetret JF, Eswar N, Cannone JJ, Gutell RR, et al. Structure of the mammalian 80S ribosome at 8.7 A resolution. Structure 2008;16:535–48.

Dammann R, Lucchini R, Koller T, Sogo JM. Chromatin structures and transcription of rDNA in yeast *Saccharomyces cerevisiae*. Nucleic Acids Res 1993;21:2331–8.

Darekar SD, Mushtaq M, Gurrapu S, Kovalevska L, Drummond C, Petruchek M, et al. Mitochondrial ribosomal protein S18-2 evokes chromosomal instability and transforms primary rat skin fibroblasts. Oncotarget 2015;25:21016–28.

De Silva D, Tu YT, Amunts A, Fontanesi F, Barrientos A. Mitochondrial ribosome assembly in health and disease. Cell Cycle 2015;14:2226–50.

De S, Brogna S. Are ribosomal proteins present at transcription sites on or off ribosomal subunits? Biochem Soc Trans 2010;38:1543–7.

de la Cruz J, Gomez-Herreros F, Rodriguez-Galan O, Begley V, de la Cruz Munoz-Centeno M, Chavez S. Feedback regulation of ribosome assembly. Curr Genet 2018;64:393–404.

DeLabre ML, Kessl J, Karamanou S, Trumpower BL. RPL29 codes for a non-essential protein of the 60S ribosomal subunit in *Saccharomyces cerevisiae* and exhibits synthetic lethality with mutations in genes for proteins required for subunit coupling. Biochim Biophys Acta 2002;1574:255–61.

Desmond E, Brochier-Armanet C, Forterre P, Gribaldo S. On the last common ancestor and early evolution of eukaryotes: reconstructing the history of mitochondrial ribosomes. Res Microbiol 2011;162:53–70.

Devi KR, Chan YL, Wool IG. The primary structure of rat ribosomal protein L18. DNA 1988;7:157–62.

Devi KR, Chan YL, Wool IG. The primary structure of rat ribosomal protein S4. Biochim Biophys Acta 1989a;1008:258–62.

Devi KR, Chan YL, Wool IG. The primary structure of rat ribosomal protein L21. Biochem Biophys Res Commun 1989b;162:364–70.

Dihazi H, Kessler R, Eschrich K. Phosphorylation and inactivation of yeast 6-phosphofructo-2-kinase contribute to the regulation of glycolysis under hypotonic stress. Biochemistry 2001;40:14669–78.

Dolezal JM, Dash AP, Prochownik EV. Diagnostic and prognostic implications of ribosomal protein transcript expression patterns in human cancers. BMC Cancer 2018;18:275.

Dvinge H, Kim E, Abdel-Wahab O, Bradley RK. RNA splicing factors as oncoproteins and tumour suppressors. Nat Rev Cancer 2016;16:413–30.

Ellis SR, Gleizes PE. Diamond Blackfan anemia: ribosomal proteins going rogue. Semin Hematol 2011;48:89–96.

Emdadul Haque M, Grasso D, Miller C, Spremulli LL, Saada A. The effect of mutated mitochondrial ribosomal proteins S16 and S22 on the assembly of the small and large ribosomal subunits in human mitochondria. Mitochondrion 2008;8:54–61.

Fatima G, Mathan G, Kumar V. The HBx protein of hepatitis B virus regulates the expression, intracellular distribution and functions of ribosomal protein S27a. J Gen Virol 2012;93:706–15.

Fearon K, Mason TL. Structure and regulation of a nuclear gene in *Saccharomyces cerevisiae* that specifies MRP7, a protein of the large subunit of the mitochondrial ribosome. Mol Cell Biol 1988;8:3636–46.

Fedyukina DV, Jennaro TS, Cavagnero S. Charge segregation and low hydrophobicity are key features of ribosomal proteins from different organisms. J Biol Chem 2014;289:6740–50.

Fermi B, Bosio MC, Dieci G. Promoter architecture and transcriptional regulation of Abf1-dependent ribosomal protein genes in *Saccharomyces cerevisiae*. Nucleic Acids Res 2016;44:6113–26.

Fonseca BD, Lahr RM, Damgaard CK, Alain TT, Berman AJ. LARP1 on TOP of ribosome production. Wiley Interdiscip Rev RNA 2018;2:e1480.

Fox GE. Origin and evolution of the ribosome. Cold Spring Harb Perspect Biol 2010;2: a003483.

Friedman JR, Nunnari J. Mitochondrial form and function. Nature 2014;505:335–43.

Gallagher JE, Dunbar DA, Granneman S, Mitchell BM, Osheim Y, Beyer AL, et al. RNA polymerase I transcription and pre-rRNA processing are linked by specific SSU processome components. Genes Dev 2004;18:2506–17.

Gallagher MJ, Chan YL, Lin A, Wool IG. Primary structure of rat ribosomal protein L36a. DNA 1988;7:269–73.

Garcia-Cantalejo J, Baladron V, Esteban PF, Santos MA, Bou G, Remacha MA, et al. The complete sequence of an 18,002 bp segment of *Saccharomyces cerevisiae* chromosome XI contains the HBS1, MRP-L20 and PRP16 genes, and six new open reading frames. Yeast 1994;10:231–45.

Ghosh A, Komar AA. Eukaryote-specific extensions in ribosomal proteins of the small subunit: structure and function. Translation (Austin) 2015;3:e999576.

Gilbert WV. Functional specialization of ribosomes? Trends Biochem Sci 2011;36:127–32.

Gluck A, Chan YL, Lin A, Wool IG. The primary structure of rat ribosomal protein S10. Eur J Biochem 1989;182:105–9.

Goldschmidt-Reisin S, Kitakawa M, Herfurth E, Wittmann-Liebold B, Grohmann L, Graack HR. Mammalian mitochondrial ribosomal proteins. N-terminal amino acid sequencing, characterization, and identification of corresponding gene sequences. J Biol Chem 1998;273:34828–36.

Gopisetty G, Thangarajan R. Mammalian mitochondrial ribosomal small subunit (MRPS) genes: a putative role in human disease. Gene 2016;589:27–35.

Graack HR, Grohmann L, Choli T. Mitochondrial ribosomes of yeast: isolation of individual proteins and N-terminal sequencing. FEBS Lett 1988;242:4–8.

Graack HR, Grohmann L, Kitakawa M. The nuclear coded mitoribosomal proteins YmL27 and YmL31 are both essential for mitochondrial function in yeast. Biochimie 1991;73:837–44.

Graack HR, Grohmann L, Kitakawa M, Schafer KL, Kruft V. YmL9, a nucleus-encoded mitochondrial ribosomal protein of yeast, is homologous to L3 ribosomal proteins from all natural kingdoms and photosynthetic organelles. Eur J Biochem 1992;187:373–80.

Graack HR, Grohmann L, Kitakawa M, Goldschmidt-Reisin S. Gene MRP-L4, encoding mitochondrial ribosomal protein YmL4, is indispensable for proper non-respiratory cell functions in yeast. Gene 1995;152:107–12.

Graack HR, Bryant ML, O'Brien TW. Identification of mammalian mitochondrial ribosomal proteins (MRPs) by N-terminal sequencing of purified bovine MRPs and

comparison to data bank sequences: the large subribosomal particle. Biochemistry 1999;38:16569–77.
Greber BJ, Boehringer D, Leibundgut M, Bieri P, Leitner A, Schmitz N, et al. The complete structure of the large subunit of the mammalian mitochondrial ribosome. Nature 2014;515:283–6.
Grohmann L, Graack HR, Kruft V, Choli T, Goldschmidt-Reisin S, Kitakawa M. Extended N-terminal sequencing of proteins of the large ribosomal subunit from yeast mitochondria. FEBS Lett 1991;284:51–6.
Grohmann L, Kitakawa M, Isono K, Goldschmidt-Reisin S, Graack HR. The yeast nuclear gene MRP-L13 codes for a protein of the large subunit of the mitochondrial ribosome. Curr Genet 1994;26:8–14.
Grou CP, Pinto MP, Mendes AV, Domingues P, Azevedo JE. The de novo synthesis of ubiquitin: identification of deubiquitinases acting on ubiquitin precursors. Sci Rep 2015;5:12836.
Grove A. Control of RNA polymerase II-transcribed genes by direct binding of TOR kinase. Curr Genet 2018;64:131–5.
Guo D, Wang H, Sun L, Liu S, Du S, Qiao W, Wang W, Hou G, Zhang K, Li C, Teng Q. Identification of key gene modules and hub genes of human mantle cell lymphoma by coexpression network analysis. PeerJ 2020;8:e8843.
Guo X, Shi Y, Gou Y, Li J, Han S, Zhang Y, et al. Human ribosomal protein S13 promotes gastric cancer growth through down-regulating p27(Kip1). J Cell Mol Med 2011;15:296–306.
Gupta V, Warner JR. Ribosome-omics of the human ribosome. RNA 2014;20:1004–13.
Haffter P, Fox TD. Suppression of carboxy-terminal truncations of the yeast mitochondrial mRNA-specific translational activator PET122 by mutations in two new genes, MRP17 and PET127. Mol Gen Genet 1992;235:64–73.
Harrer R, Schwank S, Schuller HJ, Schweizer E. Molecular cloning and analysis of the nuclear gene MRP-L6 coding for a putative mitochondrial ribosomal protein from *Saccharomyces cerevisiae*. Curr Genet 1993;24:136–40.
Hemmerich P, von Mikecz A, Neumann F, Sozeri O, Wolff-Vorbeck G, Zoebelein R, et al. Structural and functional properties of ribosomal protein L7 from humans and rodents. Nucleic Acids Res 1993;21:223–31.
Itoh T, Otaka E, Matsui KA. Primary structures of ribosomal protein YS25 from *Saccharomyces cerevisiae* and its counterparts from *Schizosaccharomyces pombe* and rat liver. Biochemistry 1985;24:7418–23.
Jacques C, Fontaine JF, Franc B, Mirebeau-Prunier D, Triau S, Savagner F, et al. Death-associated protein 3 is overexpressed in human thyroid oncocytic tumours. Br J Cancer 2009;101:132–8.
Jin H, Kelley AC, Ramakrishnan V. Crystal structure of the hybrid state of ribosome in complex with the guanosine triphosphatase release factor 3. Proc Natl Acad Sci USA 2011;108:15798–803.
Johnson JM, French SL, Osheim YN, Li M, Hall L, Beyer AL, et al. Rpd3- and spt16-mediated nucleosome assembly and transcriptional regulation on yeast ribosomal DNA genes. Mol Cell Biol 2013;33:2748–59.
Jorgensen P, Nishikawa JL, Breitkreutz BJ, Tyers M. Systematic identification of pathways that couple cell growth and division in yeast. Science 2002;297:395–400.
Jorgensen P, Rupes I, Sharom JR, Schneper L, Broach JR, Tyers M. A dynamic transcriptional network communicates growth potential to ribosome synthesis and critical cell size. Genes Dev 2004;18:2491–505.
Jourdain AA, Koppen M, Wydro M, Rodley CD, Lightowlers RN, Chrzanowska-Lightowlers ZM, et al. GRSF1 regulates RNA processing in mitochondrial RNA granules. Cell Metab 2013;17:399–410.
Juneau K, Miranda M, Hillenmeyer ME, Nislow C, Davis RW. Introns regulate RNA and protein abundance in yeast. Genetics 2006;174:511–8.

Juven-Gershon T, Hsu JY, Kadonaga JT. Perspectives on the RNA polymerase II core promoter. Biochem Soc Trans 2006;34:1047–50.
Karlin S, Brocchieri L, Campbell A, Cyert M, Mrazek J. Genomic and proteomic comparisons between bacterial and archaeal genomes and related comparisons with the yeast and fly genomes. Proc Natl Acad Sci USA 2005;102:7309–14.
Kas K, Michiels L, Merregaert J. Genomic structure and expression of the human fau gene: encoding the ribosomal protein S30 fused to a ubiquitin-like protein. Biochem Biophys Res Commun 1992;187:927–33.
Kaushal PS, Sharma MR, Agrawal RK. The 55S mammalian mitochondrial ribosome and its tRNA-exit region. Biochimie 2015;114:119–26.
Kenmochi N, Suzuki T, Uechi T, Magoori M, Kuniba M, Higa S, et al. The human mitochondrial ribosomal protein genes: mapping of 54 genes to the chromosomes and implications for human disorders. Genomics 2001;77:65–70.
Khusainov I, Vicens Q, Bochler A, Grosse F, Myasnikov A, Menetret JF, et al. Structure of the 70S ribosome from human pathogen *Staphylococcus aureus*. Nucleic Acids Res 2016;44:10491–504.
Kiely PA, Baillie GS, Barrett R, Buckley DA, Adams DR, Houslay MD, et al. Phosphorylation of RACK1 on tyrosine 52 by c-Abl is required for insulin-like growth factor I-mediated regulation of focal adhesion kinase. J Biol Chem 2009;284:20263–74.
Kitagawa M, Takasawa S, Kikuchi N, Itoh T, Teraoka H, Yamamoto H, et al. rig encodes ribosomal protein S15. The primary structure of mammalian ribosomal protein S15. FEBS Lett 1991;283:210–4.
Kitakawa M, Grohmann L, Graack HR, Isono K. Cloning and characterization of nuclear genes for two mitochondrial ribosomal proteins in *Saccharomyces cerevisiae*. Nucleic Acids Res 1990;18:1521–9.
Kitakawa M, Graack HR, Grohmann L, Goldschmidt-Reisin S, Herfurth E, Wittmann-Liebold B, et al. Identification and characterization of the genes for mitochondrial ribosomal proteins of *Saccharomyces cerevisiae*. Eur J Biochem 1997;245:449–56.
Kitaoka Y, Olvera J, Wool IG. The primary structure of rat ribosomal protein S23. Biochem Biophys Res Commun 1994;202:314–20.
Kitoh R, Oshima A, Suzuki N, Hashimoto S, Takumi Y, Usami S. Immunocytochemical localization of ubiquitin A-52 protein in the mouse inner ear. Neuroreport 2007;18:869–73.
Klinge S, Voigts-Hoffmann F, Leibundgut M, Arpagaus S, Ban N. Crystal structure of the eukaryotic 60S ribosomal subunit in complex with initiation factor 6. Science 2011;334:941–8.
Koc EC, Burkhart W, Blackburn K, Moseley A, Koc H, Spremulli LL. A proteomics approach to the identification of mammalian mitochondrial small subunit ribosomal proteins. J Biol Chem 2000;275:32585–91.
Koc EC, Burkhart W, Blackburn K, Moyer MB, Schlatzer DM, Moseley A, et al. The large subunit of the mammalian mitochondrial ribosome. Analysis of the complement of ribosomal proteins present. J Biol Chem 2001a;276:43958–69.
Koc EC, Burkhart W, Blackburn K, Koc H, Moseley A, Spremulli LL. Identification of four proteins from the small subunit of the mammalian mitochondrial ribosome using a proteomics approach. Protein Sci 2001b;10:471–81.
Komander D, Clague MJ, Urbe S. Breaking the chains: structure and function of the deubiquitinases. Nat Rev Mol Cell Biol 2009;10:550–63.
Kondrashov N, Pusic A, Stumpf CR, Shimizu K, Hsieh AC, Ishijima J, et al. Ribosome-mediated specificity in Hox mRNA translation and vertebrate tissue patterning. Cell 2011;145:383–97.
Korobeinikova AV, Garber MB, Gongadze GM. Ribosomal proteins: structure, function, and evolution. Biochemistry (Mosc) 2012;77:562–74.

Kotter P, Entian KD. Cloning and analysis of the nuclear gene MRP-S9 encoding mitochondrial ribosomal protein S9 of *Saccharomyces cerevisiae*. Curr Genet 1995;28:26–31.
Kurland CG, Andersson SG. Origin and evolution of the mitochondrial proteome. Microbiol Mol Biol Rev 2000;64:786–820.
Kuwano Y, Nakanishi O, Nabeshima Y, Tanaka T, Ogata K. Molecular cloning and nucleotide sequence of DNA complementary to rat ribosomal protein S26 messenger RNA. J Biochem 1985;97:983–92.
Kuwano Y, Olvera J, Wool IG. The primary structure of rat ribosomal protein L38. Biochem Biophys Res Commun 1991;75:551–5.
Kuwano Y, Olvera J, Wool IG. The primary structure of rat ribosomal protein S5. A ribosomal protein present in the rat genome in a single copy. J Biol Chem 1992;267:25304–8.
Kuwano Y, Wool IG. The primary structure of rat ribosomal protein L3. Biochem Biophys Res Commun 1992;187:58–64.
Kwasniak M, Majewski P, Skibior R, Adamowicz A, Czarna M, Sliwinska E, et al. Silencing of the nuclear RPS10 gene encoding mitochondrial ribosomal protein alters translation in arabidopsis mitochondria. Plant Cell 2013;25:1855–67.
Lacombe T, Garcia-Gomez JJ, de la Cruz J, Roser D, Hurt E, Linder P, et al. Linear ubiquitin fusion to Rps31 and its subsequent cleavage are required for the efficient production and functional integrity of 40S ribosomal subunits. Mol Microbiol 2009;72:69–84.
Lahr RM, Fonseca BD, Ciotti GE, Al-Ashtal HA, Jia JJ, Niklaus MR, et al. La-related protein 1 (LARP1) binds the mRNA cap, blocking eIF4F assembly on TOP mRNAs. Elife 2017;6:e24146.
Lambertsson A. The minute genes in Drosophila and their molecular functions. Adv Genet 1998;38:69–134.
Lao X, Feng Q, He G, Ji M, Zhu D, Xu P, et al. Immature colon carcinoma transcript-1 (ICT1) expression correlates with unfavorable prognosis and survival in patients with colorectal cancer. Ann Surg Oncol 2016;23:3924–33.
Lecompte O, Ripp R, Thierry JC, Moras D, Poch O. Comparative analysis of ribosomal proteins in complete genomes: an example of reductive evolution at the domain scale. Nucleic Acids Res 2002;30:5382–90.
Liko D, Slattery MG, Heideman W. Stb3 binds to ribosomal RNA processing element motifs that control transcriptional responses to growth in *Saccharomyces cerevisiae*. J Biol Chem 2007;282:26623–8.
Lin A, Chan YL, Jones R, Wool IG. The primary structure of rat ribosomal protein S12. The relationship of rat S12 to other ribosomal proteins and a correlation of the amino acid sequences of rat and yeast ribosomal proteins. J Biol Chem 1987;262:14343–51.
Lippman SI, Broach JR. Protein kinase A and TORC1 activate genes for ribosomal biogenesis by inactivating repressors encoded by Dot6 and its homolog Tod6. Proc Natl Acad Sci U S A 2009;106:19928–33.
Liu M, Spremulli L. Interaction of mammalian mitochondrial ribosomes with the inner membrane. J Biol Chem 2000;275:29400–06.
Lott BB, Wang Y, Nakazato T. A comparative study of ribosomal proteins: linkage between amino acid distribution and ribosomal assembly. BMC Biophys 2013;6:13.
Lu H, Zhu YF, Xiong J, Wang R, Jia Z. Potential extra-ribosomal functions of ribosomal proteins in *Saccharomyces cerevisiae*. Microbiol Res 2015;177:28–33.
Mager WH, Planta RJ, Ballesta JG, Lee JC, Mizuta K, Suzuki K, et al. A new nomenclature for the cytoplasmic ribosomal proteins of *Saccharomyces cerevisiae*. Nucleic Acids Res 1997;25:4872–5.
Marygold SJ, Roote J, Reuter G, Lambertsson A, Ashburner M, Millburn GH, et al. The ribosomal protein genes and minute loci of *Drosophila melanogaster*. Genome Biol 2007;8:R216.

Matsushita Y, Kitakawa M, Isono K. Cloning and analysis of the nuclear genes for two mitochondrial ribosomal proteins in yeast. Mol Gen Genet 1989;219:119–24.
Mayer C, Grummt I. Ribosome biogenesis and cell growth: mTOR coordinates transcription by all three classes of nuclear RNA polymerases. Oncogene 2006;25:6384–91.
Meyuhas O, Klein A. The mouse ribosomal protein L7 gene. Its primary structure and functional analysis of the promoter region. J Biol Chem 1990;265:11465–73.
Michiels L, van der Rauwelaert E, van Hasselt F, Kas K, Merregaert J. fau cDNA encodes a ubiquitin-like-S30 fusion protein and is expressed as an antisense sequence in the Finkel-Biskis-Reilly murine sarcoma virus. Oncogene 1993;8:2537–46.
Miller C, Saada A, Shaul N, Shabtai N, Ben-Shalom E, Shaag A, et al. Defective mitochondrial translation caused by a ribosomal protein (MRPS16) mutation. Ann Neurol 2004;56:734–8.
Molavi G, Samadi N, Hosseingholi EZ. The roles of moonlight ribosomal proteins in the development of human cancers. J Cell Physiol 2019;234:8327–41.
Monia BP, Ecker DJ, Jonnalagadda S, Marsh J. Gene synthesis, expression, and processing of human ubiquitin carboxyl extension proteins. J Biol Chem 1989;264:4093–103.
Mougey EB, O'Reilly M, Osheim Y, Miller Jr OL, Beyer A, Sollner-Webb B. The terminal balls characteristic of eukaryotic rRNA transcription units in chromatin spreads are rRNA processing complexes. Genes Dev 1993;7:1609–19.
Myers AM, Crivellone MD, Tzagoloff A. Assembly of the mitochondrial membrane system. MRP1 and MRP2, two yeast nuclear genes coding for mitochondrial ribosomal proteins. J Biol Chem 1987;262:3388–97.
Nakamura H, Tanaka T, Ishikawa K. Nucleotide sequence of cloned cDNA specific for rat ribosomal protein L7a. Nucleic Acids Res 1989;17:4875.
Nakanishi O, Oyanagi M, Kuwano Y, Tanaka T, Nakayama T, Mitsui H, et al. Molecular cloning and nucleotide sequences of cDNAs specific for rat liver ribosomal proteins S17 and L30. Gene 1985;35:289–96.
Narla A, Ebert BL. Ribosomopathies: human disorders of ribosome dysfunction. Blood 2010;115:3196–205.
Nilsson J, Sengupta J, Frank J, Nissen P. Regulation of eukaryotic translation by the RACK1 protein: a platform for signalling molecules on the ribosome. EMBO Rep 2004;5:1137–41.
O'Brien TW, Liu J, Sylvester JE, Mougey EB, Fischel-Ghodsian N, Thiede B, et al. Mammalian mitochondrial ribosomal proteins (4). Amino acid sequencing, characterization, and identification of corresponding gene sequences. J Biol Chem 2000;275:18153–9.
O'Brien TW. Properties of human mitochondrial ribosomes. IUBMB Life 2003;55:505–13.
Odintsova TI, Muller EC, Ivanov AV, Egorov TA, Bienert R, Vladimirov SN, et al. Characterization and analysis of posttranslational modifications of the human large cytoplasmic ribosomal subunit proteins by mass spectrometry and Edman sequencing. J Protein Chem 2003;22:249–58.
Olvera J, Wool IG. The primary structure of rat ribosomal protein L13. Biochem Biophys Res Commun 1994;201:102–7.
Otaka E, Kobata K. Yeast ribosomal proteins. I. Characterization of cytoplasmic ribosomal proteins by two-dimensional gel electrophoresis. Mol Gen Genet 1978;162:259–68.
Otsuka S, Tanaka M, Saito S, Yoshimoto K, Itakura M. Molecular cloning of a cDNA encoding human ribosomal protein L39. Biochim Biophys Acta 1996;1308:119–21.
Ou JH, Yen TS, Wang YF, Kam WK, Rutter WJ. Cloning and characterization of a human ribosomal protein gene with enhanced expression in fetal and neoplastic cells. Nucleic Acids Res 1987;15:8919–34.
Palade GE. A small particulate component of the cytoplasm. J Biophys Biochem Cytol 1955;1:59–68.

Parenteau J, Durand M, Veronneau S, Lacombe AA, Morin GG, Guerin VV, et al. Deletion of many yeast introns reveals a minority of genes that require splicing for function. Mol Biol Cell 2008;19:1932–41.
Partaledis JA, Mason TL. Structure and regulation of a nuclear gene in *Saccharomyces cerevisiae* that specifies MRP13, a protein of the small subunit of the mitochondrial ribosome. Mol Cell Biol 1988;8:3647–60.
Patel VB, Cunningham CC, Hantgan RR. Physiochemical properties of rat liver mitochondrial ribosomes. J Biol Chem 2001;276:6739–46.
Paz V, Chan YL, Gluck A, Wool IG. The primary structure of rat ribosomal protein S14. Nucleic Acids Res 1989a;17:9484.
Paz V, Olvera J, Chan YL, Wool IG. The primary structure of rat ribosomal protein L26. FEBS Lett 1989b;73:89–93.
Perina D, Korolija M, Roller M, Harcet M, Jelicic B, Mikoc A, et al. Over-represented localized sequence motifs in ribosomal protein gene promoters of basal metazoans. Genomics 2011;98:56–63.
Perry RP. Balanced production of ribosomal proteins. Gene 2007;401:1–3.
Petibon C, Parenteau J, Catala M, Elela SA. Introns regulate the production of ribosomal proteins by modulating splicing of duplicated ribosomal protein genes. Nucleic Acids Res 2016;44:3878–91.
Philippe L, Vasseur JJ, Debart F, Thoreen CC. La-related protein 1 (LARP1) repression of TOP mRNA translation is mediated through its cap-binding domain and controlled by an adjacent regulatory region. Nucleic Acids Res 2018;46:1457–69.
Pleiss JA, Whitworth GB, Bergkessel M, Guthrie C. Transcript specificity in yeast pre-mRNA splicing revealed by mutations in core spliceosomal components. PLoS Biol 2007;5:e90.
Prinz A, Behrens C, Rapoport TA, Hartmann E, Kalies KU. Evolutionarily conserved binding of ribosomes to the translocation channel via the large ribosomal RNA. EMBO J 2000;19:1900–6.
Rabl J, Leibundgut M, Ataide SF, Haag A, Ban N. Crystal structure of the eukaryotic 40S ribosomal subunit in complex with initiation factor 1. Science 2011;331:730–6.
Rajchel A, Chan YL, Wool IG. The primary structure of rat ribosomal protein L32. Nucleic Acids Res 1988;16:2347.
Redman KL, Rechsteiner M. Identification of the long ubiquitin extension as ribosomal protein S27a. Nature 1989;338:438–40.
Reja R, Vinayachandran V, Ghosh S, Pugh BF. Molecular mechanisms of ribosomal protein gene coregulation. Genes Dev 2015;29:1942–54.
Rossman TG, Visalli MA, Komissarova EV. fau and its ubiquitin-like domain (FUBI) transforms human osteogenic sarcoma (HOS) cells to anchorage-independence. Oncogene 2003;22:1817–21.
Russo A, Russo G. Ribosomal proteins control or bypass p53 during nucleolar stress. Int J Mol Sci 2017;18:140.
Saeboe-Larssen S, Lyamouri M, Merriam J, Oksvold MP, Lambertsson A. Ribosomal protein insufficiency and the minute syndrome in *Drosophila*: a dose-response relationship. Genetics 1998;148:1215–24.
Schmeing TM, Voorhees RM, Kelley AC, Gao YG, Murphy FV, Weir JR, et al. The crystal structure of the ribosome bound to EF-Tu and aminoacyl-tRNA. Science 2009;326:688–94.
Sengupta J, Nilsson J, Gursky R, Spahn CM, Nissen P, Frank J. Identification of the versatile scaffold protein RACK1 on the eukaryotic ribosome by cryo-EM. Nat Struct Mol Biol 2004;11:957–62.
Shah ZH, O'Dell KM, Miller SC, An X, Jacobs HT. Metazoan nuclear genes for mitoribosomal protein S12. Gene 1997;204:55–62.

Shen B, Arese M, Gualandris A, Rifkin DB. Intracellular association of FGF-2 with the ribosomal protein L6/TAXREB107. Biochem Biophys Res Commun 1998;152:524–8.

Slomnicki LP, Pietrzak M, Vashishta A, Jones J, Lynch N, Elliot S, et al. Requirement of neuronal ribosome synthesis for growth and maintenance of the dendritic tree. J Biol Chem 2016;291:5721–39.

Sonenberg N, Hinnebusch AG. Regulation of translation initiation in eukaryotes: mechanisms and biological targets. Cell 2009;136:731–45.

Stark H. Three-dimensional electron cryomicroscopy of ribosomes. Curr Protein Pept Sci 2002;3:79–91.

Stelzl U, Spahn CM, Nierhaus KH. Selecting rRNA binding sites for the ribosomal proteins L4 and L6 from randomly fragmented rRNA: application of a method called SERF. Proc Natl Acad Sci USA 2000;97:4597–602.

Stepinski D. The nucleolus, an ally, and an enemy of cancer cells. Histochem Cell Biol 2018;150:607–29.

Stoffler G, Wool IG, Lin A, Rak KH. The identification of the eukaryotic ribosomal proteins homologous with *Escherichia coli* proteins L7 and L12. Proc Natl Acad Sci U S A 1974;71:4723–6.

Su Y, Doherty T, Waring AJ, Ruchala P, Hong M. Roles of arginine and lysine residues in the translocation of a cell-penetrating peptide from ^{13}C, ^{31}P, and ^{19}F solid-state NMR. Biochemistry 2009;48:4587–95.

Suzuki K, Olvera J, Wool IG. The primary structure of rat ribosomal protein S7. FEBS Lett 1990a;51–3.

Suzuki K, Olvera J, Wool IG. The primary structure of rat ribosomal protein S13. Biochem Biophys Res Commun 1990b;519–24.

Suzuki K, Olvera J, Wool IG. The primary structure of rat ribosomal protein S19. Biochimie 1990c;299–302.

Suzuki K, Olvera J, Wool IG. The primary structure of rat ribosomal protein L35. Biochem Biophys Res Commun 1990d;1377–82.

Suzuki K, Olvera J, Wool IG. The primary structure of rat ribosomal protein L9. Gene 1990e;297–300.

Suzuki K, Olvera J, Wool IG. The primary structure of rat ribosomal protein L12. Biochem Biophys Res Commun 1990f;35–41.

Suzuki K, Olvera J, Wool IG. Primary structure of rat ribosomal protein S2. A ribosomal protein with arginine-glycine tandem repeats and RGGF motifs that are associated with nucleolar localization and binding to ribonucleic acids. J Biol Chem 1991a;266: 20007–10.

Suzuki K, Wool IG. The primary structure of rat ribosomal protein L17. Biochem Biophys Res Commun 1991b;178:322–8.

Suzuki K, Wool IG. The primary structure of rat ribosomal protein L23a. The application of homology search to the identification of genes for mammalian and yeast ribosomal proteins and a correlation of rat and yeast ribosomal proteins. J Biol Chem 1993;2755–61.

Suzuki T, Terasaki M, Takemoto-Hori C, Hanada T, Ueda T, Wada A, et al. Proteomic analysis of the mammalian mitochondrial ribosome. Identification of protein components in the 28 S small subunit. J Biol Chem 2001;276:33181–95.

Tamura S, Kuwano Y, Nakayama T, Tanaka S, Tanaka T, Ogata K. Molecular cloning and nucleotide sequence of cDNA specific for rat ribosomal protein L5. Eur J Biochem 1987;168:83–7.

Tanaka M, Tanaka T, Harata M, Suzuki T, Mitsui Y. Triplet repeat-containing ribosomal protein L14 gene in immortalized human endothelial cell line (t-HUE4). Biochem Biophys Res Commun 1998;243:531–7.

Tanaka T, Kuwano Y, Ishikawa K, Ogata K. Nucleotide sequence of cloned cDNA specific for rat ribosomal protein S11. J Biol Chem 1985;260:6329–33.

Tanaka T, Wakasugi K, Kuwano Y, Ishikawa K, Ogata K. Nucleotide sequence of cloned cDNA specific for rat ribosomal protein L35a. Eur J Biochem 1986;154:523–7.
Tanaka T, Kuwano Y, Kuzumaki T, Ishikawa K, Ogata K. Nucleotide sequence of cloned cDNA specific for rat ribosomal protein L31. Eur J Biochem 1987;162:45–8.
Tanaka T, Kuwano Y, Ishikawa K, Ogata K. Nucleotide sequence of cloned cDNA specific for rat ribosomal protein L27. Eur J Biochem 1988;173:53–6.
Tanaka T, Aoyama Y, Chan YL, Wool IG. The primary structure of rat ribosomal protein L37a. Eur J Biochem 1989;183:15–8.
Taylor DJ, Devkota B, Huang AD, Topf M, Narayanan E, Sali A, et al. Comprehensive molecular structure of the eukaryotic ribosome. Structure 2009;17:1591–604.
Teng T, Thomas G, Mercer CA. Growth control and ribosomopathies. Curr Opin Genet Dev 2013;23:63–71.
Toivonen JM, Boocock MR, Jacobs HT. Modelling in Escherichia coli of mutations in mitoribosomal protein S12: novel mutant phenotypes of rpsL. Mol Microbiol 1999;31:1735–46.
Tsang P, Gilles F, Yuan L, Kuo YH, Lupu F, Samara G, et al. A novel L23-related gene 40 kb downstream of the imprinted H19 gene is biallelically expressed in mid-fetal and adult human tissues. Hum Mol Genet 1995;4:1499–507.
Tung CS, Sanbonmatsu KY. Atomic model of the *Thermus thermophilus* 70S ribosome developed in silico. Biophys J 2004;87:2714–22.
Uechi T, Maeda N, Tanaka T, Kenmochi N. Functional second genes generated by retrotransposition of the X-linked ribosomal protein genes. Nucleic Acids Res 2002;30:5369–75.
Vahedi S, Chueh FY, Dutta S, Chandran B, Yu CL. Nuclear lymphocyte-specific protein tyrosine kinase and its interaction with CR6-interacting factor 1 promote the survival of human leukemic T cells. Oncol Rep 2015;34:43–50.
Van Riggelen J, Yetil A, Felsher DW. MYC as a regulator of ribosome biogenesis and protein synthesis. Nat Rev Cancer 2010;10:301–9.
Vandenbol M, Bolle PA, Dion C, Portetelle D, Hilger F. DNA sequencing of a 36.2 kb fragment located between the FAS1 and LAP loci of chromosome XI of *Saccharomyces cerevisiae*. Yeast 1994;10:S35–40.
Wang H, Yu J, Zhang L, Xiong Y, Chen S, Xing H, et al. RPS27a promotes proliferation, regulates cell cycle progression and inhibits apoptosis of leukemia cells. Biochem Biophys Res Commun 2014;446:1204–10.
Wang W, Nag S, Zhang X, Wang MH, Wang H, Zhou J, et al. Ribosomal proteins and human diseases: pathogenesis, molecular mechanisms, and therapeutic implications. Med Res Rev 2015;35:225–85.
Warner JR, McIntosh KB. How common are extraribosomal functions of ribosomal proteins? Mol Cell 2009;34:3–11.
Watanabe M, Furuno N, Goebl M, Go M, Miyauchi K, Sekiguchi T, et al. Molecular cloning of the human gene, CCG2, that complements the BHK-derived temperature-sensitive cell cycle mutant tsBN63: identity of CCG2 with the human X chromosomal SCAR/RPS4X gene. J Cell Sci 1991;100:35–43.
Wilson DN, Schluenzen F, Harms JM, Yoshida T, Ohkubo T, Albrecht R, et al. X-ray crystallography study on ribosome recycling: the mechanism of binding and action of RRF on the 50S ribosomal subunit. EMBO J 2005;24:251–60.
Wittmann HG, Stoffler G, Hindennach I, Kurland CG, Birge EA, Randall-Hazelbauer L, et al. Correlation of 30S ribosomal proteins of *Escherichia coli* isolated in different laboratories. Mol Gen Genet 1971;111:327–33.
Wolfe KH, Shields DC. Molecular evidence for an ancient duplication of the entire yeast genome. Nature 1997;387:708–13.

Wong QW, Li J, Ng SR, Lim SG, Yang H, Vardy LA. RPL39L is an example of a recently evolved ribosomal protein paralog that shows highly specific tissue expression patterns and is upregulated in ESCs and HCC tumors. RNA Biol 2014;11:33–41.
Wool IG, Chan YL, Paz V, Olvera J. The primary structure of rat ribosomal proteins: the amino acid sequences of L27a and L28 and corrections in the sequences of S4 and S12. Biochim Biophys Acta 1990;1050:69–73.
Wool IG, Chan YL, Gluck A, Suzuki K. The primary structure of rat ribosomal proteins P0, P1, and P2 and a proposal for a uniform nomenclature for mammalian and yeast ribosomal proteins. Biochimie 1991;73:861–70.
Wool IG, Chan YL, Gluck A. Structure and evolution of mammalian ribosomal proteins. Biochem Cell Biol 1995;73:933–47.
Xiong X, Zhao Y, He H, Sun Y. Ribosomal protein S27-like and S27 interplay with p53-MDM2 axis as a target, a substrate and a regulator. Oncogene 2011;30:1798–811.
Yusupov MM, Yusupova GZ, Baucom A, Lieberman K, Earnest TN, Cate JH, et al. Crystal structure of the ribosome at 5.5 A resolution. Science 2001;292:883–96.
Yusupova G, Yusupov M. High-resolution structure of the eukaryotic 80S ribosome. Annu Rev Biochem 2014;83:467–86.
Zanni G, Kalscheuer VM, Friedrich A, Barresi S, Alfieri P, Di Capua M, et al. A novel mutation in RPL10 (ribosomal protein L10) causes X-linked intellectual disability, cerebellar hypoplasia, and spondylo-epiphyseal dysplasia. Hum Mutat 2015;36:1155–8.
Zhang Y, Lu H. Signaling to p53: ribosomal proteins find their way. Cancer Cell 2009;16:369–77.
Zhang Z, Gerstein M. Identification and characterization of over 100 mitochondrial ribosomal protein pseudogenes in the human genome. Genomics 2003;81:468–80.
Zhou X, Liao WJ, Liao JM, Liao P, Lu H. Ribosomal proteins: functions beyond the ribosome. J Mol Cell Biol 2015;7:92–104.

CHAPTER 5

Ribosome diversity

Zaheenul Islam Siddiqui
Department of Microbiology, College of Medicine, Imam Abdulrahman Bin Faisal University, Dammam, Saudi Arabia

Introduction

All ribosomes have two subunits that are made of RNA and protein. A clear difference is consistently observed in the size and composition of ribosomes in organisms as different as prokaryotes and eukaryotes. Bacteria have relatively simple ribosomes of 2.3 MDa size whereas eukaryotes have larger and far more complex ribosomes with size ranging from 3.3 MDa in unicellular yeast to 4.3 MDa in multicellular and more complex organisms such as humans (Filipovska and Rackham, 2013). The bacterial ribosome small 30S subunit has 21 proteins and 16S ribosomal RNA (rRNA), and the large 50S subunit has 33 proteins, 23S and 5S rRNA. Contrary to this, the eukaryotic ribosome small 40S subunit is made up of 18S rRNA and 33 proteins. Whereas the eukaryotic ribosome large 60S subunit is made up of 46 proteins and 5S, 5.8S, and 25S rRNAs for yeast, and 47 proteins and 5S, 5.8S, and 28S rRNAs for humans (reviewed by Filipovska and Rackham, 2013). These are subtle differences in ribosomes found in organisms that are distantly positioned on the phylogenetic tree. However, when it comes to the ribosomal population within a single organism, the prevailing view is of a single type of ribosome with unchanged structure and composition. Recent studies suggest the existence of ribosome diversity within an organism, between different cell types within an organism, and even within a single cell. For example, subpopulations of heterogeneous and functionally specialized ribosomes may be produced when bacteria encounter environmental stress. Tissue-specific differences in rRNA and ribosomal protein (r-protein) composition are common in multicellular eukaryotes. Specialized ribosomes are also found in bacteria that lack specific r-proteins and are committed to translate only a subpopulation of mRNAs (reviewed by Briggs and Dinman, 2017). A recent review has explained the

Emerging Concepts in Ribosome Structure, Biogenesis, and Function
https://doi.org/10.1016/B978-0-12-816364-1.00001-9

best practices for describing specialized ribosomes (Ferretti and Karbstein, 2019). The precautions advised by the authors have been taken into consideration in this chapter.

Ribosome diversity arises due to differences in constituent r-proteins and rRNAs (Fig. 5.1). In both prokaryotes and eukaryotes, around 50% of the ribosomal content by mass is contributed by rRNA. The sequence of rRNA molecules of a ribosome is the guide for its assembly, structure, and catalytic activities. Therefore, any change in the rRNA either due to differences in encoding gene sequence or due to posttranscriptional modifications has a direct consequence on the structure and function of the ribosome (Lopez-Lopez et al., 2007). Besides, there are three possibilities which may result in diversity owing to variations in r-proteins—(i) differences in r-proteins complement, i.e., the number of r-proteins present in the ribosome, (ii) the presence of different versions of r-proteins with the

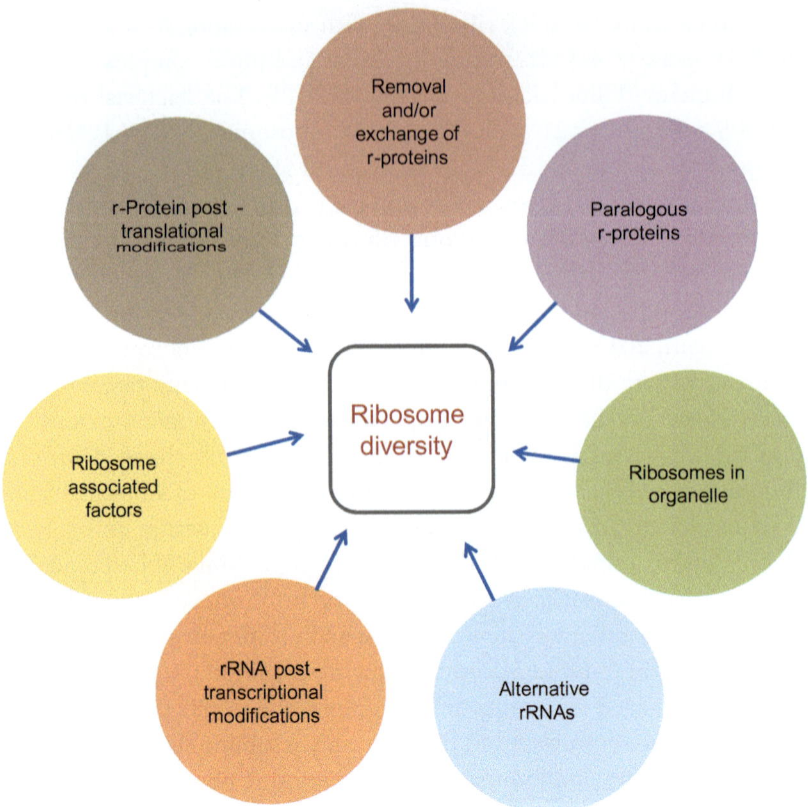

Fig. 5.1 Sources of ribosome diversity.

difference in their amino acid sequence, and (iii) differences arising due to posttranslational modifications of r-proteins.

Variation in r-protein complement

Since the discovery of ribosomes, both in prokaryotes and eukaryotes, it was presumed that it is made up of a fixed number of proteins. This perception changed after the demonstration in bacteria and slime mold that some r-proteins are dispensable, and the protein content varies based on growth conditions and differentiation (Deusser, 1972; Ramagopal and Ennis, 1981; Deusser and Wittmann, 1972). Further, it was shown in *E. coli* that there is a subpopulation of ribosomes lacking r-protein S1 which is responsible for translation of leaderless mRNA that lack 5′ untranslated region (Delvillani et al., 2011). Further in *D. discoideum*, the r-protein complement of ribosomes from ameboid cells is different from that of spores, and that this variation affects the translation specificity of these ribosomes (Ramagopal and Ennis, 1981). In a recent study on yeast and mouse models, core r-proteins complement was observed to vary based on the number of ribosomes per mRNA and the cell growth conditions (Slavov et al., 2015). Tissue-specific variations in r-protein complement were observed while studying six different human tissue samples (Gupta and Warner, 2014). Tissue-specific variations in r-proteins complement were also observed in adult mouse and their embryos, especially the differential expression of r-protein L10 in mouse embryos (Kondrashov et al., 2011; Mills et al., 1999; Wong et al., 2014). These findings suggest that some of the r-proteins are developmentally regulated during cell differentiation (Ramagopal, 1990).

The changes in r-protein complement could also be a consequence of posttranscriptional modifications affecting the stability of mRNA or the efficacy of its translation (Sauert et al., 2015). Ribosomes are found to be fully functional even after the release of several r-proteins under experimental conditions such as S1, S2, S6, S18, S12, and S21 from the mature 70S ribosome. This suggests that r-proteins complement could vary in response to environmental conditions to fine-tune ribosomal specificity and efficiency. Ample evidence is available which suggests that some ribosomes may have lesser canonical r-proteins complement, whereas others may have extra copies of specific r-proteins perhaps to the purpose of customizing ribosomes according to translational needs of the cell (reviewed by Byrgazov et al., 2013).

It is of special interest to know what happens to ribosomes (in terms of their structure and function) when one or several of their proteins become defective—partially or completely. Such ribosomes can result either from translation errors, incorrect posttranslation modifications, or changes in the physiological milieu during the process of translation (Gavrilova and Spirin, 1971; Cukras et al., 2003). Ribosomal proteins are prone to chemical changes due to the reactive oxygen species and other active radicals in the intracellular environments (Mirzaei and Regnier, 2007). Several r-proteins in the ribosome, especially those that contribute to ribosome diversity due to r-proteins complement variation, are readily diffusible with the freshly synthesized r-proteins present in the cytoplasm. This is evident from studies on *E. coli*, where it has been found that defective and inactive ribosomes become functionally active following the exchange of damaged r-proteins with undamaged r-proteins present in the surrounding environment (Byrgazov et al., 2013). This exchangeability phenomenon can further be understood by the observation that in *E. coli* ribosome, seven of the nine r-proteins encoded by spc operon are exchangeable in vivo (Pulk et al., 2010). Apart from bacteria, r-proteins exchange between the ribosome and its surrounding intracellular environment are also observed in yeast, animals, and plants (reviewed by Ballesta and Remacha, 1996). Besides the rejuvenation of defective ribosomes, the diffusion of some of the r-proteins out of the ribosome is also observed to make the already functional ribosomes specific for a subclass of mRNA. This provides a mechanism for the modulation of gene expression in response to various cellular stresses (Byrgazov et al., 2013).

Paralogues of r-proteins

Multiple versions of genes called paralogues are found to be present for several r-proteins in both prokaryotes and eukaryotes (Kellis et al., 2004; Komili et al., 2007). As an example, there are two paralogues for S14 and L31 in *B. subtilis* that differ in their zinc binding capacity. Here, the alternative r-proteins might serve the purpose of zinc storage and release under zinc-limiting conditions (Natori et al., 2007). In yeast, 70% of the paralogous proteins are observed to be expressed asymmetrically implying that these proteins may not be functional substitutes (Parenteau et al., 2011). It is proposed that there are regulatory mechanisms to ensure differential expression of these paralogues (Xue and Barna, 2012). Interestingly, *S. cerevisiae* strains lacking one or the other r-protein paralogue has distinct ribosomal populations that are functionally very different. For example, the r-protein paralogues for ribosome stalk proteins P1 and P2 form two heterodimers to bind to site A and B of P0 protein, resulting in four different combinations with nonidentical functions. The absence of any one of these heterodimers results in reduced cell growth and reduced biosynthesis of the 60S subunit (Cárdenas et al., 2012).

The presence of paralogous genes is remarkably high for the plant *A. thaliana*, where every r-protein is coded by at least two different genes and up to as much as seven genes in some cases. Proteins coded by these genes are differentially expressed during different developmental stages. Notably, several of these paralogues are formed by posttranslational modifications (Giavalisco et al., 2005). In *Arabidopsis*, r-protein S5A is highly expressed in actively dividing cells whereas r-protein S5B is expressed in cells undergoing differentiation (Weijers et al., 2001). Similarly r-protein L23A has two paralogues, where r-protein L23AA seems to be essential for the growth of the plant and maintenance of its morphological features. Further, it was found in the same study model that two paralogues of r-protein L27a differently affect ovule development, and low expression of either of them affects fertility (Zsogon et al., 2014). The presence of paralogous r-proteins is not unique to *A. thaliana*, but is common in other plants as well. For example, the rapeseed plant *B. napus* has as many as 966 genes coding for the 79 ribosomal proteins. Here the paralogues have opposite expression patterns in embryos and seedlings as compared to mature plants suggesting the possible usage of specialized ribosomes (made up of these paralogues) during the early developmental stages of the plant (Whittle and Krochko, 2009).

In the fruit fly *D. melanogaster*, ribosome heterogeneity arises due to r-protein paralogues. For example, r-protein L22 is expressed uniformly in all tissues whereas L22-like r-protein shows testis-specific expression (Kearse et al., 2011). Interestingly, r-proteins S5b, S19a, L10Aa, and L37b show enhanced expression in testis as compared to their paralogues suggesting that spermiogenesis may require specialized ribosomes to meet customized translation needs (Marygold et al., 2007).

The number of paralogous r-proteins in mammals is rather low as compared to other life forms. Till now, paralogues of only r-proteins S4, L10, L22, and L39 have been reported. In humans, r-protein S4 has three paralogues, S4X, S4Y1, and S4Y2 located on X and Y chromosomes. r-protein S4Y2 is expressed in testis and prostate, whereas r-protein S4X and S4Y1 are expressed ubiquitously suggesting that testis may have specialized ribosomes (Kearse et al., 2011; Lopes et al., 2010). The paralogues of r-proteins

S4 and L39 are highly expressed in differentiating stem cells of mouse embryonic tissues. The paralogue of r-protein L39 is specifically overexpressed in human hepatocellular carcinoma cells and some human cancer cell lines (Wong et al., 2014). Testis-specific overexpression of paralogues of r-proteins L10 and L39 were reported in studies on mouse models (Sugihara et al., 2010). In mice, r-protein L22 paralogue L22-like is not expressed in testis, but its expression is observed in the liver and mammary gland (Sugihara et al., 2010). These observations suggest that ribosomes present in different tissues may be having different paralogues of specific r-proteins which give these ribosomes special property to meet the special needs of that specific tissue (Xue and Barna, 2012).

R-protein posttranslational modifications

Posttranslational modifications of r-protein offer vast possibilities for ribosomal diversity. It has been observed in prokaryotes that ribosomes isolated from cells in stationary and exponential cell growth phases have differences in acetylation and phosphorylation levels of certain r-proteins (Macek et al., 2008; Yu et al., 2008). A good example of this is r-protien L12. Its acetylated form is called L7. In cells that are in the stationary or minimal growth phase, the acetylated form of r-protein-L7 is the predominant form. Further, the interaction of L7/12 also varies depending on its acetylation status (Gordiyenko et al., 2008). These evidence advocate protein modifications as an adaptive mechanism to fine-tune the ribosomes in accordance with the changing protein needs of bacterial cells with changing environmental conditions (Sauert et al., 2015).

In yeast, methylation, acetylation, and phosphorylation are observed in many r-proteins. These modifications affect ribosome function and cell growth (Lee et al., 2002). A study in budding yeast has demonstrated that the hydroxylation status of r-protein S23 affects the readability of ribosome nonsense (or stop) codon (Loenarz et al., 2014). r-proteins can also be modified by the addition of ubiquitin residues. It has been observed that r-protein L28 is highly ubiquitylated in S phase ribosomes as compared to those isolated from G1 phase of the cell cycle. These data strongly advocate polyubiquitylation of r-protein L28 as a mechanism for inducing the translation machinery (Spence et al., 2000).

Differences in methylation and phosphorylation patterns of r-proteins of *D. discoideum* are observed between ribosomes isolated from unicellular amoebae and multicellular slug forms of the amoeba. These modifications are believed to help the ribosome in adjusting to the translation needs of the two very different phases of *D. discoideum* life cycle (Ramagopal, 1991; Ladror et al., 2014; Lee et al., 2002). In mammalian cells, r-protein S6 is phosphorylated at multiple sites in response to external stimuli for cellular growth and proliferation. These modified ribosomes appear more specific for a subclass of mRNA harboring 5′ tract oligopyrimidine sequence (Krieg et al., 1988), whereas another study relates phosphorylation of r-protein S6 with glucose homeostasis and cell size (Ruvinsky et al., 2005). In human ribosomes, about 11 large subunit r-proteins and all small subunit r-proteins undergo posttranslational modifications (Odintsova et al., 2003; Yu et al., 2005). These include modifications at serine and threonine residues by the addition of *O*-linked β-d-*N*-acetylglucosamine (*O*-GlcNAc). This particular modification affects ribosome biogenesis and is essential for the translation machinery (Zeidan et al., 2010).

Contrary to other life forms, r-protein modifications are more common in plant ribosomes. Different types of modifications are found in at least 23 r-proteins of *A. thaliana* (Carroll et al., 2008). These modifications are involved in the induction of de novo ribosomal protein synthesis when ribosomes are damaged, for example, by ultraviolet light. Similar mechanisms of de novo protein synthesis signaled by r-protein modifications have been observed in higher plants as well as in humans (Casati and Walbot, 2004; Brenneisen et al., 2000).

Ribosome-associated factors

There are several diffusible proteins in both prokaryotes and eukaryotes termed "ribosome-associated factors" that bind to ribosomes and affect their specificity and function. At least 77 such factors have been identified in budding yeast *S. cerviceae* (Fleischer et al., 2006). Some of these factors upon binding direct the ribosomes to a subset of mRNAs. For example, the proapoptotic protein reaper found in *Drosophila* binds to a small ribosomal subunit and affects its function by interfering with the cap-dependent translation. The binding of such factors customizes the ribosome to specifically recognize proapoptotic mRNAs harboring IRES elements (Colón-Ramos et al., 2006). Another example of a ribosome-associated factor is glycogen synthase 1, which can interact with functional ribosomes and make it specific for a subpopulation of mRNAs (Fuchs et al., 2011). An interesting example of how a ribosome-associated factor can interconnect ribosomes with signaling cascades is the receptor for activated C kinase 1 (RACK1) which stabilizes the activated kinase to phosphorylate the target proteins (Adams et al., 2011; Nilsson et al., 2004). When activated, PKCβII comes in contact with ribosome-bound RACK1 resulting in phosphorylation and release of eIF6 from the large ribosomal subunit. This culminates in the joining of the two subunits to make the functional ribosome (Ceci et al., 2003). In addition, RACK1 attracts microRNA-induced silencing complex to ribosome resulting in repression of specific genes (Jannot et al., 2011). RACK1 also interacts with the integrin receptor to attach the bound ribosome to the cell membrane (Nilsson et al., 2004).

Ribosomes of higher eukaryotes have a solvent-accessible outer shell of rRNA that is exposed to the protein-rich cytoplasm giving enormous options for ribosomal interactions with nonribosomal intracellular proteins. Cytoplasmic proteins that reversibly interact with ribosomes are called ribosome-associated proteins (RAPs). Most of the RAPs are involved in cellular metabolism, cell cycle, or RNA and protein modifications. This provides a regulatory interface where the translation machinery can be modulated in accordance with the central processes of metabolism and cell division (Noeske and Cate, 2012). As an example, the RNA-binding protein FMRP binds with the preassembled 80S ribosome (Chen et al., 2014) and represses translation of specific subsets of mRNAs (Darnell et al., 2011). Similarly, the ubiquitin ligase listerin directly interacts with 60S subunit of interrupted ribosomes as a quality control pathway for the controlled degradation of nascent proteins (Shao et al., 2015). A recent murine embryo-based study identified hundreds of such RNA-binding proteins (Simsek et al., 2017). These proteins contribute to the diversity of ribosomes by altering the specificity and fidelity of ribosome and modulating gene expression as per the needs of the subcellular location.

Contribution of rRNA alternatives in ribosome diversity

The three rRNAs of prokaryotes and four rRNAs of eukaryotes are encoded by a single gene cluster called "rrn" that upon transcription make a single transcript that is later cleaved to produce the different rRNA molecules (see Chapter 3 for details). Multiple rrn copies are often present in both prokaryotic and eukaryotic genomes. This is a rich source of ribosome diversity due to the number of rrn operons, the number of genes coding for specific rRNA molecules, and the sequence variations among different rrn copies. An excellent example is a halophilic archaeon *H. marismortui* that harbors three rRNA operons rrnA, rrnB, and rrnC, where rrnB is more GC rich than rrnA and rrnC. At high temperatures, heat-stable ribosomes are generated by selective transcription of the GC-rich rrnB (Lopez-Lopez et al., 2007).

The ribosome diversity may also arise due to alternative rRNAs as in the case of protozoan parasite *P. berghei* that is known to cause malaria in certain rodents. In *P. berghei*, there are two structurally different genes encoding the small subunit rRNA. The expression of one rRNA gene predominates when the parasite is living in the mosquito whereas alternative rRNA transcript predominates when the parasite infects the mammalian host (van Spaendonk et al., 2001). In sea urchin *P. lividus*, 5S rRNA gene is present in three different clusters. Differential expression of these clusters results in structurally different 5S rRNA variants incorporated in the ribosomes contributing to rich ribosome diversity (Dimarco et al., 2012). Similarly, the genome of soil-dwelling gram-positive bacteria *S. coelicolor* has six copies of large subunit rRNA genes having differences in their sequences. These large subunit rRNA species are differentially transcribed during the morphological development of bacteria (Kim et al., 2008).

The genome of bacterium *B. subtilis* has 10 rrn operons all of which are important in defining the essential processes of bacterial life, such as generation time, sporulation, and cell motility. These operons are differentially expressed in the bacterium and are important for bacterial life (Yano et al., 2013). In *E. coli*, there are seven rrns that have considerable differences in their sequences, enough for producing rRNA of slightly different structures. Promoters of these operons are not equally turned on in response to external stimuli, resulting in the production of distinct ribosome subpopulations. Under limited nutrient conditions, a specific type of 16S rRNA predominates, resulting in functionally distinct ribosomes that are resistant to tetracycline and that are more inclined to a specific type of RNA helicase (Kurylo et al., 2018).

In humans and mice, many rrns with sequence variations are dispersed throughout the genome (Parks et al., 2018). These rrns are differentially expressed depending on the tissue type and the stage of the life cycle (Parks et al., 2018; Tseng et al., 2008). Interestingly in zebra fish, ribosomes isolated from embryonic stages are more similar to that of oocytes, but their rRNAs sequences are quite different from those isolated from the adult stages (Locati et al., 2017). Ribosome diversity that arises due to rRNAs variations is speculated to be useful for long-term adaptations to new environmental conditions (Ferretti and Karbstein, 2019).

Modifications of rRNA contribute to ribosome diversity

Some nucleotides of rRNA may be modified during the process of ribosome assembly. These modifications are acquired gradually during the assembly of r-proteins with rRNA (Siibak and Remme, 2010, reviewed by Bohne, 2014). About 2% of the total rRNA content of *S. cerviceae* undergoes transcriptional modifications; the prominent change being 2′-*O*-methylations of ribose or formation of pseudouridine by uridine isomerization. Such modifications are essential for structural stability and proper orientation of rRNA molecules (Liang et al., 2007; Polikanov et al., 2015). Most of the rRNA modifications are acquired during assembly and are irreversible, hence are not used as a short-term strategy for adaptation to environmental changes. It should be noted that rRNA modifications are guided by nucleolar RNAs called box C/D and box H/ACA snoRNAs in eukaryotes and archaea, and by proteins in bacteria (Bachellerie et al., 2002). rRNA modifications are under tight regulation and are a rich source of ribosome diversity. An exemplary of ribosome diversity due to rRNA modifications is in *T. brucei*, a parasite that causes sleeping sickness in humans. It is observed that pseudouridylation of the rRNAs is specifically upregulated when the parasite is present in the host blood stream (Chikne et al., 2016). Methylation of rRNAs is considered a common mechanism for regulation of lifespan and stress resistance in flies, worms, and yeast (Schosserer et al., 2015). Studies on *X. tropicalis*, a species of frog, have shown that the ubiquitylation status of Tcof1, a ribosome biogenesis factor, is involved in rRNA modification. This affects the translation of mRNAs that code for proteins involved in neural crest formation (Werner et al., 2015; Walker-Kopp et al., 2017; Larsen et al., 2014). Altered expression of snoRNAs is also found in various types of cancers. For example, *SNORA42* and *SNORA21* are overexpressed in lung and colorectal cancers, respectively (Mei et al., 2012; Yoshida et al., 2017). Therefore, it is tempting to speculate that altered expression of snoRNAs may be a cause of carcinogenesis via altering the modifications of rRNA (Ferretti and Karbstein, 2019).

Apart from the mainstream rRNA modifications that occur during ribosome assembly, other forms of modifications also exist in some organisms that occur directly in the assembled and fully functional ribosomes. A very good example of this is *E. coli* endoribonuclease MazF, which is activated when the cells encounter stress conditions (Aizenman et al., 1996). Activated MazF cleaves 43 nucleotides at the 3′ end of 16S rRNA that includes the anti-Shine and Dalgarno sequence essential for initiation of translation from mRNAs. This results in the production of a subpopulation of ribosomes that selectively translates leaderless mRNAs (lmRNAs) (Vesper et al., 2011). Simultaneously, MazF also catalyzes the cleavage of untranslated regions (UTR) of distinct transcripts to generate lmRNAs (Vesper et al., 2011). However, these observations have been challenged in a recent work that argues that these special ribosomes and lmRNAs are degradation intermediates of global mRNA and rRNA degradations that occur when cells are grown under stress conditions (Culviner and Laub, 2018).

Another example of rRNA modification in assembled ribosomes is that of an enzyme, erythromycin resistance methyltransferases (Erm). The enzyme dimethylates a nucleotide at position A2058 of large ribosomal subunit rRNA resulting in unusual interactions of the growing peptide with the modified rRNA in the peptide exit tunnel (Gupta et al., 2013; Starosta et al., 2010). It will be of interest to investigate the cell fate when there are no rRNA modifications or the process of modification becomes defective. Interestingly, yeast strains harboring impaired pseudouridine synthase display decreased ribosome affinities for tRNAs. Further, these ribosomes show decreased translational fidelity and IRES-dependent translational initiation (Jack et al., 2011).

Special ribosomes of mitochondria

The intracellular organelles, mitochondria and plastids, are remnants of intracellular symbiont bacteria acquired by the eukaryotic cell during the course of evolution. Nevertheless, ribosomes of these organelles are significantly different from their bacterial and cytoplasmic counterparts in terms of structure and function. Mitochondrial ribosomes are larger in size than the bacterial but smaller than the cytoplasmic counterpart (reviewed by Ferretti and Karbstein, 2019). Special proteins are found in these ribosomes at key positions that interact with mRNA, tRNA, and the growing polypeptide chain. Some of these ribosomal proteins have homologues in bacteria while others are unique to the mitochondrial ribosome. The peptide exit channel in the mitochondrial ribosome is lined with hydrophobic protein residues, in accordance with the fact that most of the mitochondrial DNA coded proteins are committed for insertion in mitochondrial membranes (Ott et al., 2016). Another example of unique mitochondria-specific ribosome structure is the r-protein L45. It can attach to the inner mitochondrial membrane and facilitate the insertion of other mitochondrial proteins into the mitochondrial membrane (Greber et al., 2014; Englmeier et al., 2017). In ribosomes of budding yeast, the mRNA exit channel is specialized to facilitate translation activators to interact with 5′ untranslated regions of the mitochondrial mRNAs (Herrmann et al., 2013; Desai et al., 2017).

Congenital anomalies and mutations of ribosomal proteins

Natural mutations in genes coding for ribosomal components may also affect the structure and function of ribosomes. This all depends on the site of mutation and the role of the affected component in the ribosome. Several human birth defects have been linked to mutations in ribosomal proteins (Gazda et al., 2008; also refer to Chapter 11). Diamond-Blackfan anemia (DBA) is a disease with congenital anomalies such as limb defects, cleft palate, abnormalities in heart development, growth failure, and vulnerability for cancer (Boria et al., 2010). Patients with DBA have mutations in genes for several r-proteins including S19, S24, S17, L5, L11, and L35A. Mutations in L5 are very common in children born with cleft palate (Gazda et al., 2008). Similarly, hereditary hypotrichosis simplex (HHS), a disease characterized by progressive hair loss early in childhood, is linked with L21 mutations (Zhou et al., 2011). These birth defects might be a consequence of tissue-specific roles of individual r-proteins in embryonic development or their role as part of the ribosome. Further, these observations suggest tissue-specific expression of r-proteins and their suspected roles during embryonic development (Xue and Barna, 2012).

Summary and conclusions

The conventional scientific view of ribosomes is that of uniform molecular factories constructed according to identical specifications. However, as the interest of the scientific community on ribosome research intensified in the previous decade, evidence started accumulating in favor of a rather diverse ribosome population. Ribosome diversity arises due to differences in the constituent proteins and rRNAs. This diversity seems to be employed by nature as a tool to adapt to environmental constraints encountered by various life forms. Besides, ribosomes lacking specific r-proteins are sometimes used by nature as a strategy to control gene expression (Briggs and Dinman, 2017). Modifications of r-proteins are involved in ribosome adaptations that are useful, for example, in adjusting to different life phases of an organism (Ladror et al., 2014), accelerate the translation process (Spence et al., 2000), promote cell growth (Sauert et al., 2015), or customize specific ribosomes for a subclass of mRNA (Krieg et al., 1988).

Exchange of defective r-proteins on the ribosome with freshly made r-proteins in the cytoplasm serves as a rejuvenation strategy for the ribosome (Byrgazov et al., 2013). Ribosome-associated factors/proteins can also diffuse in and out of the ribosome to quickly adapt to the changing environmental conditions. This process of diffusion is facilitated in higher eukaryotes due to the outer solvent-accessible rRNA shell which serves as an active interface between the proteins in the cytoplasm and the ribosomes (Simsek et al., 2017; Joazeiro, 2017; Shao et al., 2015; Chen et al., 2014).

Multiples rrns in a genome, which may have slightly different sequences, when expressed differentially, or selectively, produce diverse populations of rRNAs. This is useful in producing a diversity of ribosomes that can survive and adapt to high temperatures (Lopez-Lopez et al., 2007), antibiotics (Kurylo et al., 2018), and different physiological conditions (van Spaendonk et al., 2001). Further, some rrns are expressed in a tissue-specific manner, producing customized ribosomes for a specific tissue or a specific phase of the life cycle (Parks et al., 2018).

Similar to r-proteins modifications, the bases of rRNA can also be modified. These modifications occur either during the ribosome assembly process or in the fully functional ribosomes (reviewed by Bohne, 2014). Modifications are essential for structural stability and proper orientation of rRNA molecules (Polikanov et al., 2015). Ribosome diversity that arises due to rRNA modifications affects its fidelity, its affinity toward tRNAs, and its interactions with the exiting nascent peptide (Jack et al., 2011).

Diversity attained by rRNA alternatives or modifications is generally a strategy for long-term ribosomal adaptations. In contrast, the diversity that arises due to r-proteins complement variation, paralogues, or modifications is used for short-term adjustment. This is because r-proteins, especially those involved in diversity, are exchangeable with the pool of freshly synthesized r-proteins in the cytoplasm. In contrast, the rRNAs, being the main structural component of a ribosome, cannot be exchanged (reviewed by Bohne, 2014). Specialized ribosomes are found in organelles such as mitochondria and plastids that resemble more with the prokaryotic ribosomes than their cytoplasmic counterparts (Xue and Barna, 2012).

Some mutations in r-protein are strongly associated with human birth defects. It should be analyzed carefully before arriving at a conclusion, if the anomalies observed are due to individual effects of tissue-specific expression of r-proteins, as an extratranslational function, or because of change in ribosomal activity due to the mutated r-protein (Xue and Barna, 2012). Careful analysis is also needed while working on, or reporting a newly found ribosome variant, as what is considered as a ribosome variant may be a stable intermediate of degrading ribosomes (Ferretti and Karbstein, 2019).

In summary, ribosome diversity is the result of various types of changes in its constituent rRNA and proteins. The resulting heterogeneity offers a plethora of options for the ribosomes to adapt and adjust to the diverse environmental conditions in which they can be found.

References

Adams DR, Ron D, Kiely PA. RACK1, a multifaceted scaffolding protein: structure and function. Cell Commun Signal 2011;9:22.

Aizenman E, Engelberg-Kulka H, Glaser G. An *Escherichia coli* chromosomal "addiction module" regulated by guanosine [corrected] 30, 50-bispyrophosphate: a model for programmed bacterial cell death. PNAS 1996;93:6059–63.

Bachellerie JP, Cavaillé J, Hüttenhofer A. The expanding snoRNA world. Biochimie 2002;84:775–90.

Ballesta JP, Remacha M. The large ribosomal subunit stalk as a regulatory element of the eukaryotic translational machinery. Prog Nucleic Acid Res Mol Biol 1996;55:157–93.

Bohne AV. The nucleoid as a site of rRNA processing and ribosome assembly. Front Plant Sci 2014;5:257.

Boria I, Garelli E, Gazda HT, Aspesi A, Quarello P, Pavesi E, Ferrante D, Meerpohl JJ, Kartal M, Da Costa L, Proust A, Leblanc T, Simansour M, Dahl N, Fröjmark AS, Pospisilova D, Cmejla R, Beggs AH, Sheen MR, Landowski M, Buros CM, Clinton CM, Dobson LJ, Vlachos A, Atsidaftos E, Lipton JM, Ellis SR, Ramenghi U, Dianzani I. The ribosomal basis of diamond–Blackfan anemia: mutation and database update. Hum Mutat 2010;31:1269–79.

Brenneisen P, Wenk J, Wlaschek M, Krieg T, Scharffetter-Kochanek K. Activation of p70 ribosomal protein S6 kinase is an essential step in the DNA damage-dependent signaling pathway responsible for the ultraviolet B-mediated increase in interstitial collagenase (MMP-1) and stromelysin-1 (MMP-3) protein levels in human dermal fibroblasts. J Biol Chem 2000;275:4336–44.

Briggs JW, Dinman JD. Subtractional heterogeneity: a crucial steptoward defining specialized ribosomes. Mol Cell 2017;67:3–4.

Byrgazov K, Vesper O, Moll I. Ribosome heterogeneity: another level of complexity in bacterial translation regulation. Curr Opin Microbiol 2013;16:133–9.

Cárdenas D, Revuelta-Cervantes J, Jiménez-Díaz A, Camargo H, Remacha M, Ballesta JP. P1 and P2 protein heterodimer binding to the P0 protein of *Saccharomyces cerevisiae* is relatively non-specific and a source of ribosomal heterogeneity. Nucleic Acids Res 2012;40:4520–9.

Carroll AJ, Heazlewood JL, Ito J, Millar AH. Analysis of the *Arabidopsis* cytosolic ribosome proteome provides detailed insights into its components and their post-translational modification. Mol Cell Proteomics 2008;7:347–69.

Casati P, Walbot V. Cross linking of ribosomal proteins to RNA in maize ribosomes by UV-B and its effects on translation. Plant Physiol 2004;136:3319–32.

Ceci M, Gaviraghi C, Gorrini C, Sala LA, Offenhäuser N, Marchisio PC, Biffo S. Release of eIF6 (p27BBP) from the 60S subunit allows 80S ribosome assembly. Nature 2003;426:574–9.

Chen E, Sharma MR, Shi X, Agrawal RK, Joseph S. Fragile X mental retardation protein regulates translation by binding directly to the ribosome. Mol Cell 2014;54:407–17.
Chikne V, Doniger T, Rajan KS, Bartok O, Eliaz D, Cohen-Chalamish S, Tschudi C, Unger R, Hashem Y, Kadener S, Michaeli S. A pseudouridylation switch in rRNA is implicated in ribosome function during the life cycle of *Trypanosoma brucei*. Sci Rep 2016;6:25296.
Colón-Ramos DA, Shenvi CL, Weitzel DH, Gan EC, Matts R, Cate J, et al. Direct ribosomal binding by a cellular inhibitor of translation. Nat Struct Mol Biol 2006;13:103–11.
Cukras AR, Southworth DR, Brunelle JL, Culver GM, Green R. Ribosomal proteins S12 and S13 function as control elements for translocation of the mRNA: tRNA complex. Mol Cell 2003;12:321–8.
Culviner PH, Laub MT. Global analysis of the *E. coli* toxin MazF reveals widespread cleavage of mRNA and the inhibition of rRNA maturation and ribosome biogenesis. Mol Cell 2018;70:868–80.
Darnell JC, Van Driesche SJ, Zhang C, Hung KY, Mele A, Fraser CE, Stone EF, Chen C, Fak JJ, Chi SW, Licatalosi DD, Richter JD, Darnell RB. FMRP stalls ribosomal translocation on mRNAs linked to synaptic function and autism. Cell 2011;146:247–61.
Delvillani F, Papiani G, Deho G, Briani F. S1 ribosomal protein and the interplay between translation and mRNA decay. Nucleic Acids Res 2011;39:7702–15.
Desai N, Brown A, Amunts A, Ramakrishnan V. The structure of the yeast mitochondrial ribosome. Science 2017;355:528–31.
Deusser E. Heterogeneity of ribosomal populations in *Escherichia coli* cells grown in different media. Mol Gen Genet 1972;119:249–58.
Deusser E, Wittmann HG. Ribosomal proteins: variation of the protein composition in *Escherichia coli* ribosomes as function of growth rate. Nature 1972;238:269–70.
Dimarco E, Cascone E, Bellavia D, Caradonna F. Functional variants of 5S rRNA in the ribosomes of common sea urchin *Paracentrotus lividus*. Gene 2012;508:21–5.
Englmeier R, Pfeffer S, Förster F. Structure of the human mitochondrial ribosome studied in situ by cryoelectron tomography. Structure 2017;25:1574–1581.e2.
Ferretti MB, Karbstein K. Does functional specialization of ribosomes really exist? RNA 2019;25:521–38.
Filipovska A, Rackham O. Specialization from synthesis: how ribosome diversity can customize protein function. FEBS Lett 2013;587:1189–97.
Fleischer TC, Weaver CM, McAfee KJ, Jennings JL, Link AJ. Systematic identification and functional screens of uncharacterized proteins associated with eukaryotic ribosomal complexes. Genes Dev 2006;20:1294–307.
Fuchs G, Diges C, Kohlstaedt LA, Wehner KA, Sarnow P. Proteomic analysis of ribosomes: translational control of mRNA populations by glycogen synthase GYS1. J Mol Biol 2011;410:118–30.
Gavrilova LP, Spirin AS. Stimulation of 'non-enzymic' translocation in ribosomes by *p*-chloromercuribenzoate. FEBS Lett 1971;17:324–6.
Gazda HT, Sheen MR, Vlachos A, Choesmel V, O'Donohue MF, Schneider H, Darras N, Hasman C, Sieff CA, Newburger PE, Ball SE, Niewiadomska E, Matysiak M, Zaucha JM, Glader B, Niemeyer C, Meerpohl JJ, Atsidaftos E, Lipton JM, Gleizes PE, Beggs AH. Ribosomal protein L5 and L11 mutations are associated with cleft palate and abnormal thumbs in diamond–Blackfan anemia patients. Am J Hum Genet 2008;83:769–80.
Giavalisco P, Wilson D, Kreitler T, Lehrach H, Klose J, Gobom J, Fucini P. High heterogeneity within the ribosomal proteins of the *Arabidopsis thaliana* 80S ribosome. Plant Mol Biol 2005;57:577–91.
Gordiyenko Y, Deroo S, Zhou M, Videler H, Robinson CV. Acetylation of L12 increase interactions in the *Escherichia coli* ribosomal stalk complex. J Mol Biol 2008;380:404–14.

Greber BJ, Boehringer D, Leitner A, Bieri P, Voigts-Hoffmann F, Erzberger JP, Leibundgut M, Aebersold R, Ban N. Architecture of the large subunit of the mammalian mitochondrial ribosome. Nature 2014;505:515–9.

Gupta P, Sothiselvam S, Vazquez-Laslop N, Mankin AS. Deregulation of translation due to post-transcriptional modification of rRNA explains why erm genes are inducible. Nat Commun 2013;4:1984.

Gupta V, Warner JR. Ribosome-omics of the human ribosome. RNA 2014;20:1004–13.

Herrmann JM, Woellhaf MW, Bonnefoy N. Control of protein synthesis in yeast mitochondria: the concept of translational activators. Biochim Biophys Acta 2013;1833:286–94.

Jack K, Bellodi C, Landry DM, Niederer RO, Meskauskas A, Musalgaonkar S, Kopmar N, Krasnykh O, Dean AM, Thompson SR, Ruggero D, Dinman JD. rRNA pseudouridylation defects affect ribosomal ligand binding and translational fidelity from yeast to human cells. Mol Cell 2011;44:660–6.

Jannot G, Bajan S, Giguère NJ, Bouasker S, Banville IH, Piquet S, Hutvagner G, Simard MJ. The ribosomal protein RACK1 is required for microRNA function in both *C. elegans* and humans. EMBO Rep 2011;12:581–6.

Joazeiro CA. Ribosomal stalling during translation: providing substrates for ribosome-associated protein quality control. Annu Rev Cell Dev Biol 2017;33:343–68.

Kearse MG, Chen AS, Ware VC. Expression of ribosomal protein L22e family members in *Drosophila melanogaster*: rpL22-like is differentially expressed and alternatively spliced. Nucleic Acids Res 2011;39:2701–16.

Kellis M, Birren BW, Lander ES. Proof and evolutionary analysis of ancient genome duplication in the yeast *Saccharomyces cerevisiae*. Nature 2004;428:617–24.

Kim HL, Song WS, Kim K, Lee K. Characterization of heterogeneous LSU rRNA profiles in *Streptomyces coelicolor* under different growth stages and conditions. Curr Microbiol 2008;57:537–41.

Komili S, Farny NG, Roth FP, Silver PA. Functional specificity among ribosomal proteins regulates gene expression. Cell 2007;131:557–71.

Kondrashov N, Pusic A, Stumpf CR, Shimizu K, Hsieh AC, Ishijima J, Shiroishi T, Barna M. Ribosome mediated specificity in Hox mRNA translation and vertebrate tissue patterning. Cell 2011;145:383–97.

Krieg J, Hofsteenge J, Thomas G. Identification of the 40S ribosomal protein S6 phosphorylation sites induced by cycloheximide. J Biol Chem 1988;263:11473–7.

Kurylo CM, Parks MM, Juette MF, Zinshteyn B, Altman RB, Thibado JK, Vincent CT, Blanchard SC. Endogenous rRNA sequence variation can regulate stress response gene expression and phenotype. Cell Rep 2018;25:236–48.

Ladror DT, Frey BL, Scalf M, Levenstein ME, Artymiuk JM, Smith LM. Methylation of yeast ribosomal protein S2 is elevated during stationary phase growth conditions. Biochem Biophys Res Commun 2014;445:535–41.

Larsen DH, Hari F, Clapperton JA, Gwerder M, Gutsche K, Altmeyer M, Jungmichel S, Toledo LI, Fink D, Rask MB, Grøfte M, Lukas C, Nielsen ML, Smerdon SJ, Lukas J, Stucki M. The NBS1-treacle complex controls ribosomal RNA transcription in response to DNA damage. Nat Cell Biol 2014;16:792–803.

Lee SW, Berger SJ, Martinović S, Pasa-Tolić L, Anderson GA, Shen Y, Zhao R, Smith RD. Direct mass spectrometric analysis of intact proteins of the yeast large ribosomal subunit using capillary LC/FTICR. PNAS 2002;99:5942–7.

Liang X, Liu Q, Fournier MJ. rRNA modifications in an intersubunit bridge of the ribosome strongly affect both ribosome biogenesis and activity. Mol Cell 2007;28:965–77.

Locati MD, Pagano JFB, Girard G, Ensink WA, van Olst M, van Leeuwen S, Nehrdich U, Spaink HP, Rauwerda H, Jonker MJ, Dekker RJ, Breit TM. Expression of distinct maternal and somatic 5.8S, 18S, and 28S rRNA types during zebra fish development. RNA 2017;23:1188–99.

Loenarz C, Sekirnik R, Thalhammer A, Ge W, Spivakovsky E, Mackeen MM, McDonough MA, Cockman ME, Kessler BM, Ratcliffe PJ, Wolf A, Schofield CJ. Hydroxylation of the eukaryotic ribosomal decoding center affects translational accuracy. PNAS 2014;111:4019–24.

Lopes AM, Miguel RN, Sargent CA, Ellis PJ, Amorim A, Affara NA. The human RPS4 paralogue on Yq11.223 encodes a structurally conserved ribosomal protein and is preferentially expressed during spermatogenesis. BMC Mol Biol 2010;11:33.

Lopez-Lopez A, Benlloch S, Bonfa M, Rodriguez-Valera F, Mira A. Intragenomic 16S rDNA divergence in *Haloarcula marismortui* is an adaptation to different temperatures. J Mol Evol 2007;65:687–96.

Macek B, Gnad F, Soufi B, Kumar C, Olsen JV, Mijakovic I, Mann M. Phosphoproteome analysis of *E. coli* reveals evolutionary conservation of bacterial Ser/Thr/Tyr phosphorylation. Mol Cell Proteomics 2008;7:299–307.

Marygold SJ, Roote J, Reuter G, Lambertsson A, Ashburner M, Millburn GH, Harrison PM, Yu Z, Kenmochi N, Kaufman TC, Leevers SJ, Cook KR. The ribosomal protein genes and minute loci of *Drosophila melanogaster*. Genome Biol 2007;8:R216.

Mei YP, Liao JP, Shen J, Yu L, Liu BL, Liu L, Li RY, Ji L, Dorsey SG, Jiang ZR, Katz RL, Wang JY, Jiang F. Small nucleolar RNA 42 acts as an oncogene in lung tumorigenesis. Oncogene 2012;31:2794–804.

Mills AA, Mills MJ, Gardiner DM, Bryant SV, Stanbridge EJ. Analysis of the pattern of QM expression during mouse development. Differentiation 1999;64:161–71.

Mirzaei H, Regnier F. Identification of yeast oxidized proteins: chromatographic top-down approach for identification of carbonylated, fragmented and cross-linked proteins in yeast. J Chromatogr 2007;1141:22–31.

Natori Y, Nanamiya H, Akanuma G, Kosono S, Kudo T, Ochi K, Kawamura F. A fail-safe system for the ribosome under zinc-limiting conditions in *Bacillus subtilis*. Mol Microbiol 2007;63:294–307.

Nilsson J, Sengupta J, Frank J, Nissen P. Regulation of eukaryotic translation by the RACK1 protein: a platform for signalling molecules on the ribosome. EMBO Rep 2004;5:1137–41.

Noeske J, Cate JH. Structural basis for protein synthesis: snapshots of the ribosome in motion. Curr Opin Struct Biol 2012;22:743–9.

Odintsova TI, Müller EC, Ivanov AV, Egorov TA, Bienert R, Vladimirov SN, Kostka S, Otto A, Wittmann-Liebold B, Karpova GG. Characterization and analysis of posttranslational modifications of the human large cytoplasmic ribosomal subunit proteins by mass spectrometry and Edman sequencing. J Protein Chem 2003;22:249–58.

Ott M, Amunts A, Brown A. Organization and regulation of mitochondrial protein synthesis. Annu Rev Biochem 2016;85:77–101.

Parenteau J, Durand M, Morin G, Gagnon J, Lucier JF, Wellinger RJ, Chabot B, Elela SA. Introns within ribosomal protein genes regulate the production and function of yeast ribosomes. Cell 2011;147:320–31.

Parks MM, Kurylo CM, Dass RA, Bojmar L, Lyden D, Vincent CT, Blanchard SC. Variant ribosomal RNA alleles are conserved and exhibit tissue-specific expression. Sci Adv 2018;4:eaao0665.

Polikanov YS, Melnikov SV, Söll D, Steitz TA. Structural insights into the role of rRNA modifications in protein synthesis and ribosome assembly. Nat Struct Mol Biol 2015;22:342–4.

Pulk A, Liiv A, Peil L, Maiväli U, Nierhaus K, Remme J. Ribosome reactivation by replacement of damaged proteins. Mol Microbiol 2010;75:801–14.

Ramagopal S. Induction of cell-specific ribosomal proteins in aggregation-competent nonmorphogenetic *Dictyostelium discoideum*. Biochem Cell Biol 1990;68:1281–7.

Ramagopal S. Covalent modifications of ribosomal proteins in growing and aggregation-competent *Dictyostelium discoideum*: phosphorylation and methylation. Biochem Cell Biol 1991;69:263–8.

Ramagopal S, Ennis HL. Regulation of synthesis of cell-specific ribosomal proteins during differentiation of *Dictyostelium discoideum*. PNAS 1981;78:3083–7.

Ruvinsky I, Sharon N, Lerer T, Cohen H, Stolovich-Rain M, Nir T, Dor Y, Zisman P, Meyuhas O. Ribosomal protein S6 phosphorylation is a determinant of cell size and glucose homeostasis. Genes Dev 2005;19:2199–211.

Sauert M, Temmel H, Moll I. Heterogeneity of the translational machinery: variations on a common theme. Biochimie 2015;114:39–47.

Schosserer M, Minois N, Angerer TB, Amring M, Dellago H, Harreither E, Calle-Perez A, Pircher A, Gerstl MP, Pfeifenberger S, Brandl C, Sonntagbauer M, Kriegner A, Linder A, Weinhäusel A, Mohr T, Steiger M, Mattanovich D, Rinnerthaler M, Karl T, Sharma S, EntianKD KM, Breitenbach M, Wilson IB, Polacek N, Grillari-Voglauer R, Breitenbach-Koller L, Grillari J. Methylation of ribosomal RNA by NSUN5 is a conserved mechanism modulating organismal lifespan. Nat Commun 2015;6:6158.

Shao S, Brown A, Santhanam B, Hegde RS. Structure and assembly pathway of the ribosome quality control complex. Mol Cell 2015;57:433–44.

Siibak T, Remme J. Subribosomal particle analysis reveals the stages of bacterial ribosome assembly at which rRNA nucleotides are modified. RNA 2010;16:2023–32.

Simsek D, Tiu GC, Flynn RA, Byeon GW, Leppek K, Xu AF, Chang HY, Barna M. The mammalian ribo-interactome reveals ribosome functional diversity and heterogeneity. Cell 2017;169:1051–65.

Slavov N, Semrau S, Airoldi E, Budnik B, van Oudenaarden A. Differential stoichiometry among core ribosomal proteins. Cell Rep 2015;13:865–73.

van Spaendonk RM, Ramesar J, van Wigcheren A, Eling W, Beetsma AL, van Gemert GJ, Hooghof J, Janse CJ, Waters AP. Functional equivalence of structurally distinct ribosomes in the malaria parasite, *Plasmodium berghei*. J Biol Chem 2001;276:22638–47.

Spence J, Gali RR, Dittmar G, Sherman F, Karin M, Finley D. Cell cycle-regulated modification of the ribosome by a variant multiubiquitin chain. Cell 2000;102:67–76.

Starosta AL, Karpenko VV, Shishkina AV, Mikolajka A, Sumbatyan NV, Schluenzen F, Korshunova GA, Bogdanov AA, Wilson DN. Interplay between the ribosomal tunnel, nascent chain, and macrolides influences drug inhibition. Chem Biol 2010;17:504–14.

Sugihara Y, Honda H, Iida T, Morinaga T, Hino S, Okajima T, Matsuda T, Nadano D. Proteomic analysis of rodent ribosomes revealed heterogeneity including ribosomal proteins L10-like, L22-like 1, and L39-like. J Proteome Res 2010;9:1351–66.

Tseng H, Chou W, Wang J, Zhang X, Zhang S, Schultz RM. Mouse ribosomal RNA genes contain multiple differentially regulated variants. PLoS One 2008;3:e1843.

Vesper O, Amitai S, Belitsky M, Byrgazov K, Kaberdina AC, Engelberg-Kulka H, Moll I. Selective translation of leaderless mRNAs by specialized ribosomes generated by MazF in *Escherichia coli*. Cell 2011;147:147–57.

Walker-Kopp N, Jackobel AJ, Pannafino GN, Morocho PA, Xu X, Knutson BA. Treacher Collins syndrome mutations in *Saccharomyces cerevisiae* destabilize RNA polymerase I and III complex integrity. Hum Mol Genet 2017;26:4290–300.

Weijers D, Franke-van Dijk M, Vencken RJ, Quint A, Hooykaas P, Offringa R. An arabidopsis minute-like phenotype caused by a semi-dominant mutation in a ribosomal protein S5 gene. Development 2001;128:4289–99.

Werner A, Iwasaki S, McGourty CA, Medina-Ruiz S, Teerikorpi N, Fedrigo I, Ingolia NT, Rape M. Cell-fate determination by ubiquitin-dependent regulation of translation. Nature 2015;525:523–7.

Whittle CA, Krochko JE. Transcript profiling provides evidence of functional divergence and expression networks among ribosomal protein gene paralogs in *Brassica napus*. Plant Cell 2009;21:2203–19.

Wong QW, Li J, Ng SR, Lim SG, Yang H, Vardy LA. RPL39L is an example of a recently evolved ribosomal protein paralog that shows highly specific tissue expression patterns and is upregulated in ESCs and HCC tumors. RNA Biol 2014;11:33–41.

Xue S, Barna M. Specialized ribosomes: a new frontier in gene regulation and organismal biology. Nat Rev Mol Cell Biol 2012;13:355–69.

Yano K, Wada T, Suzuki S, Tagami K, Matsumoto T, Shiwa Y, Ishige T, Kawaguchi Y, Masuda K, Akanuma G, Nanamiya H, Niki H, Yoshikawa H, Kawamura F. Multiple rRNA operons are essential for efficient cell growth and sporulation as well as outgrowth in *Bacillus subtilis*. Microbiology 2013;159:2225–36.

Yoshida K, Toden S, Weng W, Shigeyasu K, Miyoshi J, Turner J, Nagasaka T, Ma Y, Takayama T, Fujiwara T, Goel A. SNORA21—an oncogenic small nucleolar RNA, with a prognostic biomarker potential in human colorectal cancer. EBioMedicine 2017;22:68–77.

Yu Y, Ji H, Doudna JA, Leary JA. Mass spectrometric analysis of the human 40S ribosomal subunit: native and HCV IRES-bound complexes. Protein Sci 2005;14:1438–46.

Yu BJ, Kim JA, Moon JH, Ryu SE, Pan JG. The diversity of lysine-acetylated proteins in *Escherichia coli*. J Microbiol Biotechnol 2008;18:1529–36.

Zeidan Q, Wang Z, De Maio A, Hart GW. O-GlcNAc cycling enzymes associate with the translational machinery and modify core ribosomal proteins. Mol Biol Cell 2010;21:1922–36.

Zhou C, Zang D, Jin Y, Wu H, Liu Z, Du J, Zhang J. Mutation in ribosomal protein L21 underlies hereditary hypotrichosis simplex. Hum Mutat 2011;32:710–4.

Zsogon A, Szakonyi D, Shi X, Byrne ME. Ribosomal protein RPL27a promotes female gametophyte development in a dose-dependent manner. Plant Physiol 2014;165:1133–43.

CHAPTER 6

Ribosome cycle—Assembly, degradation, and recycling

Purnima Tyagi and Vijay Kumar
Department of Molecular and Cellular Medicine, Institute of Liver and Biliary Sciences, New Delhi, India

Introduction

Translation is the process of protein biosynthesis in which mRNA sequence is copied to chain of amino acids or polypeptides with the help of ribosomes (Dever et al., 2016). Ribosomes are made up of ribosomal proteins and ribosomal RNA (rRNA). Ribosomes are composed of ribosomal proteins and ribosomal RNA (rRNA). Each ribosome has a large and a smaller subunit, which has a characteristic shape. The small and large subunits of prokaryotes are designated as 30S and 50S, respectively, whereas the eukaryotic ribosomes contain a small 40S subunit and a large 60S subunit (Petrov et al., 2015). The 40S ribosomal subunits of eukaryotes have 35 proteins and 18S rRNA, whereas in the 60S large subunits have 53 proteins and three rRNAs, viz., 5S, 5.8S, and 28S. At the time of synchronized biogenesis of eukaryotic ribosomal subunits, rRNAs are processed and assembled to subunit-specific proteins and make up the functional subunits (de la Cruz et al., 2015). The processing of rRNA (posttranscriptional modifications) takes place in a highly sequential order and is completed in the nucleolus—a specialized nuclear compartment (Tuorto and Parlato, 2019).

The ribosome assembly process can be divided into at least six distinct steps for the production of functional subunits of ribosomes: (i) transcription of pre-rRNAs; (ii) translation of ribosomal protein mRNAs to proteins in cytoplasm and their transport into the nucleolus; (iii) modification of rRNA bases; (iv) pre-rRNA folding; (v) rRNAs and ribosomal proteins assembly; and finally (vi) removal of external and internal transcribed spacers from rRNA, viz., 5-ETS, ITS1, ITS2, and 3-ETS by endonucleolytic and exonucleolytic processing (Kressler et al., 2010). To complete the ribosome assembly process, more than 200 ribosome biogenesis factors also known as assembly factors are involved along with ~80 small nucleolar RNAs (snoRNAs) (Greber, 2016;

Emerging Concepts in Ribosome Structure, Biogenesis, and Function
https://doi.org/10.1016/B978-0-12-816364-1.00005-6

Kressler et al., 2010). Besides, ribosomal assembly is an energy-consuming process requiring coordinated regulation of various cellular signaling processes (Kressler et al., 2008; Strunk and Karbstein, 2009). Moreover, in aggressively dividing cancer cells and some other diseases, the synthesis and assembly of ribosomal RNAs and proteins show a marked increase owing to increased cell proliferation (Kim et al., 2014; Wang et al., 2015).

Degradation of abrogated proteins is an inbuilt quality control mechanism in stressed cells (Amm et al., 2014; Gardner et al., 2005). Accumulation of such proteins may be harmful to the cell as this could trigger endoplasmic reticulum (ER) stress and imbalance in the ratio of rRNA: ribosomal proteins and their variation in ribosomal subunits (Fujii et al., 2009; Perry, 2007; Xue and Barna, 2012; Zhou et al., 2015). The abrogated proteins are degraded with the help of ubiquitin-proteasomal system (UPS) in two consecutive steps. In the first step, targeted proteins are modified by a set of enzymes including Ub-activating enzyme (E1), Ub-carrier protein (E2), and Ub-conjugating enzyme (E3). These enzymes accomplish chemical reactions involving activation by adenylation at the C-terminus of E1, acceptance of activated ubiquitin (Ub) by E2 at active site, and recognition E2 with activated Ub and conjugate targeted proteins by E3 enzyme (Glickman and Ciechanover, 2002; Kleiger and Mayor, 2014). In the following step, the polyubiquitinylated target proteins are chopped by the 26S proteasome machinery in small pieces of peptides (Kleiger and Mayor, 2014; Marshall and Vierstra, 2019). The ribosomes in mammalian cells can also be degraded by ribophagy, which requires a putative cargo receptor NUFIP1 (nuclear fragile X mental retardation-interacting protein 1) (Wyant et al., 2018). Under nutrient starvation conditions, NUFIP1 transports its ribosome cargo to autophagic vesicles by directly binding to microtubule-associated light chain 3B (LC3B) protein in order to promote cell survival (Wyant et al., 2018). Ribophagy is also induced under conditions of proteotoxic stress and chromosome mis-segregation in mammalian cells (An and Harper, 2018). The molecular mechanism of ribophagy and the potential for degradation of nonselective cargo during selective autophagy needs to be explored.

Recycling of ribosome must occur when polypeptide synthesis is stalled or if empty ribosomes accumulate in the cell (Moore, 2009). Ribosomes are often stalled while translating hairpins mRNAs that do not have a stop codon. The stalled ribosomes are recycled as a part of mRNA surveillance (Klauer and van Hoof, 2012; van Hoof and Wagner, 2011). The process of ribosome recycling is initiated by dissociation of the two subunits of ribosomes (also known as ribosome splitting) from mRNA followed by their reassociation to initiate a new round of mRNA translation (Hellen, 2018; Jackson et al., 2010). In addition, many ribosome splitting factors can rescue stalled ribosomes under stress conditions (Zhang et al., 2015). Ribosome splitting is regulated quite differently in eukaryotes, prokaryotes, and in archaea (Lecompte et al., 2002; Pesole et al., 1995). In eukaryotes and archaea, the ATP-binding cassette subfamily E member 1 (ABCE1) protein binds to eukaryotic translation termination factor 1 (eRF1) on the posttermination complex and ATP hydrolysis leads to a power stroke that splits the posttermination ribosome. The splitting results in separation of 60S subunit, eRF1, ABCE1, and a 40S subunit still bound to messenger RNA (mRNA) as well as deacylated transfer RNA (tRNA) (Nürenberg-Goloub et al., 2018; Pisarev et al., 2010). The empty ribosomes accumulate when cells face nutritional stress resulting in halting of protein synthesis. However, the empty ribosomes are recycled after the normal condition has been attained (Li et al., 2014).

Ribosome assembly

There is rather scanty information about the drivers of ribosome assembly. The assembly of both ribosomal subunits begins with a common large unprocessed 47S pre-rRNAs in the nucleolus (de la Cruz et al., 2015; Peña et al., 2017). The radioactive pulse labeling experiments have shown that 47S precursor rRNA are cotranscriptionally processed into 18S, 5.8S, and 28S rRNA in different subcellular compartments (Iadevaia et al., 2012; Nazar, 2004). Analysis of subcellular fractions of yeast and growing human cells after sucrose density gradient centrifugation show the presence of 90S, 66S, and 45S particles suggesting that the ribosome assembly process is conserved from yeast to human (Ebersberger et al., 2014). The 90S and 66S particles carry more proteins to RNA ratio and, therefore, are termed as "preribosome particles" (Harnpicharnchai et al., 2001; Trapman et al., 1975). The preribosome particles carry rRNA intermediates and ribosomal proteins along with some processing and assembly factors (Gallagher et al., 2004; Phipps et al., 2011). The fibrillar centers of nucleoli harbor rDNA genes and RNA polymerase I and are sites for the transcription of rRNAs (Gallagher et al., 2004). The dense fibrillar compartments which surround the fibrillar centers are considered to be the actual site for the association of nascent rRNAs and ribosomal proteins as well as different processing and assembly factors. The processing of pre-rRNA is much like pre-mRNAs for splicing but without rejoining after *exo-* and endonucleolytic cleavage. The large

45S preribosome is processed to produce mature 28S and 5.8S rRNA in the nucleolus, whereas the final processing of 18S rRNA takes place in the cytoplasmic compartment of the cell (Huang, 2002; Trumtel et al., 2000). The snoRNA U3 is mainly involved in the processing of 18S rRNA, whereas U14 and snR30 are known for the endonucleolytic cleavage at the A0, A1, and A2 sites of the pre-rRNA (Kufel et al., 2003). Based on the presence of C/D or A/ACA sequence motifs, the snoRNAs are categorized in two families. Out of nearly 80 snoRNAs, 45 have C/D box and remaining have A/ACA box that regulate methylation at 2′O and pseudouridinylation, respectively (Dupuis-Sandoval et al., 2015). Interestingly, not all snoRNAs are involved in the processing of rRNAs. For example, U3 snoRNA is essential for maintaining rRNAs in an unfolded immature state at a particular site by providing structural support during the maturation process of 90S preribosome (Dupuis-Sandoval et al., 2015). So far, nearly 130 specific rRNA modifications have been identified including methylation and pseudouridinylation and also confirmed in the three-dimensional (3D) structure of 80S ribosome by crystallography and electron microscopic studies (Sloan et al., 2017). In addition to snoRNAs and nonribosomal proteins, some helicases and acetyl transferases are also involved in the assembly of ribosomes during the initial stage of rRNAs processing. It is believed that helicases are required for the disassembly and removal of snoRNAs from the rRNA leading to proper folding, processing, and promote maturation of ribosome (Kressler et al., 2010).

The pathways and steps of ribosome assembly

Out of four rRNAs, three rRNAs, viz., 18S, 5.8S, and 25S are transcribed by RNA polymerase I (Gallagher et al., 2004), whereas 5S rRNA is transcribed by RNA polymerase III (Turowski and Tollervey, 2016). RNA pol I transcribes a 45S pre-rRNA that carries mature rRNAs sequences along with two intervening internal transcribed spacers (ITS1 and ITS2) and two external transcribed spacers (ETS1 and ETS2) one each at 5′ and 3′ ends (Fig. 6.2) (Gallagher et al., 2004). The processing of 45S pre-rRNA is completed with the help of both *exo*- and endonucleases (Henras et al., 2015). The recruitment of most ribosomal proteins occurs cotranscriptionally resulting in the formation of 90S preribosome intermediates. The assembly factors are recruited later, which is short lived. Therefore, assembly factors are rarely seen in the mature ribosomal subunit (Aubert et al., 2018). Nevertheless, different stages of preribosomes can be characterized by the presence of specific assembly factors and rRNA intermediates (Wu et al., 2016).

The ribosomal subunits are assembled in six discrete steps (Fig. 6.2): (i) the earliest step represented by 90S preribosome in which 5′ ETS1 and ITS1 regions are cleaved. The 27S pre-rRNA is cleaved by endonuclease at A2 site to generate 20S pre-rRNA. Both 20S and 27S pre-rRNAs associate with more assembly factors to form 43S and 66S structures that are precursors of 40S and 60S subunits of ribosome, respectively. Thus, at this step, the assembly of 40S and 60S subunit is independent of each other. (ii) An early step specific for 60S subunit maturation, removal of ITS1 sequences between the 27S A2 and 27S A3 sites to produce 27SB pre-rRNA and binding of more assembly factors to form 66S preribosome. (iii) Intermediate step that includes cleavage by endonuclease at C2 site in ITS2 of 27S B, generation of 7S and 25.5S rRNAs and their exit from the nucleolus. (iv) Late nuclear step in which 7S and 25.5S pre-rRNAs are processed to 5.8S and 25S rRNAs, respectively, including removal of ITS2 from 7S pre-rRNA and adjustment of the central protuberance. (v) Second last step that involves export of newly synthesized particle via the nuclear pore complex into the cytoplasm with the help of a Ran-GTPase-dependent system. (vi) Finally, maturation of subunits, processing of 5.8S rRNA, association of remaining ribosomal proteins, and removal of all assembly factors (Aubert et al., 2018; de la Cruz et al., 2015; Konikkat and Woolford Jr., 2017; Nazar, 2004; Phipps et al., 2011).

Assembly of small subunit of ribosome (40S subunit)

The 40S subunit is processed before the assembly of small subunit processome (SSU) or 90S preribosomes (Fig. 6.1). The SSUs are cotranscriptionally made with the gradual binding of chaperone complexes UTP-A, UTP-B, UTP-C, and Mpp10 (Gallagher, 2019; Henras et al., 2015). A detailed analysis of the purified SSU has revealed the existence of more than 20 assembly factors, several ribosomal proteins, U3-snoRNA, and 35S rRNA particle (Zhang et al., 2016). The assembly of eukaryotes 40S ribosome subunit is initiated by cleavage at A0, A1, and A2 sites of 5′ ETS by U3 snoRNAs and release of 20S pre-rRNA (Nazar, 2004). Several assembly factors are removed after finishing their role and additional factors are recruited along with more ribosomal proteins. The complex is now termed as 43S or late intermediate SSU (Greber, 2016). The SSU is exported into the cytoplasm in steady state from the nucleus with the help of exportins in association with Ran-GTPs (Yoneda, 2000). The nucleoplasmic 43S units carry the same structure and components as cytoplasmic 43S except for a "beak." The formation "beak" in the cytoplasmic 43S is linked to the phosphorylation/dephosphorylation

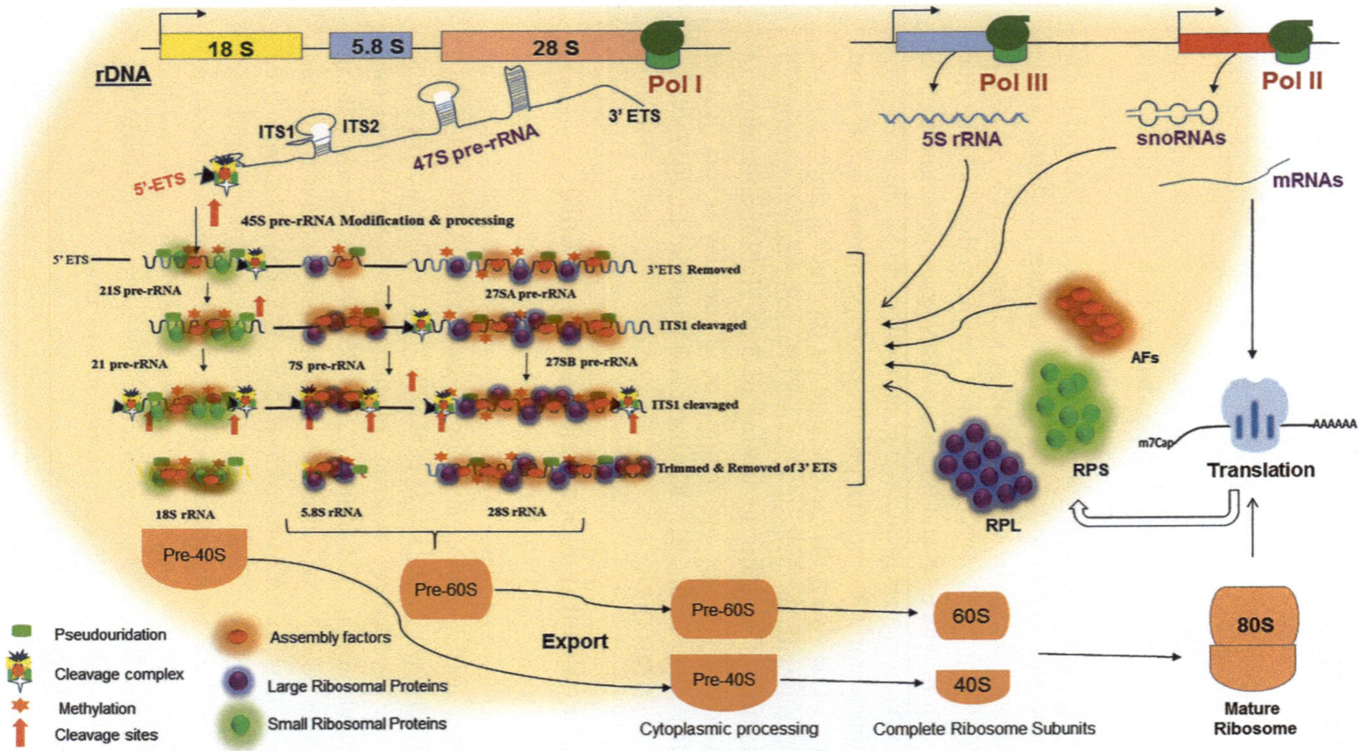

Fig. 6.1 Ribosome assembly. Eukaryotic ribosome assembly begins with RNA Pol I-driven transcription of rDNA repeats in the nucleolus to produce 47S pre-rRNA, the precursor to mature 18S, 5.8S, and 28S rRNAs. The emerging 47S pre-rRNA associates with U3, several small nucleolar RNAs (snoRNAs) transcribed by RNA Pol II and 5S rRNA transcribed by RNA Pol III. Ribosomal proteins and assembly factors translated in cytoplasm and imported to nucleus where they associate with pre-rRNAs, these associations are responsible for the processing of pre-rRNAs to mature 18S, 5.8S, and 28S. Rapidly associating assembly factors, snoRNA, and 5S rRNAs drive maturation of transcripts as they travel across the nucleoplasm and eventually translocate to the cytoplasm. This process is aided by multiple transport receptors that facilitate translocation of preribosomal through NPCs into the cytoplasm.

of a complex of small ribosomal protein 3 (RPS3) and assembly factor Enp1 (essential nuclear protein 1) by the serine/threonine protein kinase HRR25 (Panse and Johnson, 2010). The cytoplasmic 43S is processed by an endonuclease (Nob1) and assembly factors (e.g., Rio1, Rio2, Tsr1, and Fap7) to form 40S small subunit of the ribosome along with mature 18S rRNA (de la Cruz et al., 2015; Karbstein, 2011). There are two proteins which play role in the pre-40S subunits to 40S subunits S10 and S26, two of the three last r-proteins to be associated with pre-40S subunits (Ben-Shem et al., 2011). The S10 and S26 proteins are localized, respectively, on the entrance and exit sites of 40S ribosomes. S10 binding to pre-40S subunits to the beak structure is blocked by Ltv1, while Pno1/Dim2 are responsible to block the premature binding of S26. After phosphorylation via kinase HRR25, Ltv1 removed from pre-40S subunits (Strunk et al., 2011). However, the details of Pno1 release are not well understood (Ghalei et al., 2014). Nevertheless, it participates in the quality control of ribosome assembly through binding and release of assembly factors.

Assembly of large subunit of ribosome (60S subunit)

The assembly of 60S ribosome subunit from 66S preribosome is fairly well understood (Konikkat and Woolford Jr., 2017). The process starts with the cleavage at the A2 site of pre-rRNA to produce first detectable precursor termed as 66S preribosome particle (Fig. 6.1). The 66S preribosome particle is composed up of 27SA pre-rRNAs, most of the ribosomal proteins and over 30 assembly factors but no snoRNA (Nissan et al., 2002). The successive step of processing involves addition of 40 factors, eight helicases and some snoRNAs (Kos and Tollervey, 2005; Liang and Fournier, 2006). Next, the phosphorylation of Las1 endonuclease by Grc3 kinase leads to cleavage of ITS2 from the C2 site of 27SB pre-rRNA to generate 7S and 26S pre-rRNAs and the maturation of 5.8S rRNAs. Trimming of both 5.8 and 26 rRNAs at the 5′end are processed, respectively, by the nuclear exosome and Rat1 or Rai1 exonucleases (Liang and Fournier, 2006). The recruitment of 5S rRNA along with its partners (Rpl5 and Rpl11) provides specificity to 5S rRNA and allows conformational change in it required for proper assembly of 60S preribosome (Ciganda and Williams, 2011; Zhang et al., 2007). The associated ribosomal proteins are weakly bound to preribosome and thus, are unable to gain native interaction with rRNAs. Then, the early assembly factors are replaced with late assembly factors to facilitate the processing of pre-rRNAs (Espinar-Marchena et al., 2017). This hierarchical event of processing leads to the formation of peptidyl transferase center (PTC), which acts like checkpoint by dissociating Rsa4 (ribosome assembly protein 4) via ATPase (Bassler et al., 2010), and Nug2 (nucleolar GTP-binding protein 2) via GTPase (Im et al., 2011). Once passed, the 60S preribosome is exported into the cytoplasm with the help of nuclear exportin factors.

Degradation of the ribosome and ribosomal proteins

Ubiquitination—Selective degradation of ribosomal proteins

Ribosomal proteins are relatively long lived once assembled with ribosomes and short lived in their free form. The free ribosomal proteins are generally degraded by the ubiquitin-proteasome system (UPS) (Amm et al., 2014), whereas starvation-induced degradation of ribosomes occurs via autophagy (ribophagy) to promote cell survival (Fig. 6.2) (An and Harper, 2020). The quality control pathways are very restricted on misfolded and unassembled ribosomal proteins such that unnecessary excess proteins are degraded rapidly via UPS after their synthesis (Mathis et al., 2017; Sung et al., 2016a). Earlier it was believed that ubiquitin-mediated degradation is strictly focused on ribosomal proteins assembled in 40S subunits in the cytoplasm. However, till date it is unclear if all ribosomal proteins express equally in all cells. During the assembly process, misfolded and misassembled proteins require additional quality control substrate due to frequent degradation of these misassembled proteins (Amm et al., 2014; Sung et al., 2016b). It has been observed that if unassembled proteins escape proteasomal degradation process, these proteins aggregate as sedimentable particle and controlled by intranuclear quality control pathways (Takalo et al., 2013). The KEGG pathway analysis has revealed that a substantial fraction of ribosomal proteins are present near the ubiquitination sites of both yeast and human cells resulting in the accumulation of ribosomal protein-ubiquitin conjugates during the lag phase of yeast (Sung et al., 2016b).

Ribophagy—Selective degradation of ribosome

Ribophagy is a special type of catabolic process in the cell involving selective destruction of bulk of preribosomes and mature ribosomes in response to stress conditions (Kraft et al., 2008). As ribosome is considered a molecular machine with a definitive structure and function, it is important to regulate its number as well as to maintain its quality for a proper translational event under favorable environmental conditions (Beese et al., 2020).

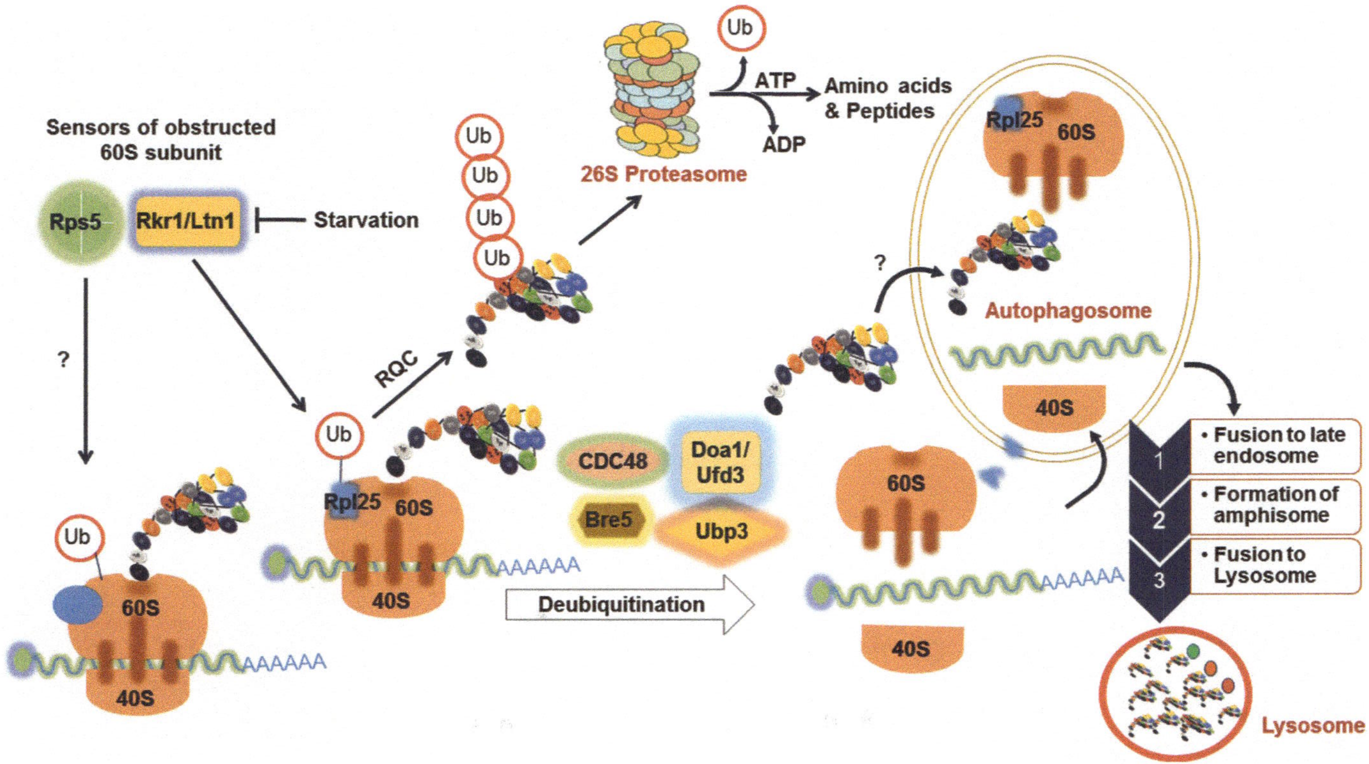

Fig. 6.2 Degradation of ribosomes and ribosomal proteins. Rps5 and Rkr1/Ltn1 are responsible to detect the obstruction in 60S subunit or cell starvation and ubiquitinate ribosomal proteins. In eukaryotes, ribosome-associated quality control (RQC) senses elongation-stalled ribosomes and promotes dissociation of ribosomal subunits. Deubiquitination of ribosomes occurs with the help of Cdc40, Doa1/Ufd3. Bre5, and UBp3 complex required for degradation of large ribosomal subunit. Antagonistic interplay between Ltn1 and Ubp3 complex for Rpl25 suggests a dynamically regulated specific ribophagy signal.

This regulation may be essential for adaption to adverse environmental condition involving recycling of the building blocks of ribosome to some other biological processes (Huang, 2002). The cellular proteins are primarily degraded by two major pathways, viz., macro-autophagy and ubiquitin-proteasome system (UPS), both being indirectly connected to each other (Fig. 6.2) (Lilienbaum, 2013; Wang et al., 2015). The ribosomal proteins are degraded by autophagy but independent of the classical pathways of autophagy that involve cytoplasm to vacuole movement. Mutational analysis of ubiquitin proteases revealed that degradation of ribosomes depends on specific or nonspecific autophagy process (Kraft et al., 2008). A comparative analysis of turnover of low- and high-abundance proteins along with ribosomal proteins suggests that ribosomal proteins have a higher turnover. Another report suggests that nonfunctional ribosome particles are ubiquitinated in a Rtt101-Mms1-dependent manner (Fujii et al., 2009).

The degradation of small and large subunit proteins is triggered by starvation and is mediated by the mTOR pathway (An and Harper, 2018). However, there is a requirement for subunit dissociation prior to protein degradation. The ribosomal proteins are degraded through interaction with the Atg8 family of proteins in the autophagy machinery (Kirkin and Rogov, 2019). A selective ribophagy receptor for ribosome has been identified from human cell lines known as nuclear fragile X mental retardation-interacting protein 1 (NUFIP1) (Wyant et al., 2018). The Ubp3 and Bre5 deubiquitinase complex is shown to be specifically required for degradation of the large ribosomal subunit (Fig. 6.2) (Kraft et al., 2008). Furthermore, in degradation of 60S subunit, mono-ubiquitination plays a significant role and works as an internalization signal. Deubiquitination of one or more ubiquitinated, misfolded, or damaged ribosomal proteins triggers the engulfment by vacuole and de novo formation of isolation membranes. This is followed by the formation of autophagosome that finally fused with autophagic body for degradation by resident hydrolases (Noda et al., 2002). The Ubp3 complex deubiquitinates lysine 74 on ribosomal protein Rpl25, the same residue that is ubiquitinated by listerin E3 ubiquitin-protein ligase 1 (Ltn1). The ubiquitination of Rpl25 serves to protect ribosomes from autophagy-mediated degradation. Thus, the antagonistic interplay between Ltn1 and the Ubp3 complex for Rpl25 suggests a dynamically regulated specific ribophagy signal (Fig. 6.2) (Ossareh-Nazari et al., 2014). Ribosomes can also be removed by NEDD4 family E3 ubiquitin-protein ligase Rsp5 (Fang et al., 2014; Sommer et al., 2014). Thus, the involvement of ubp3 in ribophagy establishes a link between the proteasomal and autophagy-mediated degradation of ribosomal proteins, RNAs, and the complete ribosome. A defect in ribophagy and proteasomal-mediated proteins often leads to the deleterious diseases and disorders (Tanaka and Matsuda, 2014).

Ribosome recycling

Ribosome recycling enables ribosomes and mRNAs to participate in multiple rounds of translation by recycling of the posttermination complex. It allows ribosomes to be reused for translation of the same transcript or a new transcript (Seshadri and Varshney, 2006). The ribosome recycling is dependent on the presence of ATP-binding cassette protein 1 (ABCE1) and the eukaryotic translation termination factor 1 (eRF1) at the A site (Fig. 6.3). ABCE1 is involved in the splitting of 80S ribosome to the 60S and 40S subunits at low concentrations of $Mg2+$ (Mancera-Martínez et al., 2017; Pisarev et al., 2010). Thus, ribosome recycling includes recycling of vacant 80S ribosomes as well as that of stalled ribosomal elongation complexes (Pisarev et al., 2010).

Splitting of 80S ribosome

The process of translational termination is associated with the binding of eRF1 and eRF3 to the entry site of translation elongating 80S and formation of ternary complex (Beißel et al., 2019). The N-terminal domain of eRF1 is responsible for the recognition of stop codon at the entry site of ribosome, whereas the GGQ sequences present in its middle domain are responsible for the release of nascent polypeptide from the peptidyl-tRNA at P site (Taylor et al., 2012). The C-terminal domain of eRF1 carries binding sites for the eRF3 and ABCE1 (Pisarev et al., 2010). eRF3 is associated with the poly(A)-binding protein (PABP) (Cosson et al., 2002) and nonsense-mediated decay factor UPF3b (Raimondeau et al., 2018). The middle domain of eRF3 has GTP-binding site for the energy stock by the hydrolysis of GTP (Kozlov and Gehring, 2010), whereas its C-terminal domain is required for binding of rescue factor Hbs1 (Hilal et al., 2016) and formation of the stable eRF1/eRF3•GTP ternary complex (Kononenko et al., 2010; Taylor et al., 2012). The eRF3 is a translational GTPase, associated with process of termination of translation with Hbs1. Hbs1 induces GTP hydrolysis to release cognate aminoacyl-tRNA from the nascent polypeptide at the peptidyl transferase center (Khade and Joseph, 2010; Simonović and Steitz, 2009).

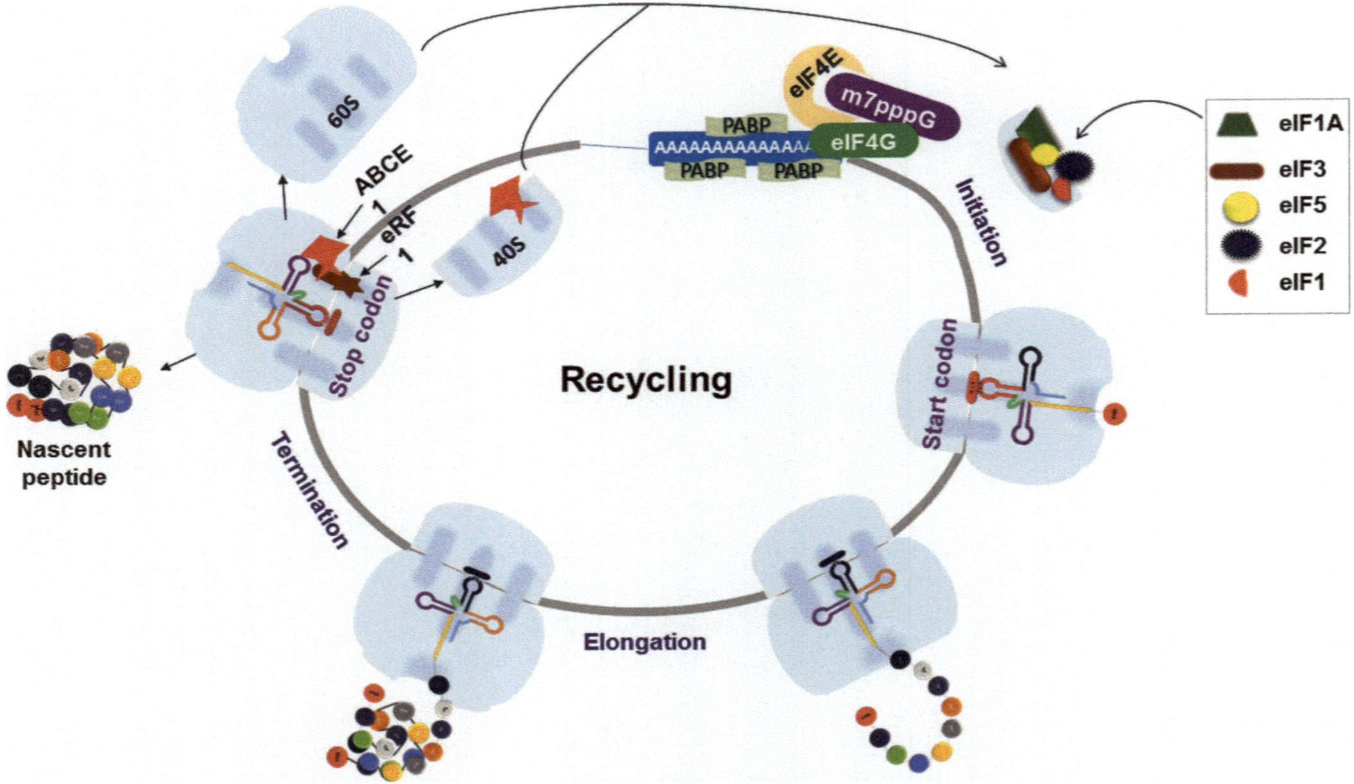

Fig. 6.3 Ribosome recycling. The ribosome recycling is mediated by ABCE1 and eRF1 at the A site. After A site recognition by the eRF1-eRF3-GTP ternary complex, eRF3 dissociates after GTP hydrolysis and allows the proper positioning of the GGQ motif in the central domain of eRF1, which facilitates peptide release and formation of termination complex with the deacyl-tRNA in the P state or P/E hybrid state. ABCE1 binds the posttermination complex, contacts eRF1, and facilitates ribosome splitting.

ABCE1-dependent 80S ribosome splitting

ABCE1 has two identical nucleotide-binding domains (NBDs) with two merged nucleotide-binding sites formed by motifs from both domains—a helix-loop-helix motif in NBD1 and an unique amino-terminal FeS domain containing two [4Fe-4S]2+ clusters, connected to the NBD core via a hinged cantilever arm (Barthelme et al., 2007, 2011; Karcher et al., 2008). The NBDs adopt an open state when bound to ADP or free nucleotide but switch to a closed form upon interaction with ATP (Heuer et al., 2017). It has been observed that ABCE1 recruited on the 80S ribosome remains in a semiclosed state (Preis et al., 2014) but must undergo domain closure to manifest its ATPase activity and participate in ribosome recycling. Many succeeding conformational changes in ABCE1 can destabilize intersubunit bridges leading to ribosomal splitting. ABCE1 binds in the ribosomal intersubunit space interacting via the HLH and hinge elements with sites on 40S subunit (h5-h15 and h8-h14), which establish contact points for ribosome-bound translational GTPases (such as Erf3) via NBD2 at a single site (rpL9) on the 60S subunit. ATP binds to the identical nucleotide-binding sites in ABC proteins (closed state). Hydrolysis of ATP and release of ADP (leading to an "open" state) make conformational changes in these proteins and associated domains (Rees et al., 2009). ATP hydrolysis by ABCE1 requires for split eRF1-bound posttermination complexes (post-TCs) and Pelota-associated ribosomal complexes (Pisarev et al., 2010; Pisareva et al., 2011; Shoemaker and Green, 2011). eIF3 is the principal factor that promotes splitting of post-termination ribosomes into 60S subunits and tRNA- and mRNA-bound 40S subunits. Its activity is enhanced by eIF3j, eIF1, and eIF1A. eIF1 also mediates release of P-site tRNA, whereas eIF3j ensures subsequent dissociation of mRNA (Pisarev et al., 2007, 2010).

Initiation of translation by post-TCs

The translation starts with the formation of 43S initiation complex including initiator tRNA-Met-eIF2-GTP and initiation factors eIF3, eIF1, eIF1A, and eIF5 with 40S subunit (Jivotovskaya et al., 2006), scanning and initiate codon-anticodon base pairing between initiator tRNAs and mRNAs result in the formation of 48S complex. Along with hydrolysis of GTP by eIF2 and eIF5, 60S subunit is recruited and finally the assembly of 80S complex is completed (Pisareva and Pisarev, 2014). During the recycling of ribosome, formation of the 48S complex is dependent on the abundance of necessary factors Met-tRNAMeti, eIFs 1, 1A, 2, 3, 4A, 4B, and 4F along with sufficient amount of free Mg2+ (Pisareva et al., 2008; Pisareva and Pisarev, 2014). After the splitting of 80S complex to 40S and 60S subunits, 40S remains associated with eIF3 in order to prevent their reassociation with 60S subunit. ABCE1 is also involved in the initiation of translation by post-TCs by facilitating the recruitment of eIF2 on the preinitiation complex (Heuer et al., 2017). Thus, ABCE1 is the heart of ribosome recycling as well as initiator of translation by engaging eIF2a, eIF2, eIF3, and eIF5 (Valásek, 2012).

Summary and conclusions

Ribosome cycle involves ribosome assembly, degradation of ribosomes, and ribosomal proteins and their recycling. Ribosome assembly is a complex process extending from the nucleolus to the cytoplasm. It involves the processing and folding of the pre-rRNA and many associated assembly factors. It is a tightly regulated activity that is intimately linked to biogenesis of ribosomes and fundamental cellular process including growth and cell division. Ribophagy is process of selective degradation of ribosomal assembled proteins under stress conditions and is essential to maintain quality for a proper translational event under favorable environmental conditions in the cell. Any defect in ribophagy often leads to the deleterious diseases and disorders.

Splitting of the 80S ribosome in ribosome recycling by the ATP-binding cassette protein 1 to release the 60S subunit. Interruptions in the ribosome recycling result in reinitiation of translation near the stop codon and accumulation of empty ribosomes during nutritional stress. A stop codon enters the A site of the ribosome mediated by release factors eRF1 and eRF3 in eukaryotes which form a ternary complex with GTP during termination of mRNA translation. eRF1 recognizes the stop codon and after hydrolysis of GTP by eRF 3 mediates release of the nascent peptide. The posttermination complex is the disassembled facilitating its constituents to participate in further rounds of translation.

References

Amm I, Sommer T, Wolf DH. Protein quality control and elimination of protein waste: the role of the ubiquitin-proteasome system. Biochim Biophys Acta 2014;1843:182–96.

An H, Harper JW. Systematic analysis of ribophagy in human cells reveals bystander flux during selective autophagy. Nat Cell Biol 2018;20:135–43.

An H, Harper JW. Ribosome abundance control via the ubiquitin-proteasome system and autophagy. J Mol Biol 2020;432:170–84.

Aubert M, O'Donohue MF, Lebaron S, Gleizes PE. Pre-ribosomal RNA processing in human cells: from mechanisms to congenital diseases. Biomolecules 2018;28. pii: E123.

Barthelme D, Scheele U, Dinkelaker S, Janoschka A, Macmillan F, Albers SV, Driessen AJ, Stagni MS, Bill E, Meyer-Klaucke W, Schünemann V, Tampé R. Structural organization of essential iron-sulfur clusters in the evolutionarily highly conserved ATP-binding cassette protein ABCE1. J Biol Chem 2007;282:14598–607.

Barthelme D, Dinkelaker S, Albers SV, Londei P, Ermler U, Tampé R. Ribosome recycling depends on a mechanistic link between the FeS cluster domain and a conformational switch of the twin-ATPase ABCE1. Proc Natl Acad Sci U S A 2011;108:3228–33.

Bassler J, Kallas M, Pertschy B, Ulbrich C, Thoms M, Hurt E. The AAA-ATPase Rea1 drives removal of biogenesis factors during multiple stages of 60S ribosome assembly. Mol Cell 2010;38:712–21.

Beese CJ, Brynjólfsdóttir SH, Frankel LB. Selective autophagy of the protein homeostasis machinery: ribophagy, proteaphagy and ER-phagy. Front Cell Dev Biol 2020;7:373.

Beißel C, Neumann B, Uhse S, Hampe I, Karki P, Krebber H. Translation termination depends on the sequential ribosomal entry of eRF1 and eRF3. Nucleic Acids Res 2019;47:4798–813.

Ben-Shem A, Garreau de Loubresse N, Melnikov S, Jenner L, Yusupova GZ, Yusupov MM. The structure of the eukaryotic ribosome at 3.0Å resolution. Science 2011;334:1524–9.

Ciganda M, Williams N. Eukaryotic 5S rRNA biogenesis. Wiley Interdiscip Rev RNA 2011;2:523–33.

Cosson B, Berkova N, Couturier A, Chabelskaya S, Philippe M, Zhouravleva G. Poly(A)-binding protein and eRF3 are associated in vivo in human and Xenopus cells. Biol Cell 2002;94:205–16.

de la Cruz J, Karbstein K, Woolford Jr JL. Functions of ribosomal proteins in assembly of eukaryotic ribosomes in vivo. Annu Rev Biochem 2015;84:93–129.

Dever TE, Kinzy TG, Pavitt GD. Mechanism and regulation of protein synthesis in *Saccharomyces cerevisiae*. Genetics 2016;203:65–107.

Dupuis-Sandoval F, Poirier M, Scott MS. The emerging landscape of small nucleolar RNAs in cell biology. Wiley Interdiscip Rev RNA 2015;6:381–97.

Ebersberger I, Simm S, Leisegang MS, Schmitzberger P, Mirus O, von Haeseler A, Bohnsack MT, Schleiff E. The evolution of the ribosome biogenesis pathway from a yeast perspective. Nucleic Acids Res 2014;42:1509–23.

Espinar-Marchena FJ, Babiano R, Cruz J. Placeholder factors in ribosome biogenesis: please, pave my way. Microb Cell 2017;4:144–68.

Fang NN, Chan GT, Zhu M, Comyn SA, Persaud A, Deshaies RJ, Rotin D, Gsponer J, Mayor T. Rsp5/Nedd4 is the main ubiquitin ligase that targets cytosolic misfolded proteins following heat stress. Nat Cell Biol 2014;16:1227–37.

Fujii K, Kitabatake M, Sakata T, Miyata A, Ohno M. A role for ubiquitin in the clearance of nonfunctional rRNAs. Genes Dev 2009;23:963–74.

Gallagher JE, Dunbar DA, Granneman S, Mitchell BM, Osheim Y, Beyer AL, Baserga SJ. RNA polymerase I transcription and pre-rRNA processing are linked by specific SSU processome components. Genes Dev 2004;18:2506–17.

Gallagher JEG. Proteins and RNA sequences required for the transition of the t-Utp complex into the SSU processome. FEMS Yeast Res 2019;19:foy120.

Gardner RG, Nelson ZW, Gottschling DE. Degradation-mediated protein quality control in the nucleus. Cell 2005;120:803–15.

Ghalei H, Schaub FX, Doherty JR, Noguchi Y, Roush WR, Cleveland JL, Stroupe ME, Karbstein K. Hrr25/CK1d-directed release of Ltv1 from pre-40S ribosomes is necessary for risobome assembly and cellular growth. J Cell Biol 2014;208:745–59.

Glickman MH, Ciechanover A. The ubiquitin-proteasome proteolytic pathway: destruction for the sake of construction. Physiol Rev 2002;82:373–428.

Greber BJ. Mechanistic insight into eukaryotic 60S ribosomal subunit biogenesis by cryo-electron microscopy. RNA 2016;22:1643–62.
Harnpicharnchai P, Jakovljevic J, Horsey E, Miles T, Roman J, Rout M, Meagher D, Imai B, Guo Y, Brame CJ, Shabanowitz J, Hunt DF, Woolford Jr JL. Composition and functional characterization of yeast 66S ribosome assembly intermediates. Mol Cell 2001;8:505–15.
Hellen CUT. Translation termination and ribosome recycling in eukaryotes. Cold Spring Harb Perspect Biol 2018;10:1–18.
Henras AK, Plisson-Chastang C, O'Donohue MF, Chakraborty A, Gleizes PE. An overview of pre-ribosomal RNA processing in eukaryotes. Wiley Interdiscip Rev RNA 2015;6:225–42.
Heuer A, Gerovac M, Schmidt C, Trowitzsch S, Preis A, Kötter P, Berninghausen O, Becker T, Beckmann R, Tampé R. Structure of the 40S-ABCE1 post-splitting complex in ribosome recycling and translation initiation. Nat Struct Mol Biol 2017;24:453–60.
Hilal T, Yamamoto H, Loerke J, Burger J, Mielke T, and Spahn C.M. Structural insights into ribosomal rescue by Dom34 and Hbs1 at near-atomic resolution. Nat Commun 2016; 7:13521.
Huang S. Building an efficient factory: where is pre-rRNA synthesized in the nucleolus? J Cell Biol 2002;157:739–41.
Iadevaia V, Zhang Z, Jan E, Proud CG. mTOR signaling regulates the processing of pre-rRNA in human cells. Nucleic Acids Res 2012;40:2527–39.
Im CH, Hwang SM, Son YS, Heo JB, Bang WY, Suwastika IN, Shiina T, Bahk JD. Nuclear/nucleolar GTPase 2 proteins as a subfamily of YlqF/YawGGTPases function in pre-60S ribosomal subunit maturation of mono- and dicotyledonous plants. J Biol Chem 2011;286:8620–32.
Jackson RJ, Hellen CU, Pestova TV. The mechanism of eukaryotic translation initiation and principles of its regulation. Nat Rev Mol Cell Biol 2010;11:113–27.
Jivotovskaya AV, Valásek L, Hinnebusch AG, Nielsen KH. Eukaryotic translation initiation factor 3 (eIF3) and eIF2 can promote mRNA binding to 40S subunits independently of eIF4G in yeast. Mol Cell Biol 2006;26:1355–72.
Karbstein K. Inside the 40S ribosome assembly machinery. Curr Opin Chem Biol 2011;15:657–63.
Karcher A, Schele A, Hopfner KP. X-ray structure of the complete ABC enzyme ABCE1 from *Pyrococcus abyssi*. J Biol Chem 2008;283:7962–71.
Khade P, Joseph S. Functional interactions by transfer RNAs in the ribosome. FEBS Lett 2010;584:420–6.
Kim TH, Leslie P, Zhang Y. Ribosomal proteins as unrevealed caretakers for cellular stress and genomic instability. Oncotarget 2014;5:860–71.
Kirkin V, Rogov VV. A diversity of selective autophagy receptors determines the specificity of the autophagy pathway. Mol Cell 2019;76:268–85.
Klauer AA, van Hoof A. Degradation of mRNAs that lack a stop codon: a decade of nonstop progress. Wiley Interdiscip Rev RNA 2012;3:649–60.
Kleiger G, Mayor T. Perilous journey: a tour of the ubiquitin-proteasome system. Trends Cell Biol 2014;24:352–9.
Konikkat S, Woolford Jr JL. Principles of 60S ribosomal subunit assemblyemerging from recent studies in yeast. Biochem J 2017;474:195–214.
Kononenko AV, Mitkevich VA, Atkinson GC, Tenson T, Dubovaya VI, Frolova LY, Makarov AA, Hauryliuk V. GTP-dependent structural rearrangement of the eRF1:eRF3 complex and eRF3 sequence motifs essential for PABP binding. Nucleic Acids Res 2010;38:548–58.
Kos M, Tollervey D. The putative RNA helicase Dbp4p is required for release of the U14 snoRNA from Preribosomes in *Saccharomyces cerevisiae*. Mol Cell 2005;20:53–64.
Kozlov G, Gehring K. Molecular basis of eRF3 recognition by the MLLE domain of poly(A)-binding protein. PLoS One 2010;5, e10169.

Kraft C, Deplazes A, Sohrmann M, Peter M. Mature ribosomes are selectively degraded upon starvation by an autophagy pathway requiring the Ubp3p/Bre5p ubiquitin protease. Nat Cell Biol 2008;10:602–10.

Kressler D, Roser D, Pertschy B, Hurt E. The AAA ATPase Rix7 powers progression of ribosome biogenesis by stripping Nsa1 from pre-60S particles. J Cell Biol 2008;181:935–44.

Kressler D, Hurt E, Bassler J. Driving ribosome assembly. Biochim Biophys Acta 2010;1803:673–83.

Kufel J, Allmang C, Verdone L, Beggs J, Tollervey D. A complex pathway for 3′ processing of the yeast U3 snoRNA. Nucleic Acids Res 2003;31:6788–97.

Lecompte O, Ripp R, Thierry JC, Moras D, Poch O. Comparative analysis of ribosomal proteins in complete genomes: an example of reductive evolution at the domain scale. Nucleic Acids Res 2002;30:5382–90.

Li L, Hong Y, Luan G, Mosel M, Malik M, Drlica K, Zhao X. Ribosomal elongation factor 4 promotes cell death associated with lethal stress. M Biol 2014;5, e01708.

Liang XH, Fournier MJ. The helicase Has1p is required for snoRNA release from pre-rRNA. Mol Cell Biol 2006;26:7437–50.

Lilienbaum A. Relationship between the proteasomal system and autophagy. Int J Biochem Mol Biol 2013;4:1–26.

Mancera-Martínez E, BritoQuerido J, Valasek LS, Simonetti A, Hashem Y. ABCE1: a special factor that orchestrates translation at the crossroad between recycling and initiation. RNA Biol 2017;14:1279–85.

Marshall RS, Vierstra RD. Dynamic regulation of the 26S proteasome: from synthesis to degradation. Fmolb 2019;6:1–31.

Mathis AD, Naylor BC, Carson RH, Evans E, Harwell J, Knecht J, Hexem E, Peelor 3rd FF, Miller BF, Hamilton KL, Transtrum MK, Bikman BT, Price JC. Mechanisms of in vivo ribosome maintenance change in response to nutrient signals. Mol Cell Proteomics 2017;16:243–54.

Moore PB. The ribosome returned. J Biol 2009;8:8. https://doi.org/10.1186/jbiol103.

Nazar RN. Ribosomal RNA processing and ribosome biogenesis in eukaryotes. IUBMB Life 2004;56:457–65.

Nissan TA, Bassler J, Petfalski E, Tollervey D, Hurt E. 60S pre-ribosome formation viewed from assembly in the nucleolus until export to the cytoplasm. EMBO J 2002;21:5539–47.

Noda T, Suzuki K, Ohsumi Y. Yeast autophagosomes: de novo formation of a membrane structure. Trends Cell Biol 2002;12:231–5.

Nürenberg-Goloub E, Heinemann H, Gerovac M, Tampé R. Ribosome recycling is coordinated by processive events in two asymmetric ATP sites of ABCE1. Life Sci Alliance 2018;1. pii: e201800095.

Ossareh-Nazari B, Niño CA, Bengtson MH, Lee JW, Joazeiro CA, Dargemont C. Ubiquitylation by the Ltn1 E3 ligase protects 60S ribosomes from starvation-induced selective autophagy. J Cell Biol 2014;204(6):909–17.

Panse VG, Johnson AW. Maturation of eukaryotic ribosomes: acquisition of functionality. Trends Biochem Sci 2010;35:260–6.

Peña C, Hurt E, Panse VG. Eukaryotic ribosome assembly, transport and quality control. Nat Struct Mol Biol 2017;24:689–99.

Perry RP. Balanced production of ribosomal proteins. Gene 2007;401:1–3.

Pesole G, Gissi C, Lanave C, Saccone C. Glutamine synthetase gene evolution in bacteria. Mol Biol Evol 1995;12:189–97.

Petrov AS, Gulen B, Norris AM, Kovacs NA, Bernier CR, Lanier KA, Fox GE, Harvey SC, Wartell RM, Hud NV, Williams LD. History of the ribosome and the origin of translation. Proc Natl Acad Sci U S A 2015;112:15396–401.

Phipps KR, Charette J, Baserga SJ. The small subunit processome in ribosome biogenesis-progress and prospects. Wiley Interdiscip Rev RNA 2011;2:1–21.

Pisarev AV, Hellen CU, Pestova TV. Recycling of eukaryotic posttermination ribosomal complexes. Cell 2007;131(2):286–99.

Pisarev AV, Skabkin MA, Pisareva VP, Skabkina OV, Rakotondrafara AM, Hentze MW, Hellen CU, Pestova TV. The role of ABCE1 in eukaryotic posttermination ribosomal recycling. Mol Cell 2010;37:196–210.

Pisareva VP, Pisarev AV, Komar AA, Hellen CU, Pestova TV. Translation initiation on mammalian mRNAs with structured 5'UTRs requires DExH-box protein DHX29. Cell 2008;135:1237–50.

Pisareva VP, Skabkin MA, Hellen CU, Pestova TV, Pisarev AV. Dissociation by Pelota, Hbs1 and ABCE1 of mammalian vacant 80S ribosomes and stalled elongation complexes. EMBO J 2011;30:1804–17.

Pisareva VP, Pisarev AV. eIF5 and eIF5B together stimulate 48S initiation complex formation during ribosomal scanning. Nucleic Acids Res 2014;42:12052–69.

Preis A, Heuer A, Barrio-Garcia C, Hauser A, Eyler DE, Berninghausen O, Green R, Becker T, Beckmann R. Cryoelectron microscopic structures of eukaryotic translation termination complexes containing eRF1-eRF3 or eRF1-ABCE1. Cell Rep 2014;8:59–65.

Raimondeau E, Bufton JC, Schaffitzel C. New insights into the interplay between the translation machinery and nonsense-mediated mRNA decay factors. Biochem Soc Trans 2018;46:503–12.

Rees DC, Johnson E, Lewinson O. ABC transporters: the power to change. Nat Rev Mol Cell Biol 2009;10:218–27.

Seshadri A, Varshney U. Mechanism of recycling of post-termination ribosomal complexes in eubacteria: a new role of initiation factor 3. J Biosci 2006;31:281–9.

Shoemaker CJ, Green R. Kinetic analysis reveals the ordered coupling of translation termination and ribosome recycling in yeast. Proc Natl Acad Sci U S A 2011;108:E1392–8.

Simonović M, Steitz TA. A structural view on the mechanism of the ribosome-catalyzed peptide bond formation. Biochim Biophys Acta 2009;1789:612–23.

Sloan KE, Warda AS, Sharma S, Entian KD, Lafontaine DLJ, Bohnsack MT. Tuning the ribosome: the influence of rRNA modification on eukaryotic ribosome biogenesis and function. RNA Biol 2017;14:1138–52.

Sommer T, Weber A, Jarosch E. Rsp5/Nedd4 clears cells of heat-damaged proteins. Nat Cell Biol 2014;16:1130–2.

Strunk BS, Karbstein K. Powering through ribosome assembly. RNA 2009;15:2083–104.

Strunk BS, Loucks CR, Su M, Vashisth H, Cheng S, Schilling J, Brooks 3rd CL, Karbstein K, Skiniotis G. Ribosome assembly factors prevent premature translation initiation by 40S assembly intermediates. Science 2011;333:1449–53.

Sung MK, Reitsma JM, Sweredoski MJ, Hess S, Deshaies RJ. Ribosomal proteins produced in excess are degraded by the ubiquitin-proteasome system. Mol Biol Cell 2016a;27:2642–52.

Sung MK, Porras-Yakushi TR, Reitsma JM, Huber FM, Sweredoski MJ, Hoelz A, Hess S, Deshaies RJ. A conserved quality-control pathway that mediates degradation of unassembled ribosomal proteins. Elife 2016b;5, e19105.

Takalo M, Salminen A, Soininen H, Hiltunen M, Haapasalo A. Protein aggregation and degradation mechanisms in neurodegenerative diseases. Am J Neurodegener Dis 2013;2:1–14.

Tanaka K, Matsuda N. Proteostasis and neurodegeneration: the roles of proteasomal degradation and autophagy. Biochim Biophys Acta 2014;1843:197–204.

Taylor D, Unbehaun A, Li W, Das S, Lei J, Liao HY, Grassucci RA, Pestova TV, Frank J. Cryo-EM structure of the mammalian eukaryotic release factoreRF1-eRF3-associated termination complex. Proc Natl Acad Sci U S A 2012;109:18413–8.

Trapman J, Retèl J, Planta RJ. Ribosomal precursor particles from yeast. Exp Cell Res 1975;90:95–104.

Trumtel S, Léger-Silvestre I, Gleizes PE, Teulières F, Gas N. Assembly and functional organization of the nucleolus: ultrastructural analysis of *Saccharomyces cerevisiae* mutants. Mol Biol Cell 2000;11:2175–89.

Tuorto F, Parlato R. rRNA and tRNA bridges to neuronal homeostasis in health and disease. J Mol Biol 2019;431:1763–79.
Turowski TW, Tollervey D. Transcription by RNA polymerase III: insights intomechanism and regulation. Biochem Soc Trans 2016;44:1367–75.
Valášek LS. 'Ribozoomin'—translation initiation from the perspective of the ribosome-bound eukaryotic initiation factors (eIFs). Curr Protein Pept Sci 2012;13:305–30.
van Hoof A, Wagner EJ. A brief survey of mRNA surveillance. Trends Biochem Sci 2011;36:585–92.
Wang W, Nag S, Zhang X, Wang MH, Wang H, Zhou J, Zhang R. Ribosomal proteins and human diseases: pathogenesis, molecular mechanisms, and therapeutic implications. Med Res Rev 2015;35:225–85.
Wyant GA, Abu-Remaileh M, Frenkel EM, Laqtom NN, Dharamdasani V, Lewis CA, Chan SH, Heinze I, Ori A, Sabatini DM. NUFIP1 is a ribosome receptor for starvation-induced ribophagy. Science 2018;360:751–75.
Wu S, Tutuncuoglu B, Yan K, Brown H, Zhang Y, Tan D, Gamalinda M, Yuan Y, Li Z, Jakovljevic J, Ma C, Lei J, Dong MQ, Woolford Jr JL, Gao N. Diverse roles of assembly factors revealed by structures of late nuclear pre-60S ribosomes. Nature 2016;534:133–7.
Xue S, Barna M. Specialized ribosomes: a new frontier in gene regulation and organismal biology. Nat Rev Mol Cell Biol 2012;13:355–69.
Yoneda Y. Nucleocytoplasmic protein traffic and its significance to cell function. Genes Cells 2000;5:777–87.
Zhang J, Harnpicharnchai P, Jakovljevic J, et al. Assembly factors Rpf2 and Rrs1 recruit 5S rRNA and ribosomal proteins rpL5 and rpL11 into nascent ribosomes. Genes Dev 2007;21:2580–92.
Zhang L, Wu C, Cai G, Chen S, Ye K. Stepwise and dynamic assembly of the earliest precursors of small ribosomal subunits in yeast. Genes Dev 2016;30:718–32.
Zhang Y, Mandava CS, Cao W, Li X, Zhang D, Li N, Zhang Y, Zhang X, Qin Y, Mi K, Lei J, Sanyal S, Gao N. HflX is a ribosome-splitting factor rescuing stalled ribosomes under stress conditions. Nat Struct Mol Biol 2015;22:906–13.
Zhou X, Liao WJ, Liao JM, Liao P, Lu H. Ribosomal proteins: functions beyond the ribosome. J Mol Cell Biol 2015;7:92–104.

CHAPTER 7

Ribosomal biogenesis in eukaryotes

Vijay Kumar
Department of Molecular and Cellular Medicine, Institute of Liver and Biliary Sciences, New Delhi, India

Introduction

In living cells, ribosomes constitute the universal translational apparatus responsible for decoding genetic information into proteins. These complex and sophisticated cellular nanomachines are inevitably linked to cellular growth and proliferation as their number fluctuates depending on the physiological state of the cell (Dai and Zhu, 2020; Farley-Barnes and Baserga, 2020). Despite morphological and functional similarities, the eukaryotic ribosomal complex is much larger and far more intricate than its prokaryotic counterpart. In eukaryotes, the 80S ribosomal machinery is composed of two universally conserved ribosomal subunits made of ribonucleoproteins. Each ribosome carries four ribosomal RNA (rRNA) molecules and roughly 80 proteins that form the 40S small and 60S large ribosomal subunits (Ben-Shem et al., 2011; Klinge and Woolford Jr., 2019). The small subunit participates in decoding the mRNA with the help of tRNAs in its decoding center (DC), whereas the large subunit facilitates the peptide bond formation at the peptidyl transferase center (PTC) (Rodnina, 2013). The 60S subunit is also characterized by the presence of P and L1 stalks, the 5S rRNA-containing central protuberance (CP), polypeptide exit tunnel, guanosine triphosphatase (GTPase)-activating center, and a binding platform for processing and folding factors (refer to Chapter 2 for details).

Synthesis of the ribosome or ribosome biogenesis is an energy-consuming biosynthetic process involving multiple factors that are under tight regulatory control to avoid unnecessary energy expenditure. There are six major well-integrated steps in ribosome biogenesis including (i) synthesis of components (rRNA, RPs, assembly factors or AFs, and small nucleolar RNAs (snoRNAs)); (ii) processing of pre-rRNAs; (iii) covalent modification of pre-RNAs, RPs, and AFs; (iv) assembly of ribosomal subunits; (v) transport

Emerging Concepts in Ribosome Structure, Biogenesis, and Function
https://doi.org/10.1016/B978-0-12-816364-1.00011-1

(nuclear import of RPs and AFs, and export of preribosomes in the cytoplasm); and (vi) quality control of ribosome maturation and its surveillance. For these reasons, ribosome biogenesis involves the coordinated action of all three DNA-dependent RNA polymerases (RNA polymerase I, II, and III) and the concerted action of more than 200 proteins (Mayer and Grummt, 2006; Thomson et al., 2013). The assembly of this two-subunit molecular machinery is a complex process spanning different cellular compartments. It is initiated in the nucleolus where RNA polymerase I (Pol I) carries out the synthesis of pre-rRNA from rDNA gene clusters in the nucleolus organizer regions (Fig. 7.1) (Goodfellow and Zomerdijk, 2013; Paule and White, 2000). The primary 35S-45S pre-rRNA transcripts include RNA regions for both the small ribosomal subunit (18S) and the large ribosomal

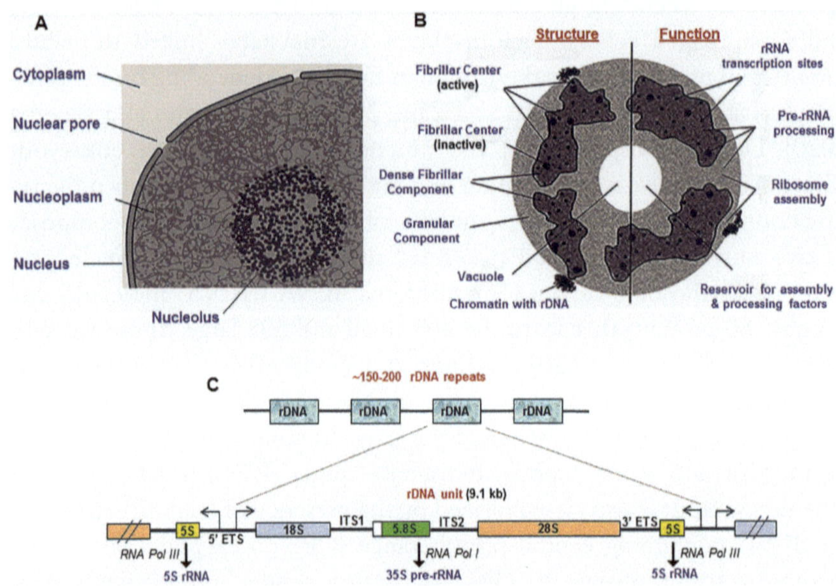

Fig. 7.1 ***Cell nucleolus and the expression of rDNA in eukaryotes.*** (A) Ribosome synthesis is initiated in the nucleolar region of the nucleus. The eukaryotic genomes carry a high copy number of rDNA or the ribosomal RNA genes clustered in the nucleolus. (B) Schematic organization of the structural components of the nucleolus and their functions. Ribosome biogenesis begins with the transcription of rDNA in the fibrillar components of the nucleoli whereas the processing of pre-rRNA and assembly of ribosomes takes place in the granular component. (C) Schematic organization of an rDNA unit in the rDNA locus (with ~150–200 units) of chromosome XII of the yeast *Saccharomyces cerevisiae*. The 5S rRNA (yellow box) is transcribed by RNA Pol III. RNA Pol I synthesize the 35S primary pre-rRNA transcripts which are processed to produce the mature 18S, 5.8S, and 25S rRNAs (filled three boxes). ETS, 5′ and 3′ external transcribed spacers; ITS, internal transcribed spacers 1 and 2.

subunit (5.8S and 25S) separated via internal transcribed spacer sequences ITS1 and ITS2. The 18S rRNA region is preceded by a 5′-external transcribed spacer sequence (5′-ETS), whereas the 25S rRNA is flanked by a 3′-ETS. The 5S rRNA gene is transcribed independently by RNA polymerase III in the opposite direction (Fig. 7.1C) (Ciganda and Williams, 2011; Paule and White, 2000). The nascent RNA Pol I transcripts associate cotranscriptionally with a subset of ribosomal as well as nonribosomal proteins including small nucleolar ribonucleoprotein particles (snoRNPs) and preribosomal factors to yield 90S ribonucleoprotein particles (RNPs) in which pre-rRNA processing and modification take place along with the concomitant assembly of the ribosomal proteins. During this process, the assembly factors orchestrate the complex series of the nucleolytic cleavages by *endo*- and *exo*-nucleases present largely in the dense fibrillar component region of the nucleolus (Fig. 7.1B), and sequential cleavage and removal of the transcribed spacers yield mature rRNA transcripts (18S, 25/28S, and 5.8S rRNA) (Lafontaine et al., 2021). The cleavage in ITS1 results in the separation of the production pathways of two ribosomal subunits. Further maturation of pre-RNA and its association with RPs for the assembly of small and large ribosomal subunits majorly occurs in the granular component region of the nucleolus (Fig. 7.1B) with additional modifications in the nucleoplasm (Lafontaine et al., 2021). These preribosomal subunits are then actively exported to the cytoplasm, where they bind to the mRNA to form functional 80S ribosomes ready to decode the genetic information by translating mRNA into proteins (Nerurkar et al., 2015; Peña et al., 2017). A proliferating HeLa cell produces approximately 7500 ribosomes per minute (Lewis and Tollervey, 2000) whereas an actively replicating yeast cell can make up to 2000 ribosomes in a minute (Warner, 1999).

Although the mature ribosomes have been well characterized at both the structural and functional levels, we still lack a good understanding of the precise molecular function of each factor involved in the assembly of this universal translating machinery. In this chapter, we discuss ribosome biogenesis in a yeast model system, which is considered as the gold standard for studying this molecular process.

Yeast as an eukaryotic model to study ribosome biogenesis

A single cell of yeast *Saccharomyces cerevisiae* carries up to 200,000 ribosomes. The rRNAs are expressed from the rDNA transcription units present as ~150–200 tandem repeats on chromosome XII (Fig. 7.1C). Apparently, the rate of ribosome biogenesis is regulated by transcription initiation (French et al., 2003), elongation (Zhang et al., 2010), and the ratio of active to inactive rDNA repeats in a cell (Sandmeier et al., 2002). Interestingly, the cell size and growth rate in yeast are intimately linked to the rate of ribosome biogenesis (Cook and Tyers, 2007). Further, the number of active genes in yeast decreases from the log to stationary phase, suggesting modulation of their numbers between growth conditions (Dammann et al., 1993). Besides the rRNAs and ribosomal proteins, nearly 75 small nucleolar RNAs and over 200 nonribosomal proteins participate in ribosome biogenesis including nucleases, RNA modifying enzymes, adenosine triphosphatases (ATPases), GTPases, kinases, and RNA helicases.

Preinitiation complex formation

Ribosome biogenesis begins with the recruitment and assembly of Pol I and other transcription factors into a preinitiation complex (PIC) at the rRNA gene promoter in the nucleolus. The yeast rRNA gene promoter contains a core element as the transcription start site (+1), an upstream control element (UCE) with modulatory function, and distal enhancer-like sequences (Fig. 7.2). In *S. cerevisiae*, the Pol I holoenzyme is made of 14 subunits (including *RPA12, RPA14, RPA34, RPA43, RPA49, RPA135, RPA190, RPB5, RPB8, RPB10, RPC10, RPC19, RPC40,* and *RPO26*) with a molecular mass of 590kDa. Four general transcription factors including upstream activity factor (UAF), composed of Rrn5, Rrn9, Rrn10, and histones (H3 and H4),

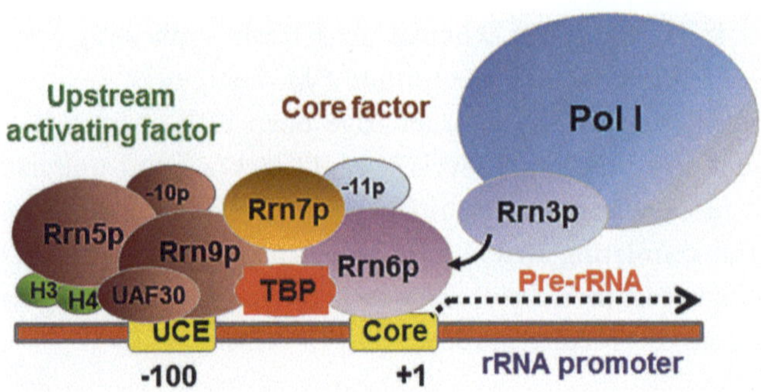

Fig. 7.2 ***The yeast Pol I preinitiation complex.*** Pol I-dependent transcription of ribosomal RNA genes in yeast involves an upstream control element (UCE localized around −100 position) and a core promoter element (core) near the transcription start site (+1). Four general transcription factors including the upstream activity factor (composed of Rrn5p, Rrn9p, Rrn10p, and histones H3 and H4), TATA-binding protein (TBP), core factor (composed of Rrn6, Rrn7, and Rrn11 proteins), and Rrn3 facilitate the recruitment of Pol I (composed of 14 subunits) to the rRNA gene promoter.

TATA-binding protein (TBP), core factor composed of Rrn6, Rrn7, and Rrn11 proteins (analogous to Selective Factor 1 or SL1 in mammals) and Rrn3 (transcription intermediary factor 1 alpha (TIF1A) in mammals) facilitate the recruitment of Pol I to the promoter by bridging Pol I and the promoter-bound core factor (Fig. 7.2). Interestingly, the initiation-competent Pol I is tightly associated with Rrn3 whereas Rrn7 helps in the formation of the preinitiation complex (Knutson et al., 2014; Sadian et al., 2017). During active transcription, the rDNA locus is also characterized by epigenetic modifications involving high histone 3′ acetylation and low histone/DNA methylation. Several epigenetic chromatin modifiers, such as histone acetyltransferase, methyltransferase, decetylase, and DNA excision repair machinery are involved in the process to facilitate transcription (O'Kane and Hyland, 2019). Besides, a number of RNA helicases also participate in ribosome biogenesis in *S. cerevisiae* including 7 for small subunit (SSU) biogenesis (including Dbp4, Dbp8, Dhr1, Dhr2, Fal1, Rok1, and Rrp3) and 10 for large subunit (LSU) biogenesis (Dbp2, Dbp3, Dbp6, Dbp7, Dbp9, Dbp10, Drs1, Mak5, Mtr4, and Sbp4). Two helicases (Has1 and Prp43) are common for both subunits (Rodríguez-Galán et al., 2013). The productive PIC formation leads to promoter opening and initiation of transcription. This is followed by promoter escape of the polymerase and elongation and synthesis of the 6.6-kb or the 35S primary transcript. The process of transcription terminates with the help of transcript-release factors bound to the transcription termination elements located at the 3′ end of the transcribed region of the rRNA gene. Subsequently, the components of PIC dissociate and are recycled.

Control of rDNA transcription

As cell growth and proliferation are intricately linked to ribosome biogenesis, the intracellular signals must coordinate the rRNA and protein synthesis in a cell. Multiple signaling pathways including PI3K-Akt-mTOR, Ras-Raf-Erk, Rb-p53, and Myc are reported to regulate rDNA transcription via overlapping but distinct and dynamic mechanisms (Fig. 7.3 and Table 7.1) (Kusnadi et al., 2015; Piazzi et al., 2019). The RAS/RAF/ERK signaling pathway stimulates rDNA transcription via ERK phosphorylation of RRN3 and UBF whereas AKT upregulates rDNA transcription at multiple steps such as including the recruitment of Pol I on the promoter, transcription elongation, and rRNA processing (Chan et al., 2011; Stefanovsky et al., 2006). The receptors for cytokines and growth factors such as the epidermal growth factor, insulin-like growth factor, and tumor necrosis factor also participate in signal transduction via the PI3K-Akt-mTOR pathway

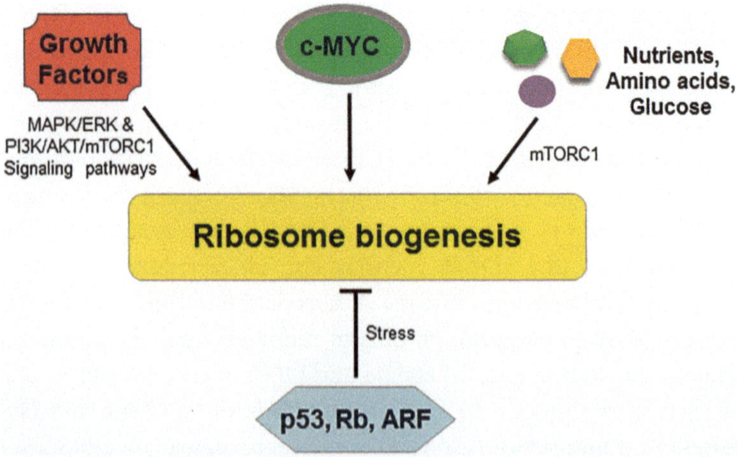

Fig. 7.3 ***Regulation of ribosome biogenesis.*** Growth factor-dependent signaling via the RAS/RAF/ERK and PI3K/AKT/mTORC1 pathways constitute an intricate control network with the transcription factor c-MYC to stimulate the rate of rDNA transcription and thus, ribosome biogenesis. mTORC1 also links the availability of nutrients, amino acids, and glucose to rDNA transcription. Under stress conditions, ribosome biogenesis can be negatively regulated as a checkpoint control aided by tumor suppressor proteins [p53, retinoblastoma protein (Rb) and the ADP-ribosylation factor (ARF)].

in order to control rDNA transcription in response to the nutrient and the energy status of the cell. Deregulation of ribosome biogenesis is sensed by quality control checkpoints leading to the stabilization of the tumor-suppressor proteins (p53, Rb, and ARF) resulting in negative regulation of ribosome biogenesis (Kusnadi et al., 2015; Piazzi et al., 2019) (Fig. 7.3).

Elevated expression of Myc is sufficient to induce the expression of rRNAs, RPs, and several cofactors necessary for ribosome biogenesis through different mechanisms (Oskarsson and Trumpp, 2005; van Riggelen et al., 2010). Myc can facilitate the recruitment of RNA Pol I cofactors to the promoter and thus, stimulate the transcription of pre-rRNA (Poortinga et al., 2004). It also stimulates the expression of the 5S rRNA by directly associating with TFIIIB beyond the direct association with rDNA promoters. Myc is well known to induce the expression of both the small and large subunit RPs and several processing factors in an RNA Pol II-dependent manner. p53, on the other hand, is known to inhibit Myc-dependent gene expression by reducing histone acetylation activity. Myc also stimulates the expression of the two most abundant nucleolar proteins, viz., nucleolin and nucleophosmin that are involved in rDNA transcription and rRNA maturation (Ginisty et al., 1999; Li and Hann, 2013). In addition, these

Table 7.1 Key activators and repressors of the Pol I transcription machinery.

Regulatory factor	Targets in Pol I promoter	References
Activators		
G1 cyclins/CDKs	Upstream binding factor (UBF)	Voit et al. (1999); Voit and Grummt (2001)
Extracellular signal-regulated kinases (ERK)	RRN3, UBF	Stefanovsky et al. (2006)
Casein kinase 2 (CK2)	UBF, RRN3	Lin et al. (2006); Panova et al. (2006)
Mechanistic Target of Rapamycin (mTOR)	UBF, RRN3	Mayer and Grummt (2006)
CREB-binding protein (CBP)	UBF	Pelletier et al. (2000); Voit and Grummt, 2001
P300/CBP-associated factor (PCAF),	Selective factor 1 (SL1)	Shen et al. (2013)
Tip60 histone acetyltransferase	UBF	Halkidou et al. (2004)
c-Myc	SL1	Grandori et al. (2005); van Riggelen et al. (2010)
Repressors		
p53	SL1	Zhai and Comai (2000)
Retinoblastoma (Rb)	UBF	Cavanaugh et al. (1995); Hannan et al. (2000); Voit et al. (1997)
Casein kinase 2 (CK2)	SL1	Panova et al. (2006)
Phosphatase and Tensin homolog (PTEN)	SL1	Zhang et al. (2005)
Glycogen synthase kinase 3 beta (GSK3β)	SL1	Vincent et al. (2008)
JNK2	RRN3	Mayer et al. (2005)

nucleolar proteins participate in the cytoplasmic-nuclear import of newly synthesized RPs and the nuclear-cytoplasmic export of the assembling ribosomal subunits. Myc can also promote the expression of translation initiation factors such as eIF2α, eIF4A-I, eIF4E, eIF4G, as well as methylation of the mRNA cap (Rosenwald et al., 1993).

Processing of pre-rRNA

The emerging 35 pre-rRNA transcripts undergo cotranscriptional folding, cleavage, and modifications and recruit RPs and assembly factors to form the earliest 90S ribosome precursors (Kornprobst et al., 2016). The initial cleavage events within the pre-rRNA release the earliest small ribosomal subunit 40S precursor (pre-40S). Pre-60S large ribosomal subunits and pre-40S particles mature independently in the nucleolus and nucleoplasm and are exported separately through the nuclear pore complexes into the cytoplasm. Cytoplasmic maturation is primarily a quality control and proof-reading step for the functional centers in ribosomes so that only correctly assembled ribosomes are available for translation (Kressler et al., 2017). Thus, preribosomes at different stages of ribosome assembly can be distinguished based on their protein composition and by the presence of distinct pre-rRNA processing intermediates.

The pre-rRNA processing involves the removal of spacer regions, such as 5′ external transcribed spacer (5′-ETS) and internal transcribed spacers 1 and 2 (ITS1 and ITS2) with the help of endonucleolytic and exonucleolytic processing reactions (Venema and Tollervey, 1995). Careful analysis of yeast chromatin spreads and EM studies have revealed cotranscriptional cleavage in the majority of nascent pre-rRNA transcripts (Osheim et al., 2004). The processing of 35S primary pre-rRNA transcripts takes place through the following steps (Fig. 7.4):

1. Generally, it begins with a cleavage reaction at the B_0 site by Rnt1p (an RNase III endonuclease) to generate the 3′ end of the pre-rRNA transcript (Kufel et al., 1999).
2. Subsequently, a cleavage may occur with the help of U3 snoRNP (see "Maturation of ribosomal subunits" section) and a cyclase-like protein Rcl1p at any one of these three sites, viz., A_0, A_1, or A_2 for the removal of 5′-ETS or ITS1. The 5′-ETS is removed in two steps, cleavage at site A_0 to generate 33S pre-rRNA and then cleavage at site A_1 to produce 32S pre-rRNA (Hughes and Ares Jr., 1991; Beltrame et al., 1994).
3. Cleavage by Rcl1p at the A_2 site in ITS1 generates the 20S and 27SA$_2$ pre-rRNAs (Udem and Warner, 1972).
4. The separating cleavage in ITS1 can also be made by RNase mitochondrial RNA processing (MRP) at the A_3 site, producing 23S and 27SA$_3$ pre-rRNAs (Lygerou et al., 1996).
5. The 20S pre-rRNA is packaged in 43S particles, and the 27SA$_2$ pre-rRNA is found in 66S pre-rRNPs (Trapman et al., 1975).
6. The 43S pre-rRNPs are exported from the nucleolus through the nucleoplasm to the cytoplasm, where 20S pre-rRNA undergoes endonucleolytic cleavage by Nob1p at site D to remove the remaining ITS1 sequences, producing mature 18S rRNA (Fatica et al., 2004; Vos and Kothe, 2020).

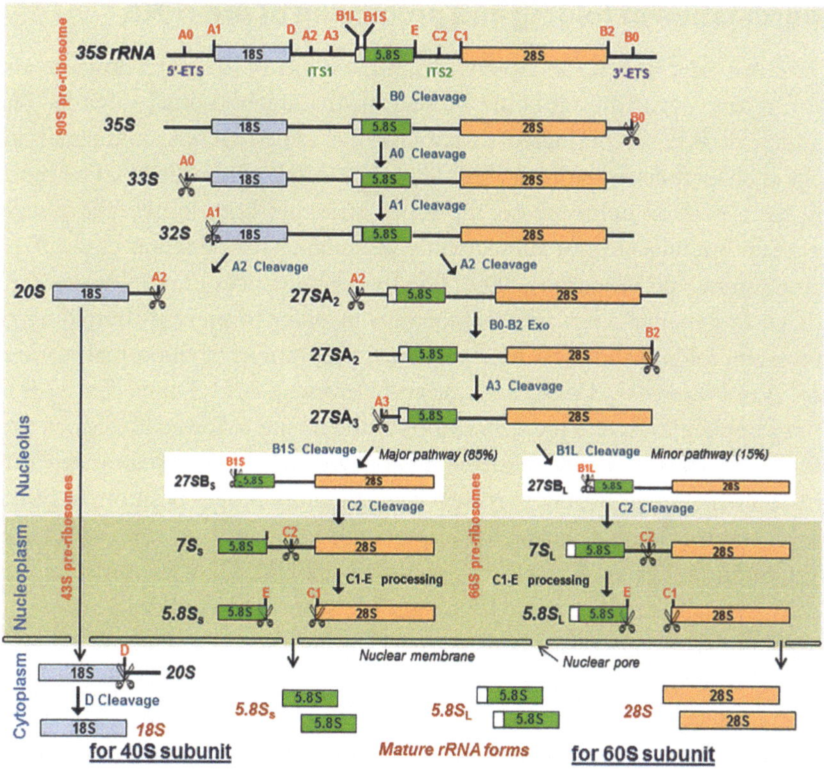

Fig. 7.4 ***Pathway for processing of yeast pre-rRNA.*** The 35S pre-rRNA is transcribed by RNA polymerase I and contains sequences for 18S, 5.8S, and 25S rRNAs (gray, green, and brown boxes) flanked and separated by external and internal transcribed spacers (ETS and ITS, solid lines). The 35S rRNA associates with ~200 assembly factors and 79 ribosomal proteins to form a 90S preribosome. The spacer sequences are removed from pre-rRNA in a sequential manner involving endonucleolytic and exonucleolytic activities in the assembling ribosomes. The majority of pre-rRNAs undergo cotranscriptional cleavage at the A0, A1, and A2 sites, which begins in the nucleolus of the cell, but later steps take place in the nucleoplasm and cytoplasm.

7. The maturation of 66S preribosomes is more complex and the $27SA_2$ pre-rRNA continues in the nucleolus by two alternative pathways with the help of exonucleases Rat1p and Rrp17p to form mature 28S rRNA (Geerlings et al., 2000).
8. The 3′ ends of 7S pre-rRNAs are processed in several steps to produce two types of mature and functional 5.8SS and 5.8SL rRNAs differing by 8 nucleotides (nt) at their 5′ ends (Henry et al., 1994; Oeffinger et al., 2009).

Coordination of folding and processing of pre-rRNA

The eukaryotic rRNA can rapidly fold into secondary structural domains and form tertiary structures as could be seen in mature ribosomal subunits. The binding of RPs is considered to stabilize the folded rRNA conformers or help as chaperones to redirect misfolded rRNA (Woodson, 2011). The binding has a marked influence on the RNA structure both locally and distally to create binding sites for subsequently assembling RPs (Henras et al., 2015; Ramaswamy and Woodson, 2009). The spacer sequences in rRNAs are considered to generate alternative conformers in order to prevent misfolding or premature folding and their removal allows progression of ribosomal maturation (Van Nues et al., 1995; Lamanna and Karbstein, 2011; Dutca et al., 2011). For example, cleavage of pre-18S rRNA at the A_2 site, followed by dissociation of the downstream ITS1 sequences (Fig. 7.4), could generate a helical structure as a conformational switch. This switch might allow proper positioning of the Nob1 endonuclease, to enable cleavage at the D site at later steps of subunit maturation (Fatica et al., 2004; Vos and Kothe, 2020). Thus, the order of two RNA cleavage reactions could be navigated by pre-rRNA folding.

A similar scenario has been reported during the early stages of pre-60S subunit biogenesis (Van Nues et al., 1995). Sequences in the 3′ end of ITS1 are predicted to base pair with nucleotides that are destined to become the 5′ end of 5.8S rRNA. In mature ribosomes, the same sequences in 5.8S rRNA base pair with nucleotides in 25S rRNA to form a helix (Woolford Jr and Baserga, 2013). Exonucleolytic removal of the 3′ end of ITS1 by Rat1 and Rrp17 during $27SA_3$ pre-rRNA processing may require unwinding this ITS1-5.8S rRNA helix, or alternatively might simply eliminate this helix (Fig. 7.4). Either way, processing of $27SA_3$ pre-rRNA may be coupled to the folding of the 5.8S/25S rRNA portion of domain I of rRNA (Granneman et al., 2011; Oeffinger et al., 2009). Likewise, the structural modeling and mutational analysis have revealed that the binding of Nop15 and Cic1 to ITS2 spacer can switch from a more flexible and open "ring" conformation to a stable, closed "hairpin" structure during subunit biogenesis (Côté et al., 2002; Granneman et al., 2011). Thus, the processing of spacer sequences can be initiated by both folding of the pre-rRNA as well as RNA-protein interaction.

Roles of the ribosomal proteins

One of the major challenges to understand ribosome assembly is to understand how pre-rRNAs fold during assembly in vivo and how its transition from secondary to proper tertiary structure takes place. Does the binding of

RPs and assembly factors affect pre-rRNA folding? If so, does pre-rRNA folding temporally regulate the interaction of these proteins? It is a general understanding that most yeast RPs are essential for growth under laboratory conditions. Importantly, 59 of the 78 yeast RPs are encoded by duplicated genes which encode proteins that differ from each other by only 1–5 amino acids, suggesting their critical role in ribosomes (Steffen et al., 2012). However, the precise functions of RPs in pre-rRNA processing and ribosome biogenesis remain elusive. It is generally considered that RPs act as chaperones to facilitate the proper folding of the nascent pre-rRNAs as nearly all RPs in mature ribosomes interact with rRNAs albeit to different extents. Some penetrate deeply into the rRNA core, while others are present on the RNA surface (Ben-Shem et al., 2011). By affecting the folding of rRNA, RPs facilitate the binding of other proteins or induce some conformational change in pre-rRNA required for the removal of spacer sequences. Besides, most RPs are also exposed on the ribosome surface, which could allow the interaction of assembly or translation factors.

Several studies on the interaction network between proteins and RNA suggest that initially the interactions are somewhat loose and dynamic but gradually it tightens during subunit assembly and eventually forms mature ribosomal subunits. The 40S ribosome subunit is sequentially assembled as follows: first, the "body" domain is formed, which contains the 5′ end and central domains of 18S rRNA and primary and secondary binding RPs, followed by the "head" domain, which is composed of the major 3′-end domain of 18S rRNA and secondary/tertiary binding RPs. The 60S subunit exhibits a hierarchical assembly starting with structural domains. The binding of early acting RPLs to domains corresponding to the 5′ part of the 25S and 5.8S rRNAs stabilizes a preribosomal structure that facilitates A2 and A3 cleavages (Fig. 7.4). Later, the assembly of middle-acting RPLs allows cleavage of the 27S-B pre-rRNA at the C2 site in ITS2. The binding of middle- and late-acting RPLs to the 3′ region of the mature 25S rRNA sequence is linked to 7S pre-rRNA processing. Finally, the 5S RNP complex (having 5S rRNA, Rpl5p, and Rpl11p) undergoes a rotation to form the central protuberance of the 60S subunit (Fig. 2.2 of Chapter 2). This step requires correct folding and processing of the 66S precursor to mature 5.8S rRNA, chaperoned by late-acting RPLs.

Role of ribosome assembly factors

In addition to RNA-mediated RNA modifications and RPs, nearly 200 different but well-conserved assembly factors are required during ribosome biogenesis in yeast (Woolford Jr and Baserga, 2013). The role of most factors is now well-defined using gene depletion and phenotype observation studies including (i) folding and processing of pre-rRNA, (ii) sequential recruitment of RPs and subunit maturation, and (iii) nuclear export. The identification of these factors could be accomplished using both classical and molecular genetics combined with affinity purification and mass spectrometric analysis of factors bound to preribosomes (Harnpicharnchai et al., 2001; Saveanu et al., 2003). The mass spectrometry data have revealed that > 70 proteins are reproducibly associated with the affinity-purified pre-40S ribosomal subunits and specifically function in their assembly. Likewise, ~90 proteins can be found in 66S precursors to mature 60S subunits and play a role in the assembly. Among these proteins, there are 19 RNA helicases, seven of which (Dbp4, Dbp8, Dhr1, Dhr2, Fal1, Rok1, and Rrp3) are involved in small-subunit assembly whereas nine helicases (Dbp2, Dbp3, Dbp6, Dbp7, Dbp9, Dbp10, Drs1, Mak5, and Spb4) are involved in large-subunit assembly. Three helicases (Has1, Mtr4, and Prp43) have been implicated in the assembly of both subunits. Additional energy-consuming enzymes include GTPases such as Bms1, Nog1, Nog2, Nug1, Lsg1, and Efl1; ATPases such as Rio1, Rio2, and Fap7; and AAA-ATPases (e.g., Mdn1, Drg1, and Rix7) (Henras et al., 2015; Kressler et al., 2017). All these factors drive ribosome assembly in a unidirectional manner aided by dedicated chaperones that help in the folding of the ribosome intermediates, escorting during nuclear maturation, and inserting them into preribosomal particles.

Maturation of ribosomal subunits

Several studies have shown that pre-40S particles undergo relatively fewer compositional changes as compared to pre-60S while transiting through the nucleoplasm and export into the cytoplasm. During a similar journey, pre-60S particles are reported to associate with nearly 100 trans-acting factors and undergo dynamic changes in their composition. However, both nascent 40S and 60S ribosomal subunits entering the cytoplasm undergo the last few rounds of processing to make mature and functional subunits (Zemp and Kutay, 2007; Panse and Johnson, 2010). It involves the final steps of pre-rRNA processing, the association of the last RPs, and the release and recycling of the assembly factors from previous cycles. It also prevents the premature association of the inactive subunits with the translational machinery and interrogates functional domains for their correct assembly.

These remodeling events are assisted by several assembly factors, which associate with the maturing preribosomal particles reviewed in (Klinge and Woolford Jr., 2019).

Initially, the pre-rRNA is synthesized and initially compacted to form 90S preribosomes. From this, the 5′-ETS is removed and ITS1 cleavage occurs. Cleavage at the A2 site in ITS1 separates the 20S pre-rRNA from 27S pre-rRNA. These two pre-rRNAs are packaged respectively into 43S and 66S precursors to mature 40S and 60S subunits. The assembly of 40S and 60S subunits, subsequent to cleavage at the A2 site, proceeds largely independently.

Maturation of the 40S subunit

The maturation of the 40S subunit precursor, also called the small subunit processome (SSU) or 90S particle, is primarily a cotranscriptional event and happens in a hierarchical fashion. The 90S particle is made up of over 30 nonribosomal protein factors, the U3 snoRNP, a few small subunit ribosomal proteins, and the 35S pre-rRNA (~5500 nucleotides long). The composition of the pre-40S particle changes drastically following the cleavage at the U3 snoRNPA-dependent sites (sites A0, A1, and A2) (Fig. 7.4). These cleavages result in the formation of 20S pre-rRNA and dissociation of ribosomal factors from the pre-40S particle. U3 is displaced from the nascent 40S by the helicase Dhr1. At this stage, the 40S preribosome contains RPs, the 20s rRNA, and a few nonribosomal factors and exhibits the "head" and "body" structures of the mature 40S subunit. The 40S preribosome is now transported from the nucleolus into the cytoplasm. The "beak" structure of the mature 40S subunit is formed after the phosphorylation and dephosphorylation step (involving the Enp1-Ltv1-Rps3 complex and the kinase, Hrr25) while a cleavage at the D-site of 20S pre-rRNA (assisted by nonribosomal factors Nob1, Rio1, Rio2, Tsr1, and Fap7) generates the mature 18S rRNA. The cytoplasmic maturation of 40S subunit is linked to the translation cycle and involves the incorporation of several RPs, formation of the 3′ end of mature 18S rRNA, and dissociation of the leftover assembly factors. These steps serve as a quality control of nascent ribosomes or SSU to ensure their translational competence.

Maturation of the pre-60S subunit

Unlike 40S subunit biogenesis, only the initial stages of the LSU assembly are cotranscriptional. Production of the 60S subunit is initiated by cleavage of nascent pre-rRNA at the A2 site. Since most RPs associate with the earliest pre-60S particles and make contact with multiple intertwined domains of rRNA secondary structure, depletion of individual RPs is able to impair the earliest steps of 27SA2 pre-rRNA processing and accumulation of 27SA2, 27SA3, or 27SB pre-rRNAs and low levels of all downstream pre-rRNAs. The assembly of LSUs proceeds in a hierarchical fashion where early acting RPs are necessary for the stable association of middle-acting RPs with pre-rRNPs and middle-acting RPs in turn are required for the assembly of late-acting RPs. Further, a similar hierarchical dependence upon RPs is also seen in the case of loading of assembly factors in preribosomes. For these reasons, pre-rRNA processing seems to happen in a sequential fashion and in concert with the binding and action of assembly factors.

The maturation and assembly of the 60S subunit involve the following steps of processing:

1. "Early steps" include removal of the ITS1 spacer sequence from 27SA2 and 27SA3 pre-rRNAs to produce 27SB pre-rRNAs in 66S preribosomes. The binding of early RPs (such as RPL3) facilitates this step.
2. "Middle steps" include cleavage at the C2 site in ITS2 of 27SB pre-rRNA to produce 7S and 25.5S pre-rRNAs, and exit of preribosomes from the nucleolus. This step is carried out with the help of the biogenesis factor Nsa2 and GTPase Nog2.
3. "Late nuclear steps" include removal of the ITS2 spacer (processing of 25.5S and 7S pre-rRNAs to 25S and 5.8S rRNAs) and remodeling of the central protuberance (CP). Several RPs and assembly factors along with Nog2 participate in this process.
4. Export of nascent particles from the nucleus into the cytoplasm is enabled by binding of export factors.
5. "Final steps" of subunit maturation occur in the cytoplasm and include the final steps of formation of 5.8S rRNA from 6S pre-RNA, assembly of the last few RPs, release of remaining AFs, and quality control checks of active sites.

Finally as for the 40S subunit, the translation-like cycle seems to be used to proofread nascent 60S subunits.

Modifications of pre-RNA

Modifications in the nucleoside of RNA were first reported in the 1950s and were subsequently observed in different types of RNAs in all organisms (Littlefield and Dunn, 1958). Two major types of covalent modifications are observed in pre-rRNAs during ribosome biogenesis, viz., (i)

2′-*O*-methylation (methylation of the 2′-hydroxyl group of sugar residues) and (ii) pseudouridylation (conversion of uridine residues to pseudouridine by base rotation). Interestingly, these modifications are well conserved between species albeit their numbers vary in different organisms. Although the precise role of such modifications in mature ribosomes is unclear, these are localized around functionally important regions such the A- and the P-site (Decatur and Fournier, 2002). Most likely the modified nucleotides allow some conformational changes of rRNAs to stabilize the overall structure and therefore the mature ribosome (King et al., 2003; Ofengand et al., 2001; Yoon et al., 2006). In eukaryotic rRNA, over 100 covalent modifications have been observed that are catalyzed by small nucleolar ribonucleoproteins (snoRNPs) (Cantara et al., 2011; Machnicka et al., 2013).

snoRNPs and rRNA modifications

Two major classes of snoRNPs exist that can be distinguished structurally and functionally. Box C/D snoRNPs are responsible for methylation reactions, whereas pseudouridylation reactions are mediated by H/ACA-box snoRNPs (Fig. 7.5A and B). The snoRNPs are composed of a snoRNA and several proteins (Table 7.2). Box C/D-associated proteins include the methyltransferase Nop1, the Nop56-Nop58 heterodimer, and Snu13. Box H/ACA-associated proteins include the pseudouridine synthase Cbf5 and Gar1, Nop10, and Nhp2. The snoRNAs present in these snoRNPs act as a guide sequence that base pair with the target rRNA and bring the processing sites into close proximity which allows cleavage in the RNA (Watkins and Bohnsack, 2012). In addition, two other snoRNPs that play key roles in eukaryotic ribosome assembly are the U3 snoRNP and RNase mitochondrial RNA processing (MRP) (reviewed in Massenet et al., 2017; Ojha et al., 2020; Reichow et al., 2007). The modifications in rRNAs are important for optimal ribosome function as their absence is linked to reduced translation efficiency (Baxter-Roshek et al., 2007; Liang et al., 2009; Sloan et al., 2017). The box H/ACA snoRNPs catalyze pseudouridylation at 44 sites on the pre-rRNA, whereas the box C/D snoRNPs catalyze 2′-*O*-ribose methylation at 67 sites (Fig. 7.5C) (Kiss et al., 2010; Watkins and Bohnsack, 2012). The methyltransferase fibrillarin/Nop1 catalyzes ribose methylation using S-adenosyl methionine as the methyl group donor, while Snu13 binds to the kink-turn in the box C/D snoRNA (Lin et al., 2011; Singh et al., 2008). However, a detailed structure of the yeast box C/D snoRNP is still not available. The snoRNPs that catalyze the pseudouridylation, carry a two-stem box H/ACA snoRNA, are characterized by a conserved H box

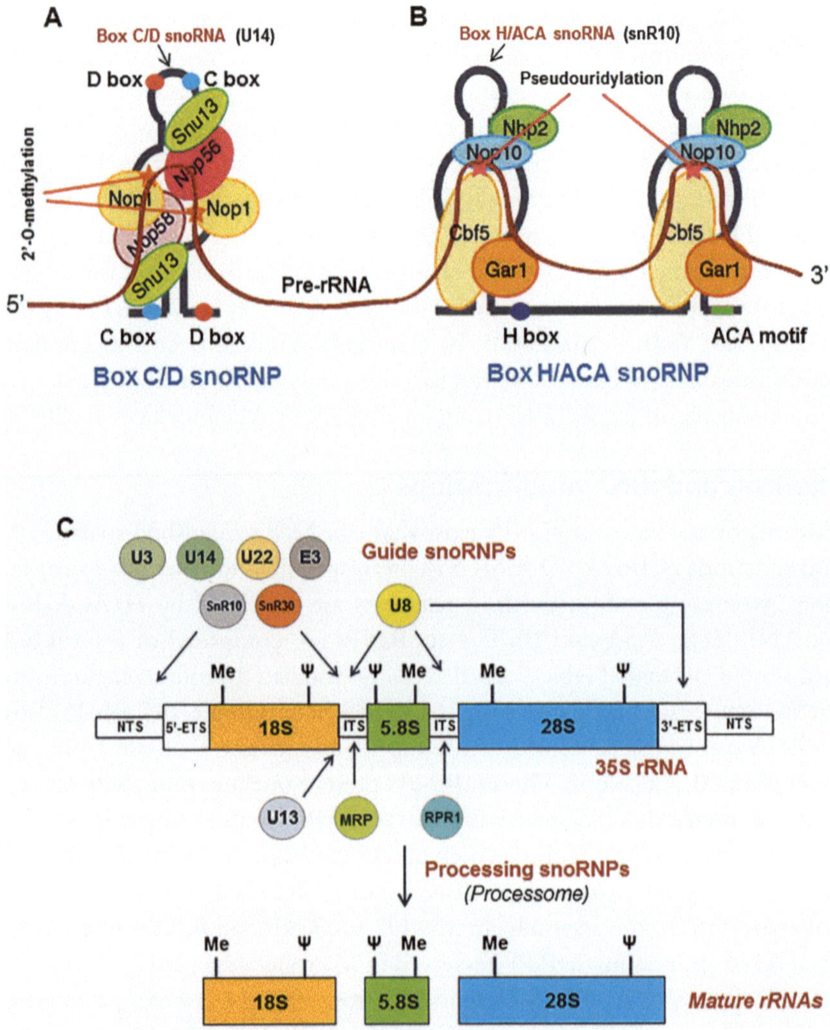

Fig. 7.5 ***SnoRNPs and their RNA and protein components.*** Based on the sequence motifs of their RNAs, snRNPs belong to two classes—The box H/ACA snoRNPs catalyze pseudouridylation (A) whereas the box C/D snoRNPs catalyze 2′-O-ribose methylation (B). Both snoRNAs target modification of pre-rRNA via base pairing. snoRNAs are shown in dark gray whereas the target rRNAs are depicted in maroon. (C) Cotranscriptional modifications/processing of rRNA. The 35S rRNA undergoes multiple RNA-mediated RNA modifications to generate mature rRNAs with the help of different classes of factors including small nucleolar ribonuclear protein particles (snoRNPs), exonucleases, endonucleases, base methylases. ETS, external transcribed spacer; ITS, internal transcribed spacer; NTS, nontranscribed spacer.

Table 7.2 Protein and RNA components of yeast snoRNPs.

	Function	Box C/D	Box H/ACA
Protein	Ribose methylation	Nop1/Fibrillarin	Cbf5
	RNA binding	Snu13	Nhp2
		Nop56	Nop10
		Nop58	Gar1
RNA	Pre-rRNA processing and pseudouridylation	–	SnR10
	Pre-rRNA processing and 2′-*O*-methylatiion	U14	–
	Pre-rRNA processing	U3	snR30

or ACA motifs, and guide sequences of 4–8 nt that base pair to the RNA target site for pseudouridylation. A single-box H/ACA snoRNA can target as many as two pseudouridylations. Further, just as for box C/D snoRNPs, the structure for the yeast box H/ACA snoRNP is not well studied.

Summary and conclusions

Ribosomes are macromolecular machines of the cell responsible for translating the genetic code into proteins. Eukaryotic ribosomes are 80S ribosomes, consisting of 60S large subunit and 40S small subunit. In the model organism *S. cerevisiae*, the mature 40S subunit consists of 18S rRNA and 33 RPs, whereas the 60S subunit consists of 28S, 5.8S, and 5S runes and 46 RPs. Ribosome biogenesis is a dynamic and tightly regulated process that is initiated around the nucleolus having multiple copies of rDNA genes. The process involves the expression of rRNAs, their processing, modification, proper folding and association with RPs, and finally the export of maturing ribosomal subunits to the cytoplasm. Over 200 nonribosomal proteins and 75 snoRNAs transiently associate with assembling ribosomes during its biogenesis. However, we still lack complete information about the interacting partners of individual proteins vis-à-vis their composite role in preribosomal particles. Recent progress in the structural understanding of nascent preribosomal particles has greatly enhanced our knowledge about ribosome biogenesis but remains far from complete. Solving the puzzle of ribosome assembly requires a combination of functional and structural approaches, which will be instrumental for future advances.

References

Baxter-Roshek JL, Petrov AN, Dinman JD. Optimization of ribosome structure and function by rRNA base modification. PLoS One 2007;2, e174.

Beltrame M, Henry Y, Tollervey D. Mutational analysis of an essential binding site for the U3 snoRNA in the 5′ external transcribed spacer of yeast pre-rRNA. Nucleic Acids Res 1994;22:5139–47.

Ben-Shem A, Garreau de Loubresse N, Melnikov S, Jenner L, Yusupova G, Yusupov M. The structure of the eukaryotic ribosome at 3.0Å resolution. Science 2011;334:1524–9.

Cantara WA, Crain PF, Rozenski J, McCloskey JA, Harris KA, Zhang X, Vendeix FA, Fabris D, Agris PF. The RNA modification database, RNAMDB: 2011 update. Nucleic Acids Res 2011;39:D195–201.

Cavanaugh AH, Hempel WM, Taylor LJ, Rogalsky V, Todorov G, Rothblum LI. Activity of RNA polymerase I transcription factor UBF blocked by Rb gene product. Nature 1995;374:177–80.

Chan JC, Hannan KM, Riddell K, Ng PY, Peck A, Lee RS, Hung S, Astle MV, Bywater M, Wall M, Poortinga G, Jastrzebski K, Sheppard KE, Hemmings BA, Hall MN, Johnstone RW, McArthur GA, Hannan RD, Pearson RB. AKT promotes rRNA synthesis and cooperates with c-MYC to stimulate ribosome biogenesis in cancer. Sci Signal 2011;4:ra56.

Ciganda M, Williams N. Eukaryotic 5S rRNA biogenesis. Wiley Interdiscip Rev RNA 2011;2:523–33.

Cook M, Tyers M. Size control goes global. Curr Opin Biotechnol 2007;18:341–50.

Côté CA, Greer CL, Peculis BA. Dynamic conformational model for the role of ITS2 in pre-rRNA processing in yeast. RNA 2002;8:786–97.

Dai X, Zhu M. Coupling of ribosome synthesis and translational capacity with cell growth. Trends Biochem Sci 2020;45:681–92.

Dammann R, Lucchini R, Koller T, Sogo JM. Chromatin structures and transcription of rDNA in yeast *Saccharomyces cerevisiae*. Nucleic Acids Res 1993;21:2331–8.

Decatur WA, Fournier MJ. rRNA modifications and ribosome function. Trends Biochem Sci 2002;27:344–51.

Dutca LM, Gallagher JE, Baserga SJ. The initial U3 snoRNA: pre-rRNA base pairing interaction required for pre-18S rRNA folding revealed by in vivo chemical probing. Nucleic Acids Res 2011;39:5164–80.

Farley-Barnes KI, Baserga SJ. Ribosome biogenesis and its role in cell growth and proliferation in the liver. In: Arias IM, Alter HJ, Boyer JL, Cohen DE, Shafritz DA, Thorgeirsson SS, et al., editors. The liver: biology and pathobiology. 6th ed. John Wiley & Sons Ltd; 2020. p. 174–82.

Fatica A, Tollervey D, Dlakić M. PIN domain of Nob1p is required for D-site cleavage in 20S pre-rRNA. RNA 2004;10:1698–701.

French SL, Osheim YN, Cioci F, Nomura M, Beyer AL. In exponentially growing *Saccharomyces cerevisiae* cells, rRNA synthesis is determined by the summed RNA polymerase I loading rate rather than by the number of active genes. Mol Cell Biol 2003;23:1558–68.

Geerlings TH, Vos JC, Raué HA. The final step in the formation of 25S rRNA in *Saccharomyces cerevisiae* is performed by 5′- ->3′ exonucleases. RNA 2000;6:1698–703.

Ginisty H, Sicard H, Roger B, Bouvet P. Structure and functions of nucleolin. J Cell Sci 1999;112:761–72.

Goodfellow SJ, Zomerdijk JC. Basic mechanisms in RNA polymerase I transcription of the ribosomal RNA genes. Subcell Biochem 2013;61:211–36.

Grandori C, Gomez-Roman N, Felton-Edkins ZA, Ngouenet C, Galloway DA, Eisenman RN, White RJ. c-Myc binds to human ribosomal DNA and stimulates transcription of rRNA genes by RNA polymerase I. Nat Cell Biol 2005;7:311–8.

Granneman S, Petfalski E, Tollervey D. A cluster of ribosome synthesis factors regulate pre-rRNA folding and 5.8S rRNA maturation by the Rat1 exonuclease. EMBO J 2011;30:4006–19.

Halkidou K, Logan IR, Cook S, Neal DE, Robson CN. Putative involvement of the histone acetyltransferase Tip60 in ribosomal gene transcription. Nucleic Acids Res 2004;32:1654–65.

Hannan KM, Hannan RD, Smith SD, Jefferson LS, Lun M, Rothblum LI. Rb and p130 regulate RNA polymerase I transcription: Rb disrupts the interaction between UBF and SL-1. Oncogene 2000;19:4988–99.

Harnpicharnchai P, Jakovljevic J, Horsey E, Miles T, Roman J, Rout M, Meagher D, Imai B, Guo Y, Brame CJ, Shabanowitz J, Hunt DF, Woolford Jr JL. Composition and functional characterization of yeast 66S ribosome assembly intermediates. Mol Cell 2001;8:505–15.

Henras AK, Plisson-Chastang C, O'Donohue MF, Chakraborty A, Gleizes PE. An overview of pre-ribosomal RNA processing in eukaryotes. Wiley Interdiscip Rev RNA 2015;6:225–42.

Henry Y, Wood H, Morrissey JP, Petfalski E, Kearsey S, Tollervey D. The 5′ end of yeast 5.8S rRNA is generated by exonucleases from an upstream cleavage site. EMBO J 1994;13:2452–63.

Hughes JM, Ares Jr M. Depletion of U3 small nucleolar RNA inhibits cleavage in the 5′ external transcribed spacer of yeast pre-ribosomal RNA and impairs formation of 18S ribosomal RNA. EMBO J 1991;10:4231–9.

King TH, Liu B, McCully RR, Fournier MJ. Ribosome structure and activity are altered in cells lacking snoRNPs that form pseudouridines in the peptidyl transferase center. Mol Cell 2003;11:425–35.

Kiss T, Fayet-Lebaron E, Jády BE. Box H/ACA small ribonucleoproteins. Mol Cell 2010;37:597–606.

Klinge S, Woolford Jr JL. Ribosome assembly coming into focus. Nat Rev Mol Cell Biol 2019;20:116–31.

Knutson BA, Luo J, Ranish J, Hahn S. Architecture of the *Saccharomyces cerevisiae* RNA polymerase I Core factor complex. Nat Struct Mol Biol 2014;21:810–6.

Kornprobst M, Turk M, Kellner N, Cheng J, Flemming D, Koš-Braun I, Koš M, Thoms M, Berninghausen O, Beckmann R, Hurt E. Architecture of the 90S pre-ribosome: a structural view on the birth of the eukaryotic ribosome. Cell 2016;166:380–93.

Kressler D, Hurt E, Baßler J. A puzzle of life: crafting ribosomal subunits. Trends Biochem Sci 2017;42:640–54.

Kufel J, Dichtl B, Tollervey D. Yeast Rnt1p is required for cleavage of the pre-ribosomal RNA in the 3' ETS but not the 5' ETS. RNA 1999;5:909–17.

Kusnadi EP, Hannan KM, Hicks RJ, Hannan RD, Pearson RB, Kang J. Regulation of rDNA transcription in response to growth factors, nutrients and energy. Gene 2015;556:27–34.

Lafontaine DLJ, Riback JA, Bascetin R, Brangwynne CP. The nucleolus as a multiphase liquid condensate. Nat Rev Mol Cell Biol 2021;22:165–82.

Lamanna AC, Karbstein K. An RNA conformational switch regulates pre-18S rRNA cleavage. J Mol Biol 2011;405:3–17.

Lewis JD, Tollervey D. Like attracts like: getting RNA processing together in the nucleus. Science 2000;288:1385–9.

Li Z, Hann SR. Nucleophosmin is essential for c-Myc nucleolar localization and c-Myc-mediated rDNA transcription. Oncogene 2013;32:1988–94.

Liang XH, Liu Q, Fournier MJ. Loss of rRNA modifications in the decoding center of the ribosome impairs translation and strongly delays pre-rRNA processing. RNA 2009;15:1716–28.

Lin CY, Navarro S, Reddy S, Comai L. CK2-mediated stimulation of Pol I transcription by stabilization of UBF-SL1 interaction. Nucleic Acids Res 2006;34:4752–66.

Lin J, Lai S, Jia R, Xu A, Zhang L, Lu J, Ye K. Structural basis for site-specific ribose methylation by box C/D RNA protein complexes. Nature 2011;469:559–63.
Littlefield JW, Dunn DB. Natural occurrence of thymine and three methylated adenine bases in several ribonucleic acids. Nature 1958;181:254–5.
Lygerou Z, Allmang C, Tollervey D, Séraphin B. Accurate processing of a eukaryotic precursor ribosomal RNA by ribonuclease MRP in vitro. Science 1996;272:268–70.
Machnicka MA, Milanowska K, Osman Oglou O, Purta E, Kurkowska M, Olchowik A, Januszewski W, Kalinowski S, Dunin-Horkawicz S, Rother KM, Helm M, Bujnicki JM, Grosjean H. MODOMICS: a database of RNA modification pathways—2013 update. Nucleic Acids Res 2013;41:D262–7.
Massenet S, Bertrand E, Verheggen C. Assembly and trafficking of box C/D and H/ACA snoRNPs. RNA Biol 2017;14:680–92.
Mayer C, Grummt I. Ribosome biogenesis and cell growth: mTOR coordinates transcription by all three classes of nuclear RNA polymerases. Oncogene 2006;25:6384–91.
Mayer C, Bierhoff H, Grummt I. The nucleolus as a stress sensor: JNK2 inactivates the transcription factor TIF-IA and down-regulates rRNA synthesis. Genes Dev 2005;19:933–41.
Nerurkar P, Altvater M, Gerhardy S, Schütz S, Fischer U, Weirich C, Panse VG. Eukaryotic ribosome assembly and nuclear export. Int Rev Cell Mol Biol 2015;319:1–34.
Oeffinger M, Zenklusen D, Ferguson A, Wei KE, El Hage A, Tollervey D, Chait BT, Singer RH, Rout MP. Rrp17p is a eukaryotic exonuclease required for 5′ end processing of Pre-60S ribosomal RNA. Mol Cell 2009;36:768–81.
Ofengand J, Del Campo M, Kaya Y. Mapping pseudouridines in RNA molecules. Methods 2001;25:365–73.
Ojha S, Malla S, Lyons SM. snoRNPs: functions in ribosome biogenesis. Biomolecules 2020;10:783.
O'Kane CJ, Hyland EM. Yeast epigenetics: the inheritance of histone modification states. Biosci Rep 2019;39, BSR20182006.
Osheim YN, French SL, Keck KM, Champion EA, Spasov K, Dragon F, et al. Pre-18S ribosomal RNA is structurally compacted into the SSU processome prior to being cleaved from nascent transcripts in *Saccharomyces cerevisiae*. Mol Cell 2004;16:943–54.
Oskarsson T, Trumpp A. The Myc trilogy: lord of RNA polymerases. Nat Cell Biol 2005;7:215–7.
Panova TB, Panov KI, Russell J, Zomerdijk JC. Casein kinase 2 associates with initiation-competent RNA polymerase I and has multiple roles in ribosomal DNA transcription. Mol Cell Biol 2006;26:5957–68.
Panse VG, Johnson AW. Maturation of eukaryotic ribosomes: acquisition of functionality. Trends Biochem Sci 2010;35:260–6.
Paule MR, White RJ. Survey and summary: transcription by RNA polymerases I and III. Nucleic Acids Res 2000;28:1283–98.
Pelletier G, Stefanovsky VY, Faubladier M, Hirschler-Laszkiewicz I, Savard J, Rothblum LI, Côté J, Moss T. Competitive recruitment of CBP and Rb-HDAC regulates UBF acetylation and ribosomal transcription. Mol Cell 2000;6:1059–66.
Peña C, Hurt E, Panse VG. Eukaryotic ribosome assembly, transport and quality control. Nat Struct Mol Biol 2017;24:689–99.
Piazzi M, Bavelloni A, Gallo A, Faenza I, Blalock WL. Signal transduction in ribosome biogenesis: a recipe to avoid disaster. Int J Mol Sci 2019;20:2718.
Poortinga G, Hannan KM, Snelling H, Walkley CR, Jenkins A, Sharkey K, Wall M, Brandenburger Y, Palatsides M, Pearson RB, McArthur GA, Hannan RD. MAD1 and c-MYC regulate UBF and rDNA transcription during granulocyte differentiation. EMBO J 2004;23:3325–35.
Ramaswamy P, Woodson SA. Global stabilization of rRNA structure by ribosomal proteins S4, S17, and S20. J Mol Biol 2009;392:666–77.

Reichow SL, Hamma T, Ferré-D'Amaré AR, Varani G. The structure and function of small nucleolar ribonucleoproteins. Nucleic Acids Res 2007;35:1452–64.
Rodnina MV. The ribosome as a versatile catalyst: reactions at the peptidyl transferase center. Curr Opin Struct Biol 2013;23:595–602.
Rodríguez-Galán O, García-Gómez JJ, de la Cruz J. Yeast and human RNA helicases involved in ribosome biogenesis: current status and perspectives. Biochim Biophys Acta 2013;1829:775–90.
Rosenwald IB, Rhoads DB, Callanan LD, Isselbacher KJ, Schmidt EV. Increased expression of eukaryotic translation initiation factors eIF-4E and eIF-2 alpha in response to growth induction by c-myc. Proc Natl Acad Sci U S A 1993;90:6175–8.
Sadian Y, Tafur L, Kosinski J, Jakobi AJ, Wetzel R, Buczak K, Hagen WJ, Beck M, Sachse C, Müller CW. Structural insights into transcription initiation by yeast RNA polymerase I. EMBO J 2017;36:2698–709.
Sandmeier JJ, French S, Osheim Y, Cheung WL, Gallo CM, Beyer AL, Smith JS. RPD3 is required for the inactivation of yeast ribosomal DNA genes in stationary phase. EMBO J 2002;21:4959–68.
Saveanu C, Namane A, Gleizes PE, Lebreton A, Rousselle JC, Noaillac-Depeyre J, Gas N, Jacquier A, Fromont-Racine M. Sequential protein association with nascent 60S ribosomal particles. Mol Cell Biol 2003;23:4449–60.
Shen M, Zhou T, Xie W, Ling T, Zhu Q, Zong L, Lyu G, Gao Q, Zhang F, Tao W. The chromatin remodeling factor CSB recruits histone acetyltransferase PCAF to rRNA gene promoters in active state for transcription initiation. PLoS One 2013;8, e62668.
Singh SK, Gurha P, Gupta R. Dynamic guide-target interactions contribute to sequential 2'-O-methylation by a unique archaeal dual guide box C/D sRNP. RNA 2008;14:1411–23.
Sloan KE, Warda AS, Sharma S, Entian KD, Lafontaine DLJ, Bohnsack MT. Tuning the ribosome: the influence of rRNA modification on eukaryotic ribosome biogenesis and function. RNA Biol 2017;14:1138–52.
Stefanovsky VY, Langlois F, Bazett-Jones D, Pelletier G, Moss T. ERK modulates DNA bending and enhancesome structure by phosphorylating HMG1-boxes 1 and 2 of the RNA polymerase I transcription factor UBF. Biochemistry 2006;45:3626–34.
Steffen KK, McCormick MA, Pham KM, MacKay VL, Delaney JR, Murakami CJ, et al. Ribosome deficiency protects against ER stress in *Saccharomyces cerevisiae*. Genetics 2012;191:107–18.
Thomson E, Ferreira-Cerca S, Hurt E. Eukaryotic ribosome biogenesis at a glance. J Cell Sci 2013;126:4815–21.
Trapman J, Retèl J, Planta RJ. Ribosomal precursor particles from yeast. Exp Cell Res 1975;90:95–104.
Udem SA, Warner JR. Ribosomal RNA synthesis in *Saccharomyces cerevisiae*. J Mol Biol 1972;65:227–42.
Van Nues RW, Venema J, Rientjes JM, Dirks-Mulder A, Raue HA. Processing of eukaryotic pre-rRNA: the role of the transcribed spacers. Biochem Cell Biol 1995;73:789–801.
van Riggelen J, Yetil A, Felsher DW. MYC as a regulator of ribosome biogenesis and protein synthesis. Nat Rev Cancer 2010;10:301–9.
Venema J, Tollervey D. Processing of pre-ribosomal RNA in *Saccharomyces cerevisiae*. Yeast 1995;11:1629–50.
Vincent T, Kukalev A, Andäng M, Pettersson R, Percipalle P. The glycogen synthase kinase (GSK) 3beta represses RNA polymerase I transcription. Oncogene 2008;27:5254–9.
Voit R, Grummt I. Phosphorylation of UBF at serine 388 is required for interaction with RNA polymerase I and activation of rDNA transcription. Proc Natl Acad Sci U S A 2001;98:13631–6.
Voit R, Schäfer K, Grummt I. Mechanism of repression of RNA polymerase I transcription by the retinoblastoma protein. Mol Cell Biol 1997;17:4230–7.

Voit R, Hoffmann M, Grummt I. Phosphorylation by G1-specific cdk-cyclin complexes activates the nucleolar transcription factor UBF. EMBO J 1999;18:1891–9.

Vos TJ, Kothe U. snR30/U17 small nucleolar ribonucleoprotein: a critical player during ribosome biogenesis. Cell 2020;9:2195.

Warner JR. The economics of ribosome biosynthesis in yeast. Trends Biochem Sci 1999;24:437–40.

Watkins NJ, Bohnsack MT. The box C/D and H/ACA snoRNPs: key players in the modification, processing and the dynamic folding of ribosomal RNA. Wiley Interdiscip Rev RNA 2012;3:397–414.

Woodson SA. RNA folding pathways and the self-assembly of ribosomes. Acc Chem Res 2011;44:1312–9.

Woolford Jr JL, Baserga SJ. Ribosome biogenesis in the yeast *Saccharomyces cerevisiae*. Genetics 2013;195:643–81.

Yoon A, Peng G, Brandenburger Y, Zollo O, Xu W, Rego E, et al. Impaired control of IRES-mediated translation in X-linked dyskeratosis congenita. Science 2006;312:902–6.

Zemp I, Kutay U. Nuclear export and cytoplasmic maturation of ribosomal subunits. FEBS Lett 2007;581:2783–93.

Zhai W, Comai L. Repression of RNA polymerase I transcription by the tumor suppressor p53. Mol Cell Biol 2000;20:5930–8.

Zhang C, Comai L, Johnson DL. PTEN represses RNA polymerase I transcription by disrupting the SL1 complex. Mol Cell Biol 2005;25:6899–911.

Zhang Y, Smith 4th AD, Renfrow MB, Schneider DA. The RNA polymerase-associated factor 1 complex (Paf1C) directly increases the elongation rate of RNA polymerase I and is required for efficient regulation of rRNA synthesis. J Biol Chem 2010;285:14152–9.

Further reading

Borer RA, Lehner CF, Eppenberger HM, Nigg EA. Major nucleolar proteins shuttle between nucleus and cytoplasm. Cell 1989;56:379–90.

Claypool JA, French SL, Johzuka K, Eliason K, Vu L, Dodd JA, Beyer AL, Nomura M. Tor pathway regulates Rrn3p-dependent recruitment of yeast RNA polymerase I to the promoter but does not participate in alteration of the number of active genes. Mol Biol Cell 2004;15:946–56.

Reeder RH. Regulation of RNA polymerase I transcription in yeast and vertebrates. Prog Nucleic Acid Res Mol Biol 1999;62:293–327.

CHAPTER 8

Ribosome biogenesis in prokaryotes

Zaheenul Islam Siddiqui
Department of Microbiology, College of Medicine, Imam Abdulrahman Bin Faisal University, Dammam, Saudi Arabia

Introduction

Ribosomes are minute energy-driven decoding biological machines that decode the message encrypted in the base sequence of mRNAs to synthesize proteins. The prokaryotic ribosome is the 70S (Svedberg unit) functional unit, made of 50S large and 30S small subunits, named according to their sedimentation coefficient (Yusupov, 2001). The function of the small subunit is to recognize and bind with translation start site on mRNA, whereas the function of the large subunit is to form peptide bonds between incoming amino acids, to accommodate the nascent polypeptide chain and to give it an exit tunnel (Shajani et al., 2011). The 50S particle is made up of 33 ribosomal proteins (RPs) and two ribosomal RNA (rRNA) molecules—23S (2904 nucleotides in length) and 5S (120 nt) (Cerretti et al., 1983). The 30S subunit is composed of 21 RPs and a 16S rRNA of 1542 nt (Wimberly, 2000). The RPs are numbered from L1 to L36 and from S1 to S21 for the large and small subunits, respectively, in order of their increasing electrophoretic mobilities (Cerretti et al., 1983). Ribosome biogenesis is a multistep process, which involves rRNA and RP synthesis, and their assembly to make the mature ribosomes. Since the bacterial cell is a single noncompartmentalized membrane-bound biological chamber, all the steps of ribosome biogenesis are carried out in the cytosol. The whole process of ribosome biogenesis is fast and efficient requiring only 2 minutes for the production of the mature and active ribosome.

Emerging Concepts in Ribosome Structure, Biogenesis, and Function
https://doi.org/10.1016/B978-0-12-816364-1.00009-3

Escherichia coli as a model organism to study ribosome biogenesis

The earliest work on protein biosynthetic machinery was reported by G. Gamier, who performed microscopic studies during 1897–99 on basophilic components of glandular cell cytoplasm (Siekevitz and Zamecnik, 1981). Between 1920 and 1940, various researchers including Bordet and Ciuca (1921), Bronfenbrenner and Korb (1925), Bronfenbrenner (1932), Werkman (1927), Wollman (1925), and Wollman and Wollman (1937) used *E. coli* to study ribosome because of its availability, fast growth, small genome, and easy to culture under laboratory conditions. The success of these experiments established *E. coli* as a model organism to study the structure and function of the ribosome (Blount, 2015). Many questions on ribosome biogenesis remain to be answered even for the simplest study model, i.e., *E. coli* and very few researchers tried to move to other prokaryotes to explore the basics of ribosomal biogenesis (Nomura, 2009). Therefore, most ribosomal biogenesis details given here are based on the reports published on *E. coli* as a study model unless otherwise indicated. A separate section on ribosome biogenesis in archaea is included toward the end of this chapter.

r-RNA gene structure, transcription, and posttranscriptional processing

The three rRNAs—16S, 23S, and 5S are encoded by a single ribosomal RNA operon (rrn) in prokaryotes. Multiple copies of rrn are generally present in prokaryotes, which varies from species to species, with one to 15 copies in bacteria and one to four copies in archea (Stoddard et al., 2015). The sequence of coding regions for the three rRNAs is 16S-23S-5S (Fig. 8.1A), connected by "internally transcribed spacer" regions coding for one or more tRNA and some conserved adjacent regions (Espejo and Plaza, 2018). In *E. coli* there are seven copies of rrns scattered throughout the genome (Fig. 8.1B). The length of a single rrn of *E. coli* is 5.5 kb (Schmidt, 1998). A single type of RNA polymerase (RNAP) is responsible for the synthesis of all types of transcripts in prokaryotes including rRNA, tRNA, mRNA, and noncoding RNA. So RNAP transcribes the rrn to form a single precursor rRNA, which is controlled by promoter and terminator regions (Fig. 8.1A). The presynthesized RPs present in the cytoplasm start assembling on the nascent transcript cotranscriptionally (Jin et al., 2017).

During transcription, the 16S and 23S regions of nascent pre-rRNA come together forming loops, on which a ribonuclease called RNase III

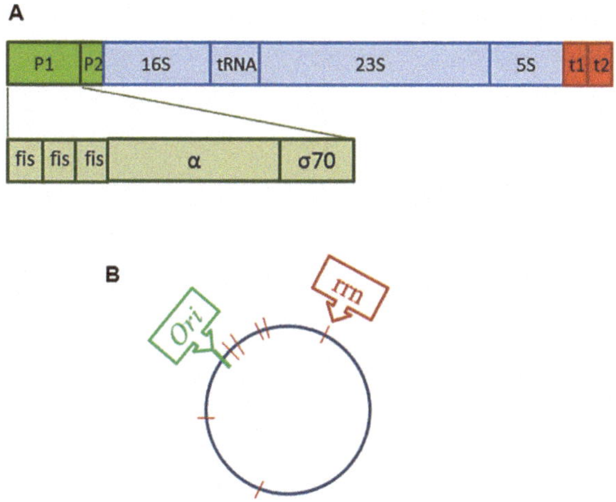

Fig. 8.1 ***(A) Schematic diagram of rrn operon in*** *E. coli. P1 and P2*: promoters for transcription. *t1* and *t2*: terminators of transcription. *tRNA*: region coding for one or multiple tRNAs. *16S*, *23S*, and *5S*: coding regions of rRNA. *Fis*: Transcription factor Fis binding sites. α: site for binding of two α subunits of RNA polymerase (RNAP). σ: site for binding of σ70 subunit of RNAP. (B) Locations of *rrn* on *E. coli* genome map (represented here as circle). The green mark on the map represents the origin of replication (*ori*). Red marks on the map are the approximate locations of rrn operons in the genome with respect to the *ori* site. (Adapted and modified from (A) Paul B.J., Ross W., Gaal T., Gourse R.L. rRNA transcription in *Escherichia coli*. Annu Rev Genet 2004b;38:749–770 and (B) Gaal T, Bratton BP, Sanchez-Vazquez P, Sliwicki A, Sliwicki K, Vegel A, Pannu R and Gourse RL. Colocalization of distant chromosomal loci in space in *E. coli*: a bacterial nucleolus. Genes Dev 2016;30:2272–2285.)

act to form three precursor (pre) rRNA—17S, 23S, and 9S (Fig. 8.2). These precursors are cleaved by various RNases to form mature 16S, 23S and 5S rRNA, respectively. Extra 115 nt from the 5′ end and 33 nt at the 3′ end of 17S rRNA are removed to generate the 16S rRNA (Deutscher, 2009; Kaczanowska and Rydén-Aulin, 2007; Shetty and Varshney, 2016). The 23S pre-rRNA formed during initial nucleolytic excision by RNase III has extra 3 or 7 nt at the 5′ end and extra 7 or 9 nt at the 3′ end (Fig. 8.2). The 5′ end is processed by an unknown enzyme, and the 3′ end processing requires a mature 5′ end along with RNase T. The final processing step of 23S is delayed till the first round of translation (Deutscher, 2009; Kaczanowska and Rydén-Aulin, 2007; Shetty and Varshney, 2016). The third fragment of the initial nucleolytic excision by RNase III is 9S pre-rRNA. It is cleaved by RNase P to release the tRNA part of 9S precursor (Fig. 8.2). The resulting 9S fragment has extra 84 nt at the 5′ end and extra 42 nt at the 3′ end. This fragment is further excised by RNase E to produce a pre-5S rRNA with

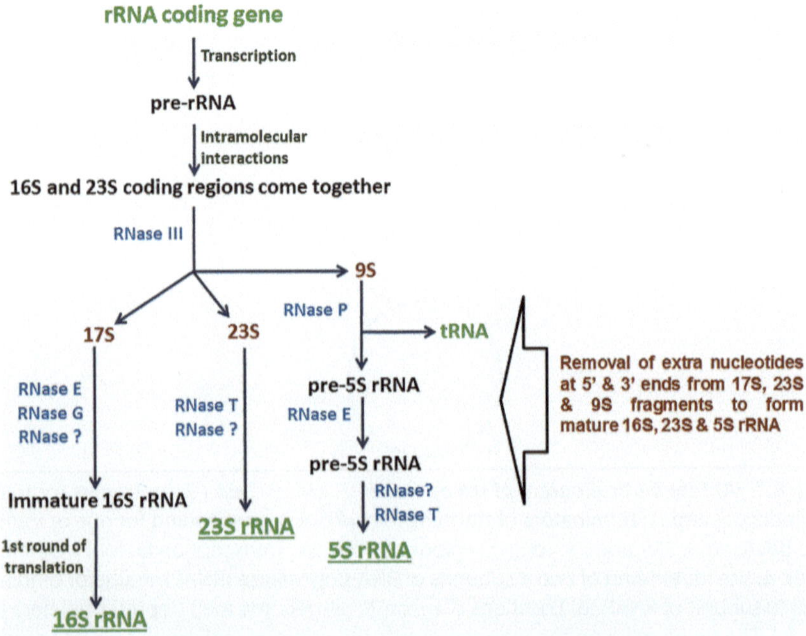

Fig. 8.2 ***Posttranscriptional processing of precursor rRNA to form mature 16S, 23S, and 5S rRNA.*** YbeY, an endonuclease, controls the maturation of all the three rRNAs. Pre-rRNA are precursor rRNA; and RNases (?) are those ribonucleases on which more research is required to define identity.

three extra nucleotides at both the 5′ and 3′ ends. These extra nucleotides are removed by an unknown enzyme at the 5′ end and RNase T at the 3′ end (Deutscher, 2009; Kaczanowska and Rydén-Aulin, 2007; Shetty and Varshney, 2016).

It is reported that an endonuclease-YbeY is also necessary for the maturation of the three rRNAs, and its deletion in *E. coli* affects the maturation of the 5′ and 3′ termini of 16S rRNA, as well as the maturation of the 5′-termini of 23S and 5S rRNAs (Davies et al., 2010). It is recently reported that participation of initiator tRNA via its universally conserved three consecutive GC base pairs in anticodon stem, in the first round of the initiation complex formation (Fig. 8.2), licenses the final steps of ribosome biogenesis by signaling RNases to trim the immature 16S rRNAs (Shetty and Varshney, 2016).

Some nucleotides of rRNA are modified by methylations and pseudouridylations at several positions during the assembly process. The nucleotide modifications are acquired gradually during the association of rRNA with r-proteins (reviewed by Bohne, 2014; Siibak and Remme, 2010). Currently, the 16S rRNA is known to undergo 11 modifications (1 pseudouridylation and 10 methylations), and the 23S rRNA undergoes 25 modifications (1 unknown, 14 methylations, 9 pseudouridylation, and 1 methylated pseudouridylation) (Shajani et al., 2011). Most of these rRNA modifications are at nucleotide positions that are functionally important, such as decoding and peptidyl transferase centers, and the interface of the 30S and 50S subunits, so it is assumed that the modifications might have some role in the structure and function of the ribosome. For example, methylation adds to the hydrophobicity of nucleotides and can influence base stacking. Similarly, pseudouridine formation increases the hydrogen bonding potential of the nucleotides and affects stacking interactions. These modifications also contribute to the stability of rRNA. Some modifications may function as checkpoints of assembly, and as quality control for intermediate preribosomal subunit structures. These modifications may also be involved in antibiotic resistance/sensitivity, or may function as molecular glues to differentiate the functionally active rRNA conformation from other pre-rRNA intermediate conformations and misfolded rRNA populations (Sergeeva et al., 2014a, 2014b).

Ribosome biogenesis is an ordered series of events comprising sequential addition of RPs to rRNA, cleavage and trimming of rRNA precursors and rRNA modifications. Each step facilitates the accuracy and efficiency of the succeeding step until the ribosome subunits have fully assembled. Multiple factors are involved in rRNA maturation, such as r-proteins, RNA helicases, rRNA modification enzymes, chaperones, and GTPases (Kaczanowska and Rydén-Aulin, 2007).

It has been observed that the rrn operons in a bacterial genome come in close proximity to form structures similar to eukaryotic nucleoli. The promoter regions of rrn are involved in this clustering process (Gaal et al., 2016). In fast-growing bacteria, there may be as many as 50 rrns in a single dividing cell. Clustering of these operons has been observed to optimize the usage of RNAP and maximize the synthesis of rRNA to form new ribosomes and meet the increasing demand for protein biosynthesis (Jin et al., 2017). rRNA accounts for more than 80% of RNA synthesized, and as much as 65% of RNAP is allocated for rRNA synthesis in fast-growing prokaryotic cells (Jin et al., 2015). When the cells return to the slow growth/stationary phase, or become stressed due to limited/starved nutrient supply, the RNAP is redistributed to other gene loci for the production of gene products needed for slow growth or stress conditions (Jin et al., 2015).

Regulation of rrn expression

The transcription of rrn operon in *E. coli* is controlled by promoters, *P1* and *P2* and transcription terminators *t1* and *t2* (Fig.8.1A). The expression is regulated primarily at the level of initiation of transcription by these promoters and is considered as the rate-limiting step for ribosome biogenesis. During rapid bacterial growth, the rrn transcription is controlled by *P1*, whereas during slow growth and stationary phases, expression is mediated by *P2*. Although transcription factor "Fis" helps RNAP binding to *P1*, the main regulators of *P1* are initiating concentration of NTP (iNTP) and guanosine tetra- and pentaphosphates (ppGpp). A high NTP concentration increases rrn expression by *P1*. Low levels of amino acid during limited nutrient conditions decrease the level of charged tRNA. The presence of uncharged tRNA at the A site of ribosome triggers ribosome-associated protein RelA to produce ppGpp. This is followed by the interaction of ppGpp with RNAP and a subsequent decrease in transcription via a decrease in the half-life of RNAP open complexes (transcription initiation complexes). This effect is enhanced specifically for rrn operon due to another transcription factor "DksA" which act by directly interacting with RNAP transcription initiation complexes. Such mechanisms are specifically involved in the modulation of rrn promoter activity during changes in levels of NTP and ppGpp under varying nutrient conditions (Paul et al., 2004a, 2004b). Apart from *E. coli*, the involvement of ppGpp in the regulation of gene expression has been observed throughout the bacterial kingdom. For example, in the case of *Bacillus subtilis,* ppGpp regulates rrn promoters via reducing the intracellular GTP levels (Krasny and Gourse, 2004; Kriel et al., 2012).

When the bacterial cells grow under optimal nutrient conditions, the rrn operons come in close proximity to form clusters. These clusters are concentrated with RNAP molecules to form eukaryotic nucleoli like structures that segregate the transcription machinery from the replication machinery. This maximizes rRNA expression and ribosome biogenesis under favorable growth conditions to meet the growing needs of protein biosynthesis (Jin et al., 2017). In fast-growing cells, the majority of RNAP molecules are allocated to rrn transcription. When cell growth slows down or cells become stressed due to nutrient exhaustion or other stress conditions, the rrn expression is turned off and RNAP molecules are redistributed throughout the genome to turn on the genes involved in slow growth, stationary phase or the stress responses (reviewed by Jin et al., 2015).

Structure of the ribosomal protein-coding genes, the operons and regulation of expression

There are 54 ribosomal protein genes (RPGs) that encode *L1-L36* proteins of the large ribosomal subunit and *S1-S21* proteins of the small ribosomal subunit. These genes are organized in 19 transcriptional units or operons in the *E. coli* genome (Fig. 8.3) (Cerretti et al., 1983). These are *S20 (S20), S2 (S2), S1 (S1), L32 (L32), thrS/L20/pheS (L35, L20), L25 (L25), S16 (S16, L19), S21 (S21), S15 (S15), L13 (L13, S9), str (S12, S7), S10 (S10, L3, L4, L23, L2, S19, L22, S3, L16, L29, S17), spc (L14, L24, L5, S14, S8, L6, L18, S5, L30, L15, L36), alpha (S13, S11, S4, L17), L28 (L28, L33), L34 (L34), L11 (L11,L1), L10 (L10,L12), and S6 (S6, S18, L9)*. Note that the operon names are written outside the brackets and gene(s) in that respective operon is/are inside. One copy of each RP coded by these genes is present in the mature bacterial ribosome, except L7/12, four copies of which are present per ribosome. In the naming system, we find that there is no gene locus for L7, L8, and L26. This is because L7 is the name given to the N-terminal acetylated form of L12 protein. L8 is the name given to a complex made of two L12 dimers complexed with a molecule of L10 (Mandava, 2011). Because a yeast RP is named L26, and its homologous protein in bacteria is called L24, no protein in the bacterial ribosome was named L26 to avoid ambiguity (Babiano et al., 2012). So in total, there are 33 genes of the large ribosomal subunit and 21 genes of the small ribosomal subunit (Fig. 8.3B).

Sometimes we come across with genes/operons named different from the normally used LX and SX convention. This is because, mutations in some of the r-protein coding regions give bacterial resistance against specific antibiotics, and these genes/operons are historically named after that specific antibiotic. For example, the genes for S12 and S5, and the β subunit of RNAP are referred to as the *str* (streptomycin resistance), *spc* (spectinomycin resistance), and *rif* (rifampicin resistance) genes, respectively. In the same way, the operon containing the S12 gene is referred to as *str* operon and the S5 encoding operon is called *spc* operon (reviewed by Zengel and Lindahl, 1994). Some operons include coding regions for nonribosomal proteins such as β and β′ subunits of RNAP. For the simpler illustration of operons, the details of nonribosomal products have been omitted (Fig. 8.3B). Transcription of all the operons is coregulated and synthesized in equimolar amounts except L12/L7. Most of the r-protein operons have promoters and terminators at their 5′ and 3′ ends, respectively (Fig. 8.3B).

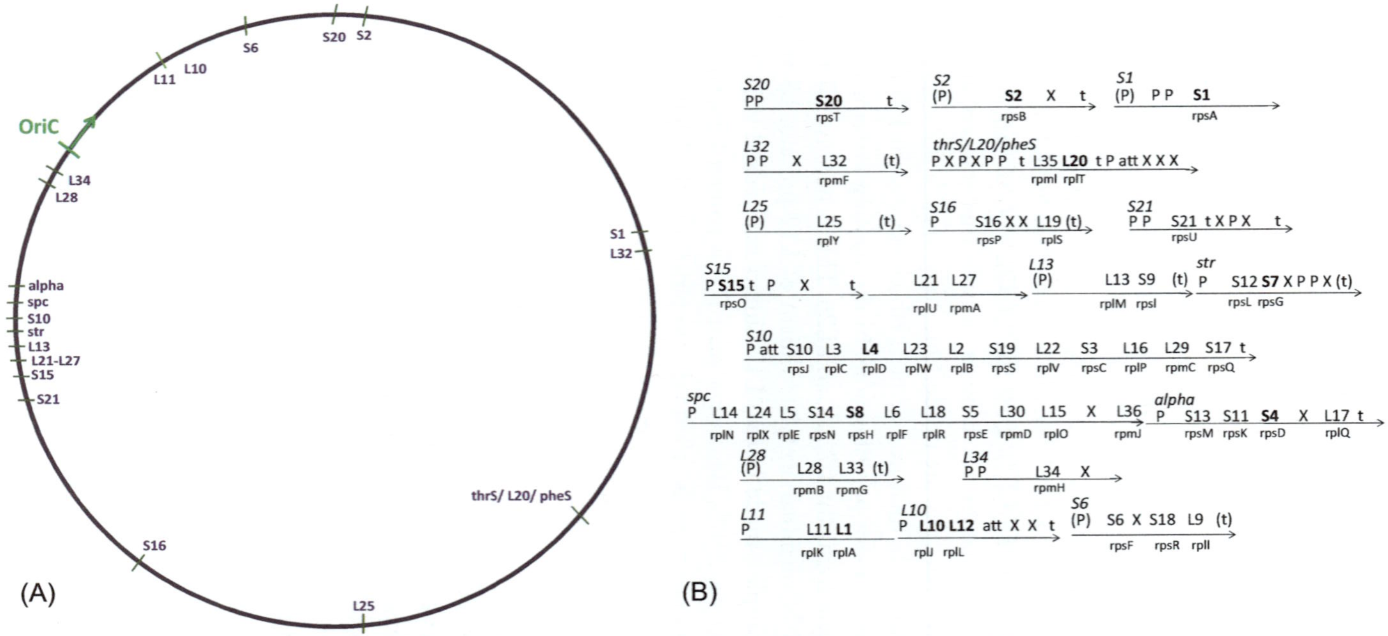

Fig. 8.3 ***Schematic diagrams of genes coding ribosomal proteins (r-proteins).*** (A) The approximate position of 19 operons of r-proteins on the *E. coli* genome with respect to the *Ori* site. *Ori* is the origin of replication. (B) Details of r-protein operons, showing the regulatory and coding regions for each gene. The details of nonribosomal proteins encoded by r-protein operons are omitted and marked X. Genes represented in bold letters encode regulatory proteins. Names of genes coding r-proteins are mentioned below its coding region. *P*: promoter, (*P*): tentative promoter, *t*: transcription terminator, (*t*): tentative transcription terminator. (The map and gene/operon order is drawn with the help of data from Babiano R, Gamalinda M, Woolford JL, Cruza J. Saccharomyces cerevisiae ribosomal protein L26 is not essential for ribosome assembly and function. Mol Cell Biol 2012; 32:3228–41; Fu Y, Deiorio-Haggar K, Anthony J, Meyer MM. Most RNAs regulating ribosomal protein biosynthesis in *Escherichia coli* are narrowly distributed to Gammaproteobacteria. Nucleic Acids Res 2013;41:3491–503; Lindahl L, Zengel JM. Ribosomal genes in *Escherichia coli*. Annu Rev Genet 1986;20:297–326; Mandava CS. Ribosomal stalk protein L12: structure, function and application. Acta Universitatis Upsaliensis, 2011; p. 49; Maouche R, Burgos HL, My L, Viala JP, Gourse RL, Bouveret E. Coexpression of *Escherichia coli* obgE, encoding the evolutionarily conserved Obg GTPase, with ribosomal proteins L21 and L27. J Bacteriol 2016;198:1857–67; Zengel JM, Lindahl L. Diverse mechanisms for regulating ribosomal protein synthesis in *Escherichia coli*. Prog Nucleic Acid Res Mol Biol 1994;47:331–70.)

But some operons have two promoters (e.g., for spc and alpha), while some do not have a terminator (e.g., for L11), and some have weak terminators (e.g., for spc). Internal promoters (e.g., for S21, *str*, and L20) and internal terminators or attenuators (e.g., for S10, L10, and L20) are also present. The attenuators allow partial termination of transcripts coded by these operons. Some operons have leaky terminators (e.g., for S10 operon and S15 gene), allowing read-throughs into neighboring operons (reviewed by Lindahl and Zengel, 1986; Zengel and Lindahl, 1994).

The production of RPs is coordinated with the production of rRNA in such a way that all the produced RPs are used up to assemble with rRNA for making mature ribosome. The RPs that bind first to rRNA are called primary binding RPs. The conformation of rRNA changes after binding with these proteins. On these rRNA-primary protein structures, secondary binding RPs bind. In the same way, tertiary binding RPs interact with new tertiary conformations adapted after secondary RP binding events. Some of the RPGs are shown highlighted in bold letters in Fig. 8.3B. The product of these genes regulates the whole operon, belonging to that gene, autogenously. All these regulatory RPs are primary rRNA binding proteins. When the production of a specific primary binding RP is higher than the available site for it on rRNA, the excess protein represses its own production and the production of other proteins coded by that operon through feedback inhibition (Fig. 8.4). These excess regulatory proteins directly bind to the mRNA of that operon halting translation (Fig. 8.4). For example, L10 and L12 bind to the L10 operon mRNA, S4 to the alpha operon mRNA, S8 to *spc* mRNA, S15 to S15 mRNA, and S7 to str mRNA. These bindings result in a hindrance of ribosome binding to

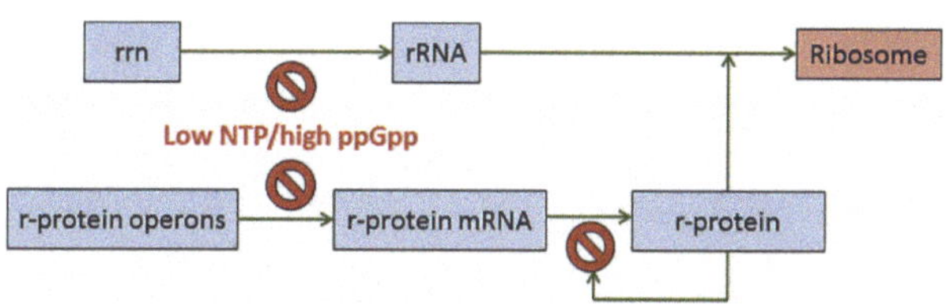

Fig. 8.4 ***rRNA and r-proteins expression controlled by autoregulation and by intracellular levels of ppGpp and NTP.*** rrn: rRNA gene; rRNA: Ribosomal RNA; r-protein: Ribosomal protein; NTP: Nucleoside triphosphate; ppGpp: Guanosine tetra-/pentaphosphates. (Adapted and modified from Lemke JJ, Sanchez-Vazquez P, Burgos HL, Hedberg G, Ross W, Gourse RL. Direct regulation of Escherichia coli ribosomal protein promoters by the transcription factors ppGpp and DksA. Proc Natl Acad Sci U S A 2011;108:5712–5717.)

the nascent mRNA. These regulatory protein-mRNA interactions may induce a conformational change in mRNA that leaves Shine-Dalgarno (SD) sequence inaccessible for the ribosome. The binding may also disrupt the pairing of the initiation codon and the formation of the initiation complex. Direct steric hindrance due to blockage of binding site by regulatory protein interactions may also be possible. Generally, translation of most of the proximal genes in autogenously regulated operons is directly inhibited. Translation of other genes is inhibited by translation inhibition coupling, i.e., when the ribosome will not read the very first gene, other genes that follow in that operon will not be translated (reviewed by Zengel and Lindahl, 1994).

It is shown by some researchers that for some operons, such as str and spc, binding of regulatory RPS to mRNA result in destabilization of mRNA and its degradation (Mattheakis et al., 1989). The binding of regulatory RPs to mRNA sometimes also regulates the elongation of that mRNA. This is the case for the S10 operon where L4 binding results in premature attenuation of growing transcript (Zengel and Lindahl, 1992).

Ribosome synthesis is the most energy-consuming process in a bacterial cell. When the nutrients are exhausted in the environment where the cells are growing, the immediate effect is the blockage of ribosomal biogenesis. This is mediated by a cellular response to amino acid exhaustion in the surrounding media, known as "stringent response." There are several mechanisms for the repression of the production of ribosomal components during the stringent response. The main effector is intracellular ppGpp. The level of ppGpp increases sharply in response to amino acid starvation, which results in repression of ribosome synthesis and upregulation of amino acid biosynthesis. The stringent response involves widespread changes in gene expression resulting in the diversion of resources from growth and division to preparation for stationary or stress phases. ppGpp binds directly to RNAP and inhibits transcription from rrn and RP gene promoters (Fig. 8.4). dksA, a protein, acts as a cofactor of ppGpp in this process (reviewed by Aseev et al., 2016). The pool of nucleoside triphosphates (NTPs) also reflects the nutritional state of a cell. Low levels of NTPs, meaning nutritional starvation, cause a halt in rRNA and RP synthesis at the transcriptional level (Fig. 8.4). The ribosome biosynthesis is resumed when the nutritional condition improves and increased NTP levels induce transcription of rrn and RPGs (Burgos et al., 2017).

Modification of RPs

In *E. coli*, a number of RPs undergo posttranslational modifications (PTM). RPs S11, L3, L7/L12, L16, and L33 are mono-methylated, L11 is methylated at nine positions, three proteins are acetylated (S5, S18, and L12), the protein S12 is methyl-thiolated, additional glutamic acid residues are added to the C-terminus of protein S6, the C-terminal 8 amino acid residues are partially removed from L31. The protein L12 is also present in acetylated form, which is known as L7. The ratio of these two forms, i.e., L7:L12, in a cell is important as it varies with the cell growth rate. Functional roles of these protein modifications are not clear, as the ribosome remains fairly functional in the absence of these modifications, with exception of L3 methylation (Nesterchuk et al., 2011). *E. coli* mutant with impaired L3 methylation shows poor growth, cold sensitivity, and accumulation of precursors for the ribosomal subunits (Kaczanowska and Rydén-Aulin, 2007). The posttranslational modifications apparently modifies the protein structurally or functionally, sometimes activating the protein catalytic site, sometimes modifying its efficiency, or in some cases inactivating the protein. It is also possible that the modifications enhance binding efficiencies and specificities of RPs with rRNA, and may optimize the translation machinery. It is speculated that these modifications might also have roles in the regulation of gene expression, owing to the fact that the central role of ribosomes is to read the genetic information of a cell (Nesterchuk et al., 2011; Kaczanowska and Rydén-Aulin, 2007).

Assembly of ribosome

Early studies by Miller and his colleagues showed that RPs assemble over rRNA cotranscriptionally (Gotta et al., 1991; Hofmann and Miller Jr, 1977; Miller Jr and Beatty, 1969; Miller Jr et al., 1970). The assembly process progresses through alternating series of RNA conformational changes and protein-binding events. The binding of RPs on rRNA occurs in an interdependent fashion so that some binders come first and the succeeding binders depend on early protein bindings (Herold and Nierhaus, 1987; Mizushima and Nomura, 1970). Each of the primary binders binds to rRNA independent of other protein binding events. These interactions induce conformational changes in rRNA, which assist further binding of secondary binding proteins with new conformations. The tertiary binding proteins bind toward the end to complete the assembly process (Sykes, 2009).

Assembly of 30S subunit

The 16S rRNA structure can be divided into three parts—5′, central and 3′ domains (Table 8.1). This division is based on the early works of Mizushima and Nomura (1970) on 30S subunit reconstruction. The primary binding proteins for the 16S rRNA are S4, S7, S8, S15, S17, and S20. Though the binding of these r-proteins on the rRNA is independent of each other, they mutually affect each other's binding. The assembly of proteins proceeds from 5′ to 3′ direction (Table 8.1) (Bunner et al., 2010; Held, 1973).

Thermodynamic, kinetic, and cryo-electron microscopic data from in vivo studies on ribosomal assembly made it possible to understand the whole process and also identify intermediates that are formed during the assembly process. In vivo studies revealed 14 intermediary stages in the 30S subunit assembly, which were grouped as early, intermediate, and late. Assembly starts with the formation of early intermediates using 16S rRNA, S4, S17, and S20, as well as S16, which were grouped as secondary binding RPs by Nomura (Mizushima and Nomura, 1970). Early intermediates are formed by rapid binding of primary and secondary RPs to the 5′ and the central domain of rRNA. The second round of intermediates is formed which utilize the early intermediates upon which the primary and secondary proteins bind slowly in the central and 3′ domains of 16S rRNA. Finally, the tertiary binding proteins bind gradually to complete the assembly. During the whole process, cofactors involved in ribosome assembly minimize the formation of erroneous intermediates (Mulder, 2010).

In vivo studies using radioactive pulse labeling of ribosomes combined with chromate mass spectrometry have revealed fine details of the assembly process which are in agreement with earlier electron microscopic studies. These studies divided the RPs into four groups based on the order of binding (Fig. 8.5). RPs S4, S6, S8, S15, S16, S17, S18, and S20 are the initial

Table 8.1 Order of protein binding in Nomura in vitro experiments for reconstitution of 30S subunit in 5′ to 3′ direction (Bunner et al., 2010).

	Domains	Primary binding proteins	Secondary binding proteins	Tertiary binding proteins
30S subunit	5′ Domain	S17,S4, S20	S16	S5, S12
	Central domain	S8, S15	S18, S6, S11	S21
	3′ Domain	S7	S9, S13, S19	S10, S14, S3, S2, S1

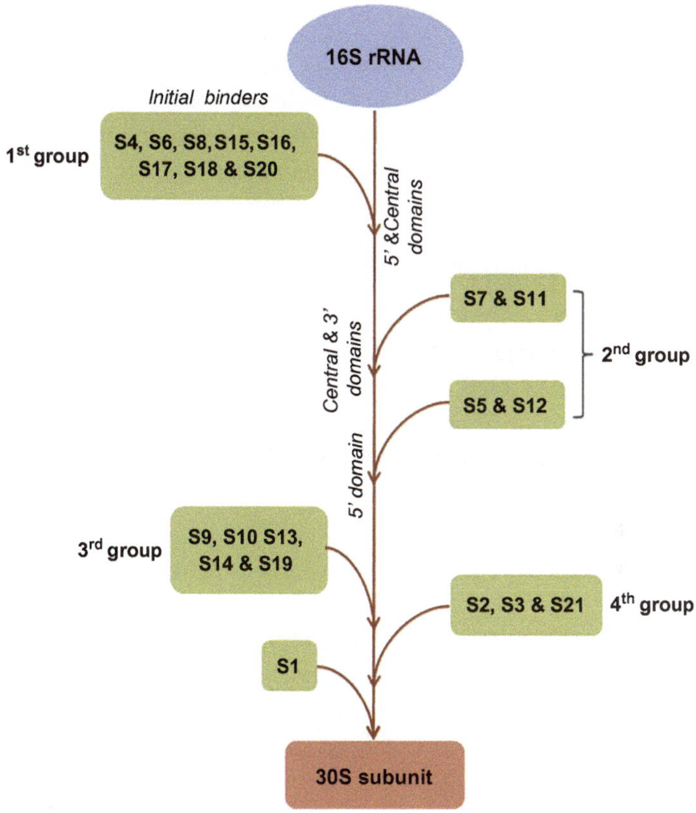

Fig. 8.5 ***Assembly of 30S subunit.***

binders to the 5′ and central domains of 16S rRNA. Subsequently, S7 and S11 bind to the central and 3′ domains, and the secondary binding proteins S5 and S12 bind to the 5′ domain. The binding of S5 and S12 is slower than the binding of other proteins and do not follow the 5′ to 3′ direction. This slow binding may be because of the fact that these proteins (S5 and S12) bind at the intercross of all the domains of 16S rRNA. The third group of proteins that bind to the complex are S9, S10, S13, S14, and S19. The fourth group of binders consists of S2, S3, and S21. The last RP that interacts with the 16S rRNA-protein complex is S1, which completes the 30S subunit assembly (Fig. 8.5). Ribosome assembly factors, viz., Era, RimM, and RimP are thought to be involved in the assembly of the 30S subunit. These assembly factors may be responsible for the fast and efficient assembly of the small subunit (Bunner et al., 2010; Bunner, 2010; Chen and Williamson, 2013; Moll et al., 2002).

Assembly of 50S subunit

The 23S rRNA can also be divided into three domains—5′, central and 3′, similar to the 16S rRNA. These domains are sometimes designated as 13S, 8S, and 12S, respectively, according to the sedimentation coefficients of separated domains. The more commonly used domain division, generally referred to in the ribosomal assembly is the six domain system (I–VI). These six domains are well defined double-stranded RNA helices around a single-stranded central region (Noller et al., 1981; Petrov et al., 2013). The assembly of the 50S subunit is more intricate than that of the 30S subunit. This is because of bigger size of 23S as compared to 16S, more number of proteins to be assembled, and involvement of 5S RNA. While the 30S subunit is assembled in 5′ to 3′direction of 16S rRNA, the 50S subunit assembly does not follow this rule. Nierhaus for the first time successfully reconstituted the 50S subunit in a cell-free system and mapped the order of RP binding (Nierhaus, 1974).

The functional 50S subunits can be reconstructed in vitro in four steps: First, 23S rRNA, 5S rRNA, and 22 RPs are incubated at 0°C in the presence of 4 mM Mg^{2+} to produce a 33S intermediate RI50(1). Further incubation at 44°C results in the formation of a 41S subparticle called RI*50 (1). Adding the remaining proteins at the same temperature and Mg^{2+} concentration results in the formation of 48S subparticle RI50(2). Heating to 50°C and increasing Mg^{2+} concentration to 20 mM results in the formation of the functional 50S subunit. The sedimentation coefficients of intermediates of the 50S subunit observed during in vivo assembly studies are similar to the intermediates observed during in vitro studies. Further, more time is required for the assembly of 50S subunit under in vitro conditions as compared to in vivo conditions. Also, the time required for in vitro assembly of the 50S is observed to be more than that required for 30S assembly (Shajani et al., 2011).

Recent studies utilizing a combination of radioactive pulse labeling of ribosomes with chromato-mass spectrometry showed more insights into the details of 50S ribosomal assembly under in vivo conditions. The assembly process involves the formation of six intermediates (Fig. 8.6). The first intermediate contains L20, L21, L22, and L24, which bind to the 5′ domain of the 23S rRNA. The second intermediate contains additional proteins L1, L3, L4, L13, L15, L17, and L23. The third intermediate is formed with the addition of L5, L18, L29, and L34. The fourth intermediate is formed with the addition of L14, L2, L19, and L32. Addition of L6, L9, L11, L28, and L33 forms the fifth intermediate. Finally, the addition of

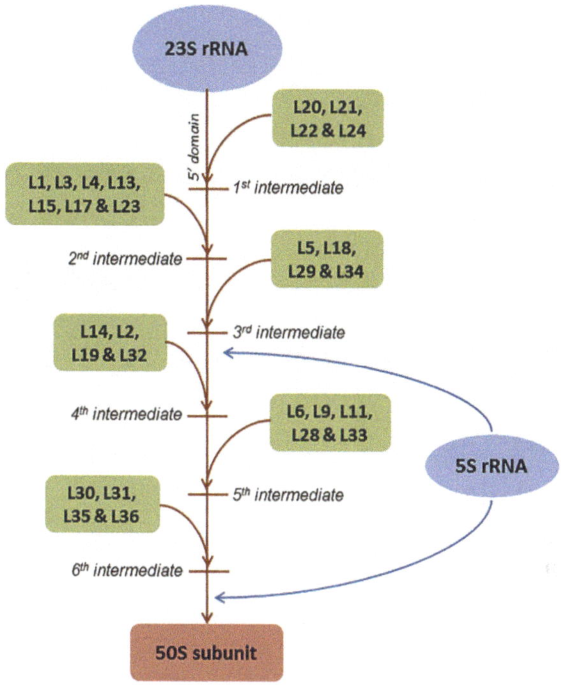

Fig. 8.6 ***Assembly of 50S subunit.*** Ribosomal RNA 5S interacts with 23S rRNA-protein complex somewhere in between the two arrows (in blue).

L30, L31, L35, and L36 forms the 50S subunit. The 5S rRNA comes into the assembly process during later stages because it requires prebinding of L5, L15, L18, and L25 (Fig. 8.6). These r-proteins mediate the principal interactions between 5S and 23S rRNA, and are necessary for the correct formation of the central protuberance of the 50S subunit (Chen and Williamson, 2013; Kaczanowska and Rydén-Aulin, 2007; Korepanov et al., 2007).

Differences in protein composition of the ribosome from exponential and stationary bacterial growth phases

There is a slight difference in RP composition of ribosomes in cells under the stationary phase as compared to cells under the exponential growth phase. Stationary phase ribosomes have an extra RP S22 (isoform of L31), and the amount of acetylated L7 is higher in the protein complex L7/L12 (Nanamiya et al., 2004). The functional significance of such differences is not understood.

Role of ions in ribosome assembly

The divalent magnesium ion (Mg^{2+}) is required for the correct folding of ribosomes (Fig. 8.7). It stabilizes the compact tertiary structures of rRNA. In vivo and in vitro experiments show that lower than the optimal concentration of Mg^{2+} results in low functionality of 70S ribosomes and a higher amount of dissociated 30S and 50S subunits. This shows an active role of Mg^{2+} in the formation of the 70S functional ribosome during translation. Further, the ribosome-bound Mg^{2+} neutralizes the negative charge of rRNA, as the bound RPs are not sufficient for this task. An optimum concentration of K^+ is also necessary (Fig. 8.7) for the ribosome subunits to form the 70S particle (Nierhaus, 2014; Pontes et al., 2016).

Proteins facilitating ribosome assembly

Fully functional ribosomes can assemble in a cell-free system from the building blocks, i.e., the RPs and the three rRNAs, but the process is slower as compared to in vivo, and it needs nonphysiological temperature and ionic concentration. So the question is what assists the fast and efficient ribosomal assembly, to such an extent of perfection that almost all the ribosomes produced in the cell are perfectly folded and fully functional. A set of around 20 known protein factors, that include helicases, chaperones, and GTPases (Fig. 8.7), and some unknown factors, yet to be discovered, help in ribosome assembly (Sergeeva et al., 2014a; Shajani et al., 2011).

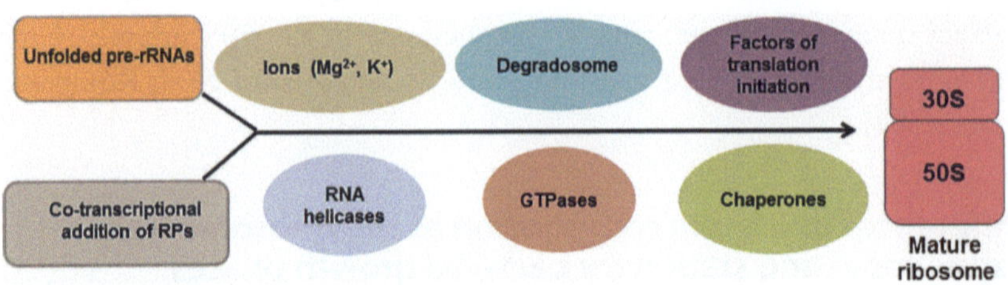

Fig. 8.7 ***Ribosome assembly components in prokaryotes.*** Under in vivo conditions, ribosome assembly is a spontaneous process where the sequential addition of r-proteins to rRNAs leads to the formation of the mature ribosome. Under physiological conditions, the cellular components—ions, chaperons, RNA helicases, GTPases, degradosome, and factors involved in translation initiation, are required for successful assembly of the mature ribosome. These components hasten the assembly process, lower the chances of protein misfolding and assembly errors, and maintain the quality of the ribosome.

RNA helicases

RNA helicases exhibit RNA-dependent ATPase activity. They perform local unwinding of RNA secondary structures, assist in RNA-protein interactions, and sometimes function as chaperons (Elles, 2009). They have a conserved 350 amino acid domain harboring nine conserved motifs, one of which being Asp-Glu-Ala-Asp (or DEAD), so these helicases are also referred to as Dead box helicases (Tanner, 2003). RNA helicases SrmB, CsdA, DbpA, RhlE, and RhlB are identified in *E. coli* (Elles, 2009). SrmB directly interacts with L4 and L24, which are components of the first intermediate formed during cell-free assembly experiments performed for the large ribosome subunit. It assists in the early stages of ribosome biogenesis (Fig. 8.7). *srmB* deletion in *E. coli* results in the accumulation of free 30S and 50S subunits and their precursors. CsdA acts downstream of SrmB in the biogenesis and is associated with the precursor of the 50S subunit. Deletion of the gene coding for CsdA in *E. coli* increases the proportion of free subunits and decreases polysome formation (reviewed by Sergeeva et al., 2014a). DbpA is activated by direct interaction with 23S rRNA and is involved during the late stages of ribosome assembly. Overexpression of DbpA in *E. coli* results in the accumulation of free subunits and affects 50S subunit composition (Elles, 2009).

Chaperones

Chaperons play important roles in rRNA folding (Fig. 8.7). They provide checkpoints for ribosome assembly, separating the whole process into distinct steps. Maturation of 45S precursor to form 50S subunit requires the well-known chaperon GroEL, but it is not involved in 30S subunit assembly (El Hage et al., 2001). Another chaperon RbfA directly interacts with the 5′ end of 16S rRNA and is associated with 30S subunits, 70S ribosome, and polysomes. Deletion of *RbfA* in *E. coli* results in altered ribosome composition and accumulation of unused precursors of 16S rRNA. This effect is reversed by overexpression of Era GTPase (Olins, 1981).

The assembly factor RimM is involved in the processing of 16S rRNA and is directly involved in the maturation of the 30S subunit. RimM possibly expedites the interactions of S13 and S19 with two helices of 16S rRNA (Bunner, 2010). The RimP protein is involved in the maturation of the 30S subunit and the deletion of *rimP* results in the accumulation of free subunits. Under in vitro conditions, preincubation of 16S rRNA with RimP facilitates its binding with S5, S12, S9, S3, S7, and S10. On the other hand, RimJ interacts with S5, as well as with a precursor of the 30S subunit. It is speculated that RimJ binding to the precursor of 30S subunit occurs after S4 interaction with the 5′ domain and before S3 interaction with the 3′ domain of 16S rRNA (Bunner, 2010).

GTPases

Era is a GTPase that under in vitro conditions, directly binds with the small ribosomal subunit as well as its RNA component (Fig. 8.7). The presence of Era increases the rate of binding of S9, S11, S5, and S12 to 16S rRNA by twofold. This enhanced binding of RPs to rRNA may be due to Era bound to the 16S rRNA. The binding of S13, S14, and S19 to the 16S rRNA is also enhanced but to a lesser extent. Further, deletion of *Era* causes accumulation of 16S rRNA precursor and reduced levels of polysomes (Britton, 2009).

The activity of RsgA, a GTPase, is enhanced in the presence of 30S subunit and 70S ribosome. It is speculated to play roles in ribosome disassembly by interacting with the 30S subunit. Deletion of *rsgA* results in accumulation of unbound ribosomal subunits, and 16S rRNA precursor. These effects are reversed following overexpression of *Era* (Himeno, 2004).

Downregulation of Der, another GTPase, increases the amount of dissociated ribosomal subunits and the precursors of 16S and 23S rRNA (Hwang, 2001). YihI, a binding partner protein of Der, when bound, stimulates the function of Der. Overexpression of YihI inhibits Der function and the formation of the 50S subunit, along with the accumulation of precursors for 16S, 23S, and 5S rRNAs. This is also accompanied by decreased interaction of L9, L18, and L25 with 23S rRNA. Der regulates the assembly of 50S subunit during lag and log phases of cell growth, with the help of YihI (Hwang, 2006, 2010).

CgtAE, yet another GTPase, binds to all types of rRNAs in the presence of guanidylimido diphosphate, and is shown to coprecipitate with 16S and 23S rRNAs in vitro in the presence of GTP. It directly interacts with S3, S4, S5, S13, S16, L2, L4, L16, and L17 (Jiang, 2006). Deletion of *cgtAE* increases the level of unbound 30S and 50S subunits and also increases the error rate during 16S and 23S rRNA processing. This deletion also leads to the formation of 50S subunits devoid of L16, L33, and L34. Under in vitro conditions, CgtAE interacts with mature 50S subunit, but not with its precursors, suggesting a role for CgtAE in later stages of 50S subunit assembly (Woodson, 2008).

Association and dissociation of ribosomal subunits

The 3′ end of 16S rRNA of small ribosome subunit carries anti-SD sequence that interacts with the SD sequence upstream of the initiation codon (AUG) of mRNA and allows the association of mRNA with the ribosome small subunit. Initiation factors IF-1, IF2 and IF3 help in this process. At this point the IFs help in the accommodation of the start codon at the P-site of the small subunit. The initiator tRNA (fMet-tRNA) along with IF2 recognizes the AUG codon and binds to the P-site of the small subunit of the ribosome. This macromolecular aggregation is known as initiation complex (IC) (Kaczanowska and Rydén-Aulin, 2007). The formation of the 70S complex is mediated by IF2, requiring i-tRNA for stable binding (Stokes et al., 2014). There are three GC pairs in the anticodon stem of i-tRNA. These three GC pairs are involved in the binding of mRNA bound 30S subunit with the 50S subunit to make the elongation competent 70S ribosome (Shetty et al., 2014). This association of 50S with 30S signals the RNases to trim the 3′ end of immature 16S rRNA. The interaction between i-tRNA and the three conserved GC pairs of 16S rRNA are very important for this final processing. Now the ribosome is fully functional and the ribosome biogenesis can be said as completed (Shetty and Varshney, 2016).

The termination of translation commences when the ribosome encounters a stop codon in the mRNA. The termination codon is recognized by release factor 1 (RF1 for UAA/UAG) or RF2 (for UAA/UGA). Binding of RF1/RF2 to the ribosome activates hydrolysis and release of the peptide chain from the P site tRNA. After that, RF3, a GTPase, binds to the ribosome and helps the release of RF1/RF2 from the A site. Now the ribosome is left with a bound mRNA and a deacylated tRNA at the P site. The subunits are disassembled with the help of ribosome recycling factor and EFG. Subsequently, the deacylated tRNA at the P site is replaced by IF3. Finally, mRNA is either detached from the 30S subunit complex or forms another stable SD–anti-SD interaction with a downstream translation initiation site (Kaczanowska and Rydén-Aulin, 2007).

Ribosome degradation, modifications, and recycling

Ribosomes are stable under normal growth conditions. However, their degradation is required under certain conditions to favor cell growth. Errors may occur during the ribosome assembly process rendering the ribosome nonfunctional. Because of the long half-life of ribosomes, chemical modifications may accumulate during the course of time, rendering the ribosome nonfunctional. During stress conditions, or slow or nil growth, the surplus of ribosomes that were otherwise required during fast growth, are degraded to supply the urgently required biomolecules for nucleic acid and protein synthesis under the new growth conditions (Maiväli et al., 2013).

Bacterial cell divides exponentially under favorable nutrient conditions, but when subjected to nutrient exhaustion, it enters stationary phase. More than half of the ribosomal population of a cell is degraded before it enters the stationary growth phase. The remaining population of the ribosome is stable for many hours in the stationary phase (Piir et al., 2011). Interestingly, a considerable proportion of the ribosome pool of a bacterial cell of the stationary phase is stored as inactivated 100S dimers, 70S monomers or dissociated subunits. These inactivated ribosomes become functional and ready for usage once the growth conditions become favorable (Agafonov et al., 1999; El-Sharoud, 2004). As much as half of the ribosomal population of a cell can be present as 100S particles at the time of entry into the stationary phase (Wada et al., 2000). The 100S particles are formed when two ribosomes join together by interaction of their small subunits. When the stationary phase is prolonged, these dimers dissociate after few hours of association and again reengage as dimers. The ribosome is then stored in this dimeric form for several days. If the growth condition is not resumed, and the cell continues in the stationary phase, the dimmers dissociate and the surplus ribosome is degraded (Wada et al., 2000). Proteins—RMF and HPF are required for 100S dimer formation. RMF blocks anti-SD region whereas HPF blocks the binding sites for tRNA, IF1, and IF3 on the small subunit of the ribosome (Polikanov et al., 2012).

As mentioned above, ribosomes, when not in use, are stored in inactive form as 70S particles, dimers or free subunits. Bacterial cells adopt several ways for this inactivation. For example, the binding of YfiA protein to 70S ribosome makes it inactive (Polikanov et al., 2012). Similarly, YqjD protein can also inactivate ribosomes by sequestering them near the cell membrane (Yoshida et al., 2012). Protein RsfA (also called YbeB) binds to the large ribosomal subunit causing dissociation of the 70S particle into subunits (Häuser et al., 2012).

Cells under stress conditions decrease the pool of ribosome by their degradation (Deutscher, 2003, 2009; El-Sharoud, 2004). The stress conditions may be due to starvation (Zundel et al., 2009); the presence of antibiotics (Silvers and Champney, 2005); or protein overexpression

(Dong et al., 1995) among several other reasons. A reason for the degradation of a mature ribosome is the exposure of RNA-rich intersubunit surfaces of misassembled subunits to cellular RNases. It has been reported that misassembled 70S ribosomes are sometimes formed when a structurally defective 30S subunit is assembled with a properly folded 50S subunit. These defective ribosomes are targeted by cellular endonuclease YbeY, degrading both normal and misassembled subunits (Jacob et al., 2013). Though the trigger may be a misassembled ribosome, after the onset of the degradation process, even the functional ribosome population is subjected to degradation. This type of degradation occurs only in the actively dividing cells, but absent in stationary phase cells (Paier et al., 2015). This also draws attention to the fact that there are different mechanisms for ribosome degradation under different circumstances each with distinct effectors.

Under various stress conditions, such as stringent response, oxidative stress, and heat shock, the 3′ end of 16S rRNA including the anti-SD sequence is cleaved by an endonuclease MazF (Moll and Engelberg-Kulka, 2012; Vesper et al., 2011). The resulting ribosome population, also called stress ribosome, translates only leaderless mRNA that does not require SD sequence. The products of these genes are required in stress conditions. Interestingly, sequences of several mRNAs are also cleaved at their 5′ end to make them leaderless and suitable for the stress ribosomes (Vesper et al., 2011).

Ribosomal rejuvenation

Some RPs in the fully folded 70S ribosomes can be exchanged with newly synthesized RPs in the cytoplasm. As much as 10 RPs are observed to undergo such exchanges in in vivo studies on cells under stationary conditions. Such swap over is also observed under in vitro conditions for more than 10 RPs. It is speculated that the purpose of this exchange is to replace old RPs harboring undesirable random chemical modifications with newly synthesized RPs to rejuvenate the ribosome (Pulk et al., 2010).

Ribosomal quality control

As mentioned above, the final step of 3′ end of 16S rRNA processing is delayed till the initiation of the first round of translation and 70S ribosome formation. This serves as the main quality control strategy by the bacterial system. The absence of properly processed 3′ end of 16S rRNA in 70S ribosome triggers ribosomal degradation. Similar degradation fate is met for ribosomes that are structurally perturbed in cells growing in the presence of antibiotics. Most of these ribosomal degradations are catalyzed by the metallo-endoribonuclease YbeY in *E. coli* (Frazier and Champney, 2012; Jacob et al., 2013; Maiväli et al., 2013).

Degradosome, a multiprotein complex in bacteria, involved in RNA metabolism and posttranscriptional control of gene expression, makes a close association with ribosomes (Tsai et al., 2012). Degradosome is composed of RNase E, RNA helicase RhlB, 3′–5′ exonuclease polynucleotide phosphorylase (PNPase) and glycolytic enzyme enolase. The function of RNase E is to degrade misfolded rRNA; Rh1B unwinds double-stranded RNA and mediates RNA-protein interactions; and PNPase is involved in mRNA metabolism and degradation. Keeping in view the roles of the components of degradosome, the interaction between mature ribosome and degradosome is most likely a part of the quality control machinery for ribosome biogenesis (Basturea et al., 2011; Maes et al., 2011; Maiväli et al., 2013).

Rescue of stalled ribosomes on mRNAs with defective or no stop codon

In general, the mRNAs produced in a cell have "stop" codon which is read by the ribosome to stop protein synthesis. mRNAs that do not have a stop codon are called truncated or nonstop mRNAs. Such mRNAs are produced under many circumstances that include premature transcription termination, mRNA damage, suppression of stop codon, or frameshifting events. Creation of nonstop mRNAs is a common event in *E. coli*. Translation of these mRNAs is unsuccessful, and the ribosome translating it is sequestered on the mRNA as it does not encounter a stop codon. Accumulation of such stalled ribosomes results in the shortage of ribosomes in the cell. Trapped ribosomes are rescued by a set of factors including trans-translation system mediated by tmRNA and small protein B (SmpB). tmRNA is a stable 300–400 nt RNA that rescues the stalled ribosome on a nonstop mRNA through its actions as a tRNA and an mRNA. SmpB binds with tmRNA and is required in the ribosome release process. It has been found recently that two other systems, ArfA (alternative rescue factor) in combination with release factor RF2; and ArfB (also called YaeJ), a class I release factor homologue, operate to relieve the stalled ribosome in *E. coli*. Ribosome rescue systems are ubiquitous in the bacterial system but are not essential for all species of bacteria. Further, many bacterial species have multiple redundant ribosome rescue systems (Himeno et al., 2015; Huter et al., 2017).

Ribosome biogenesis in archaebacteria

It is widely acknowledged that the fundamental processes of life, such as replication, transcription, and translation, are similar in archaea and eukarya (Eschrich et al., 2002). So it is not surprising to observe more likeliness of archaeal ribosome biogenesis to eukarya than eubacteria. Archaeal ribosome is structurally similar to that of bacteria with the size of 70S made up 30S and 50S subunits. However, the primary structures of archaeal rRNA and RPs are more similar to eukaryotes. The archaeal ribosome is composed of 3 rRNA molecules—16S, 23S, and 5S, and up to 68 RPs (Londei, 2010).

All the archaeal RPs are represented in eukaryotes, but none is shared with bacteria. The protein composition of archaeal ribosome varies from species to species. The archaebacteria under the phylum Crenarchaeota have more protein content in their ribosomes as compared to archaebacteria under the umbrella of halobacteriales and thermoplasmatales (Londei, 2010). The bacterial RPs are present one copy per ribosome, except L7/L12. Contrary to this, the ribosomes of archaeal species *P. furiosus* and *T. Kodakaraensis* have two copies of S24e and L14e and three copies of L8e. Further, all archaeal ribosomes have at least two copies of L8e (Armache et al., 2013).

The basic details of RPs and the coding genes for various archaeal species is still quite preliminary and incomplete. Based on the available scanty knowledge, it can be speculated that the regulation of RP synthesis in archaea is similar to that in bacteria (Dönhöfer, 2012; Ramirez et al., 1993). Among the three domains of life, viz., eukarya, bacteria, and archaea, 17 RPs of the large subunit and 19 RPs of the small subunit are universal. The archaea and eukarya share 11 RPs of the large subunit and 20 RPs of the small subunit, none of which are common between archaea and bacteria except those that are universal. Archaeal RPGs are clustered to form operons similar to that in bacteria. Examples are *S10*, *str*, *spc*, and *S13* clusters. The clusters unique to archaea are L18, L30, S24-S27, L31-L39-LX-S19, S28-L24, L14-L34, and L44-S27. The RPGs that are present alone in combination with non-RPGs are L21, L37, S6, and L15. Regulation of expression of these genes/operons exploit similar strategies as detailed for bacteria (reviewed by Wang et al., 2009).

Details of posttranscriptional regulation of RPs for methanogenic archaea are recently reported (Qi et al., 2017). In these life forms, 5′ untranslated regions (UTRs) and associated processing sites are observed. Further, polycistronic regions coding for RPs also have intergenic regions (IGRs) that harbor genes coding for enzymes that regulate the expression of mRNAs from that polycistron. Untranslated regions, processing sites and IGRs are involved in the regulation of mRNA translation and fine-tuning of RP expression. Further, posttranscriptional processing of mRNA regulates RP synthesis, cell growth, and growth-dependent ribosome synthesis (reviewed by Qi et al., 2017).

The number of genes for 16S, 23S, and 5S rRNA (rrn) and their organization varies among different archaeal species. One to four copies of 16S rRNA genes are present in archaeal genomes depending on the species. For example, the genomes of microbes under euryarchaea have only one copy of 16S rRNA gene, while others have more than one copy. The rrn genes are mostly polycistronic with the gene orientation—16S, 23S, and 5S. *Haloferax volcanii* has two full copies of rrn, whereas *H. cutirubrum* has only one copy. As in bacteria, coding regions for tRNAs are present within archaeal polycistronic rrn. However, polycistronic nature is not a rule for archaeal rrn, and 16S, 23S, or 5S may also be present separately elsewhere in the genome. For example, in *D. mobilis*, 16S and 23S are present in one operon, and 5S is present separately, whereas, in *T. acidophilum*, 16S, 23S, and 5S are present separately under their own promoters. Further, archaea specific excision mechanisms exist for processing pre-rRNA transcript into mature 16S, 23S, and 5S rRNA (Yip et al., 2013).

Unlike bacteria, introns are present in rRNA genes of some species of archaea. For example, in *D. mobilis*, the intron is found in 23S rRNA gene (Kjems and Garrett, 1985), while in *P. aerophilum*, the intron is found in 16S rRNA coding region (Burggraf et al., 1993). Multiple introns of variable lengths are reported in the 16S rRNA gene of *T. namibiensis* (Salman et al., 2012). Excision of these introns is carried out by splicing mechanisms unique to archaea (Lykke-Andersen et al., 1997).

In contrast to bacteria, chemical modifications of archaeal rRNA are extensive and comparable to eukaryotes. As an example, there are 70 modifications in *S. solfataricus*, as compared to bacteria with around 40 modifications. While most of the modifications in bacterial rRNA are enzyme guided, archaeal rRNA modifications are ribonucleoprotein (RNP) guided similar to eukaryotes. Most of the modifications are 2′OH methylation of ribose and pseudouridylation, catalyzed by box C/Ds(*no*)RNPs and box H/ACA s(*no*)RNPs, respectively (Yip et al., 2013). Protein Nep1p, conserved in archaea and eukaryotes, is involved directly in pre-rRNA processing and indirectly in its methylation (Eschrich et al., 2002). The precise function of rRNA modifications is unclear, but as they are mostly present in functionally important regions of rRNA and their functional roles in archaeal ribosome are expected. It is speculated that these modifications might have roles in stabilizing the archaeal ribosome in divergently extreme environmental conditions (reviewed by Yip et al., 2013).

Though the RPs and ribosome assembly factors of archaea share more homologies with eukarya than eubacteria, the ribosome assembly process in archaea shares more similarities with eubacteria. This is because archaea and bacteria are unicellular, and in both cases, RPs and rRNAs are assembled in the same compartment where they are synthesized. For this reason, RPs assembly on rRNA is also expected to be cotranscriptional similar to that in bacteria (Ramirez et al., 1993). Around 200 protein factors are known to be involved in eukaryotic ribosome assembly. In comparison to this, fewer proteins are employed by archaea for ribosome assembly (Blombach et al., 2011). In archaea, about 50 small RNAs (these are modification guiding RNAs) and 40 protein transacting factors participate in ribosome biogenesis (Ebersberger et al., 2013). Several GTPases are involved in ribosome assembly in a similar way as in bacteria. An example of such a GTPase is MJ1464, which is present in most of the archaea species (Blombach et al., 2011). The large and small subunits of assembled archaea ribosome show more similarities in structure/function with eukaryotes than bacteria. This has been demonstrated by successfully making completely functional ribosome hybrids using one subunit from archaea and the other from yeast (Lott et al., 2013).

It is reported that more protein factors are required for initiation of translation in archaea than in bacteria, and most of them have homologues in eukaryotes (Benelli and Londei, 2007). Further, there are two to three distinct mechanisms of mRNA-ribosome interaction for the initiation of translation. Examples of archaeal ribosome initiation factors are aIF2 and aIF6 (Benelli and Londei, 2011). IF2 regulates the formation of 70S ribosome by binding the 30S interacting face of the 50S subunit and preventing the formation of the functional heterodimer (Benelli et al., 2009). IF6 functions similar to IF2, but other functions are also attributed to this protein. These include roles in checking assembly of premature ribosome subunits, recycling of subunits, termination of translation, and as a dissociator or anti-association factor for the functional 70S particle (Benelli and Londei, 2011).

In archaea, when a stop codon in mRNA is encountered by a translating ribosome, termination initiates with the recruitment of a class I release factor and a GTPase with GTP at A site of the ribosome. Here the release factor is aRF1, and the GTPase is aEF1α. Hydrolysis of GTP by aEF1α and subsequent rearrangement of aRF1 triggers peptide release from the ribosome, dissociation of GDP-aEF1α and dissociation of 70S ribosome into subunits. Recently, it is reported that another release factor—ABCE1 is also required for dissociation of eukaryotic and archaeal ribosomes (Kobayashi et al., 2012; Nürenberg and Tampé, 2013).

Summary and conclusions

The biogenesis of prokaryotic ribosome is distinct from eukaryotes, as there is no physical barrier separating transcription of rRNAs and translation of RPs for ribosome assembly, and the whole process is to be completed in a single compartment. To solve this spatial crisis, RPs are assembled on rRNA cotranscriptionally. This also helps in regulating, integrating, and making a balance between the synthesis of rRNAs and RPs. Because of varying/unreliable energy source availability and also because ribosome biogenesis is an energy-consuming process, prokaryotes evolved energy-efficient and adaptable regulatory systems to speed up, slow down or halt ribosome biogenesis depending on the demands of the translation machinery. Though believed as a less evolved life form, ribosome biogenesis in prokaryotes is considered one of the most efficient and fast biological processes. Most of the details of ribosome biogenesis that we know today are based on various studies done on *E. coli*, and active research is going on to explore the details of the biogenesis of ribosomes in other prokaryotes. The biogenesis of archaeal ribosome has differences from bacteria because of their extreme and divergent habitat.

References

Agafonov DE, Kolb VA, Nazimov IV, Spirin AS. A protein residing at the subunit interface of the bacterial ribosome. Proc Natl Acad Sci U S A 1999;96:12345–9.

Armache JP, Anger AM, Márquez V, Franckenberg S, Fröhlich T, Villa E, Berninghausen O, Thomm M, Arnold GJ, Beckmann R, Wilson DN. Promiscuous behaviour of archaeal ribosomal proteins: implications for eukaryotic ribosome evolution. Nucleic Acids Res 2013;41:1284–93.

Aseev LV, Koledinskaya LS, Boni IV. Regulation of ribosomal protein operons rplM-rpsI, rpmB-rpmG, and rplU-rpmA at the transcriptional and translational levels. J Bacteriol 2016;198:2494–502.

Babiano R, Gamalinda M, Woolford JL, Cruza J. *Saccharomyces cerevisiae* ribosomal protein L26 is not essential for ribosome assembly and function. Mol Cell Biol 2012;32:3228–41.

Basturea GN, Zundel MA, Deutscher MP. Degradation of ribosomal RNA during starvation: comparison to quality control during steady-state growth and a role for RNase PH. RNA 2011;17:338–45.

Benelli D, Londei P. In vitro studies of archaeal translational initiation. Methods Enzymol 2007;430:79–109.

Benelli D, Londei P. Translation initiation in archaea: conserved and domain-specific features. Biochem Soc Trans 2011;39:89–93.

Benelli D, Marzi S, Mancone C, Alonzi T, la Teana A, Londei P. Function and ribosomal localization of aIF6, a translational regulator shared by archaea and eukarya. Nucleic Acids Res 2009;37:256–67.

Blombach F, Brouns SJJ, Oost J. Assembling the archaeal ribosome: roles for translation-factor-related GTPases. Biochem Soc Trans 2011;39:45–50.

Blount ZD. The natural history of model organisms: the unexhausted potential of *E. coli*. Elife 2015;4, e05826.

Bohne AV. The nucleoid as a site of rRNA processing and ribosome assembly. Front Plant Sci 2014;5:257.

Bordet J, Ciuca M. Remarques sur l'historique des recherches, concernant la lyse microbienne transmissible. C R Biol 1921;84:745–7.

Britton R. Role of GTPases in bacterial ribosome assembly. Annu Rev Microbiol 2009;63:155–76.

Bronfenbrenner JJ. The heat inactivation of bacteriophages. Proc Soc Exp Biol Med 1932;29:802–4.

Bronfenbrenner JJ, Korb C. Studies on the bacteriophage of d'Herelle: II. Effect of alcohol on the bactiophage of d'Herelle. J Exp Med 1925;42:419–29.

Bunner A. The effect of ribosome assembly cofactors on in vitro 30S subunit reconstitution. J Mol Biol 2010;398:1–7.

Bunner AE, Beck AH, Williamson JR. Kinetic cooperativity in *Escherichia coli* 30S ribosomal subunit reconstitution reveals additional complexity in the assembly landscape. Proc Natl Acad Sci U S A 2010;107:5417–22.

Burggraf S, Larsen N, Woese CR, Stetter KO. An intron within the 16S ribosomal RNA gene of the archaeon *Pyrobaculum aerophilum*. Proc Natl Acad Sci U S A 1993;90:2547–50.

Burgos HL, O'Connor K, Sanchez-Vazquez P, Gourse RL. Roles of transcriptional and translational control mechanisms in regulation of ribosomal protein synthesis in *Escherichia coli*. J Bacteriol 2017;199:e00407–17.

Cerretti DP, Dean D, Davis GR, Bedwell DM, Nomura M. The spc ribosomal protein operon of *Escherichia coli*: sequence and cotransciption of the ribosomial protein genes and a protein export gene. Nucleic Acids Res 1983;11:2599–616.

Chen SS, Williamson JR. Characterization of the ribosome biogenesis landscape in *E. coli* using quantitative mass spectrometry. J Mol Biol 2013;425:767–79.

Davies BW, Köhrer C, Jacob AI, Simmons LA, Zhu J, Aleman LM, Rajbhandary UL, Walker GC. Role of *Escherichia coli* YbeY, a highly conserved protein, in rRNA processing. Mol Microbiol 2010;78:506–18.

Deutscher MP. Degradation of stable RNA in bacteria. J Biol Chem 2003;278:45041–4.

Deutscher MP. Maturation and degradation of ribosomal RNA in bacteria. Prog Mol Biol Transl Sci 2009;85:369–91.

Dong H, Nilsson L, Kurland CG. Gratuitous overexpression of genes in *Escherichia coli* leads to growth inhibition and ribosome destruction. J Bacteriol 1995;177:1497–504.

Dönhöfer A. Auxiliary factors regulating ribosome function: Roles in biogenesis, antibiotic resistance and ribosome recycling. Dissertation, LMU München: Faculty of Chemistry and Pharmacy. URN: urn:nbn:de:bvb:19-156364; 2012.

Ebersberger I, Simm S, Leisegang MS, Schmitzberger P, Mirus O, von Haeseler A, Bohnsack MT, Schleiff E. The evolution of the ribosome biogenesis pathway from a yeast perspective. Nucleic Acids Res 2013;42:1509–23.

El Hage A, Sbaï M, Alix JH. The chaperonin GroEL and other heat-shock proteins, besides DnaK, participate in ribosome biogenesis in *Escherichia coli*. Mol Gen Genet 2001;264:796–808.

Elles L. A dominant negative mutant of the *E. coli* RNA helicase DbpA blocks assembly of the 50S ribosomal subunit. Nucleic Acids Res 2009;37:6503–14.

El-Sharoud WM. Ribosome inactivation for preservation: concepts and reservations. Sci Prog 2004;87:137–52.

Eschrich D, Buchhaupt M, Kötter P, Entian KD. Nep1p (Emg1p), a novel protein conserved in eukaryotes and archaea, is involved in ribosome biogenesis. Curr Genet 2002;40:326–38.

Espejo RT, Plaza N. Multiple ribosomal RNA operons in bacteria; their concerted evolution and potential consequences on the rate of evolution of their 16S rRNA. Front Microbiol 2018;9:1232.

Frazier AD, Champney WS. Impairment of ribosomal subunit synthesis in aminoglycoside-treated ribonuclease mutants of *Escherichia coli*. Arch Microbiol 2012;194:1033–41.

Gaal T, Bratton BP, Sanchez-Vazquez P, Sliwicki A, Sliwicki K, Vegel A, Pannu R, Gourse RL. Colocalization of distant chromosomal loci in space in *E. coli*: a bacterial nucleolus. Genes Dev 2016;30:2272–85.

Gotta SL, Miller Jr OL, French SL. rRNA transcription rate in *Escherichia coli*. J Bacteriol 1991;173:6647–9.

Häuser R, Pech M, Kijek J, Yamamoto H, Titz B, Naeve F, Tovchigrechko A, Yamamoto K, Szaflarski W, Takeuchi N, Stellberger T, Diefenbacher ME, Nierhaus KH, Uetz P. RsfA (YbeB) proteins are conserved ribosomal silencing factors. PLoS Genet 2012;8, e1002815.

Held W. Reconstitution of *Escherichia coli* 30S ribosomal subunits from purified molecular components. J Biol Chem 1973;248:5720–30.

Herold M, Nierhaus KH. Incorporation of six additional proteins to complete the assembly map of the 50S subunit from *Escherichia coli* ribosomes. J Biol Chem 1987;262:8826–33.

Himeno H. A novel GTPase activated by the small subunit of ribosome. Nucleic Acids Res 2004;32:5303–9.

Himeno H, Nameki N, Kurita D, Muto A, Abo T. Ribosome rescue systems in bacteria. Biochimie 2015;114:102–12.

Hofmann S, Miller Jr OL. Visualization of ribosomal ribonucleic acid synthesis in a ribonuclease III-deficient strain of *Escherichia coli*. J Bacteriol 1977;132:718–22.

Huter P, Müller C, Arenz S, Beckert B, Wilson DN. Structural basis for ribosome rescue in bacteria. Trends Biochem Sci 2017;42:669–80.

Hwang J. An essential GTPase, Der, containing double GTP binding domains from *Escherichia coli* and *Thermotoga maritima*. J Biol Chem 2001;276:31415–21.

Hwang J. The tandem GTPase, Der, is essential for the biogenesis of 50S ribosomal subunits in *Escherichia coli*. Mol Microbiol 2006;61:1660–72.

Hwang J. A bacterial GAP like protein, YihI, regulating the GTPase of Der, an essential GTP binding protein in *Escherichia coli*. J Mol Biol 2010;399:759–72.

Jacob AI, Köhrer C, Davies BW, Raj Bhandary UL, Walker GC. Conserved bacterial RNase YbeY plays key roles in 70S ribosome quality control and 16S rRNA maturation. Mol Cell 2013;49:427–38.

Jiang M. The *Escherichia coli* GTPase CgtAE is involved in late steps of large ribosome assembly. J Bacteriol 2006;188:6757–70.

Jin DJ, Cagliero C, Martin CM, Izard J, Zhou YN. The dynamic nature and territory of transcriptional machinery in the bacterial chromosome. Front Microbiol 2015;6:497.

Jin DJ, Martin CM, Sun Z, Cagliero C, Zhou YN. Nucleolus-like compartmentalization of the transcription machinery in fast-growing bacterial cells. Crit Rev Biochem Mol Biol 2017;52:96–106.

Kaczanowska M, Rydén-Aulin M. Ribosome biogenesis and the translation process in *Escherichia coli*. Microbiol Mol Biol Rev 2007;71:477–94.

Kjems J, Garrett RA. An intron in the 23S ribosomal RNA gene of the archaebacterium *Desulfurococcus mobilis*. Nature 1985;318:675–7.
Kobayashi K, Saito K, Ishitani R, Ito K, Nureki O. Structural basis for translation termination by archaeal RF1 and GTP-bound EF1a complex. Nucleic Acids Res 2012;40:9319–28.
Korepanov A, Gongadze GM, Graber MB, Court DL, Bubunenko M. Importance of the 5S rRNA-binding ribosomal proteins for cell viability and translation *Escherichia coli*. J Mol Biol 2007;366:1199–208.
Krasny L, Gourse RL. An alternative strategy for bacterial ribosome synthesis: *Bacillus subtilis* rRNA transcription regulation. EMBO J 2004;23:4473–83.
Kriel A, Bittner AN, Kim SH, Liu K, Tehranchi AK, Zou WY, Rendon S, Chen R, Tu BP, Wang JD. Direct regulation of GTP homeostasis by (p)ppGpp: a critical component of viability and stress resistance. Mol Cell 2012;48:231–41.
Lindahl L, Zengel JM. Ribosomal genes in *Escherichia coli*. Annu Rev Genet 1986;20:297–326.
Londei P. Archaeal ribosomes. John Wiley & Sons, Ltd; 2010.
Lott BB, Wang Y, Nakazato T. A comparative study of ribosomal proteins: linkage between amino acid distribution and ribosomal assembly. BMC Biophys 2013;6:13.
Lykke-Andersen J, Aagaard C, Semionenkov M, Garrett RA. Archaeal introns: splicing, intercellular mobility and evolution. Trends Biochem Sci 1997;22:326–31.
Maes A, Gracia C, Hajnsdorf E, Régnier P. Search for poly(A) polymerase targets in *E. coli* reveals its implication in surveillance of Glu tRNA processing and degradation of stable RNAs. Mol Microbiol 2011;83:436–51.
Maiväli U, Paier A, Tenson T. When stable RNA becomes unstable: the degradation of ribosomes in bacteria and beyond. Biol Chem 2013;394:845–55.
Mandava CS. Ribosomal stalk protein L12: structure, function and application. Acta Universitatis Upsaliensis 2011;49.
Mattheakis L, Vu L, Sor F, Nomura M. Retroregulation of the synthesis of ribosomal proteins L14 and L24 by feedback repressor S8 in *Escherichia coli*. Proc Natl Acad Sci U S A 1989;86:448–52.
Miller Jr OL, Beatty BR. Visualization of nucleolar genes. Science 1969;164:955–7.
Miller Jr OL, Hamkalo BA, Thomas Jr CA. Visualization of bacterial genes in action. Science 1970;169:392–5.
Mizushima S, Nomura M. Assembly mapping of 30S ribosomal proteins from *E. coli*. Nature 1970;226:1214.
Moll I, Engelberg-Kulka H. Selective translation during stress in *Escherichia coli*. Trends Biochem Sci 2012;37:493–8.
Moll I, Grill S, Gründling A, Bläsi U. Effects of ribosomal proteins S1, S2 and the DeaD/CsdA DEAD-box helicase on translation of leaderless and canonical mRNAs in *Escherichia coli*. Mol Microbiol 2002;44:1387–96.
Mulder A. Visualizing ribosome biogenesis: parallel assembly pathways for 30S subunit. Science 2010;330:673–7.
Nanamiya H, Akanuma G, Natori Y, Murayama R, Kosono S, Kudo T, Kobayashi K, Ogasawara N, Park SM, Ochi K, Kawamura F. Zinc is a key factor in controlling alternation of two types of L31 protein in the *Bacillus subtilis* ribosome. Mol Microbiol 2004;52:273–83.
Nesterchuk MV, Sergiev PV, Dontsova OA. Posttranslational modifications of ribosomal proteins in *Escherichia coli*. Acta Naturae 2011;3:22–33.
Nierhaus K. Total reconstitution of functionally active 50S ribosomal subunits from *Escherichia coli*. Proc Natl Acad Sci U S A 1974;71:4713–7.
Nierhaus KH. Mg^{2+}, K^{+}, and the ribosome. J Bacteriol 2014;196:3817–9.
Noller HF, Kop J, Wheaton V, Brosius J, Gutell RR, Kopylov AM, Dohme F, Herr W, Stahl DA, Gupta R, Waese CR. Secondary structure model for 23S ribosomal RNA. Nucleic Acids Res 1981;9:6167–89.

Nomura M. Switching from prokaryotic molecular biology to eukaryotic molecular biology. J Biol Chem 2009;284:9625–35.
Nürenberg E, Tampé R. Tying up loose ends: ribosome recycling in eukaryotes and archaea. Trends Biochem Sci 2013;38:64–74.
Olins PO. Translational regulation by ribosomal protein S8 in *Escherichia coli*: structural homology between rRNA binding site and feedback target on mRNA. Nucleic Acids Res 1981;9:1757–64.
Paier A, Leppik M, Soosaar A, Tenson T, Maivali U. The effects of disruptions in ribosomal active sites and in intersubunit contacts on ribosomal degradation in *Escherichia coli*. Sci Rep 2015;5:7712.
Paul BJ, Barker MM, Ross W, Schneider DA, Webb C, Foster JW, Gourse RL. DksA: a critical component of the transcription initiation machinery that potentiates the regulation of rRNA promoters by ppGpp and the initiating NTP. Cell 2004a;118:311–22.
Paul BJ, Ross W, Gaal T, Gourse RL. rRNA transcription in *Escherichia coli*. Annu Rev Genet 2004b;38:749–70.
Petrov AS, Bernier CR, Hershkovits E, Xue Y, Waterbury CC, Hsiao C, Stepanov VG, Gaucher EA, Grover MA, Harvey SC, Hud NV, Wartell RM, Fox GE, Williams LD. Secondary structure and domain architecture of the 23S and 5S rRNAs. Nucleic Acids Res 2013;41:7522–35.
Piir K, Paier A, Liiv A, Tenson T, Maivali U. Ribosome degradation in growing bacteria. EMBO Rep 2011;12:458–62.
Polikanov YS, Blaha GM, Steitz TA. How hibernation factors RMF, HPF, and YfiA turn off protein synthesis. Science 2012;336:915–8.
Pontes MH, Yeom J, Groisman EA. Reducing ribosome biosynthesis promotes translation during low Mg^{2+} stress. Mol Cell 2016;64:480–92.
Pulk A, Liiv A, Peil L, Maivali U, Nierhaus K, Remme J. Ribosome reactivation by replacement of damaged proteins. Mol Microbiol 2010;75:801–14.
Qi L, Yue L, Feng D, Qi F, Li J, Dong X. Genome-wide mRNA processing in methanogenic archaea reveals post-transcriptional regulation of ribosomal protein synthesis. Nucleic Acids Res 2017;45:7285–98.
Ramirez C, Kopke AKE, Yang DC, Boeckh T, Matheson AT. The structure, function and evolution of archaeal ribosomes. In: The Biochemistry of Archaea (Archaebacteria); 1993. p. 439–66. Chapter 14.
Salman V, Amann R, Shub DA, Schulz-Vogt HN. Multiple self-splicing introns in the 16S rRNA genes of giant sulfur bacteria. Proc Natl Acad Sci U S A 2012;109:4203–8.
Schmidt TM. Multiplicity of ribosomal RNA operons in prokaryotic genomes. Bacterial Genomes: Physical Structure and Analysis 1998;221–30.
Sergeeva OV, Sergiev PV, Bogdanov AA, Dontsova OA. Ribosome: lessons of a molecular factory construction. Mol Biol 2014a;48:468–84.
Sergeeva OV, Bogdanov AA, Sergiev PV. What do we know about ribosomal RNA methylation in *Escherichia coli*? Biochimie 2014b;1–9.
Shajani Z, Sykes MT, Williamson JR. Assembly of bacterial ribosomes. Annu Rev Biochem 2011;80:501–26.
Shetty S, Varshney U. An evolutionarily conserved element in initiator tRNAs prompts ultimate steps in ribosome maturation. Proc Natl Acad Sci U S A 2016;E6126–34.
Shetty S, Nadimpalli HP, Shah RA, Arora S, Das G, Varshney U. An extended Shine-Dalgarno sequence in mRNA functionally bypasses a vital defect in initiator tRNA. Proc Natl Acad Sci U S A 2014;111:E4224–33.
Siekevitz P, Zamecnik PC. Ribosomes and protein synthesis. J Cell Biol 1981;91:53s–65s.
Siibak T, Remme J. Subribosomal particle analysis reveals the stages of bacterial ribosome assembly at which rRNA nucleotides are modified. RNA 2010;16:2023–32.

Silvers JA, Champney WS. Accumulation and turnover of 23S ribosomal RNA in azithromycin-inhibited ribonuclease mutant strains of *Escherichia coli*. Arch Microbiol 2005;184:66–77.
Stoddard SF, Smith BJ, Hein R, Roller BRK, Schmidt TM. rrnDB: improved tools for interpreting rRNA gene abundance in bacteria and archaea and a new foundation for future development. Nucleic Acids Res 2015;43:D593–8.
Stokes JM, Davis JH, Mangat CS, Williamson JR, Brown ED. Discovery of a small molecule that inhibits bacterial ribosome biogenesis. Elife 2014;3, e03574.
Sykes M. Complex assembly landscape for the 30S ribosomal subunit. Annu Rev Biophys 2009;38:197–215.
Tanner N. The Q motif: A newly identified motif in DEAD box helicases may regulate ATP binding and hydrolysis. Mol Cell 2003;11:127–38.
Tsai YC, Du D, Dominguez-Malfavon L, Dimastrogiovanni D, Cross J, Callaghan AJ, Garcia-Mena J, Luisi BF. Recognition of the 70S ribosome and polysome by the RNA degradosome in *Escherichia coli*. Nucleic Acids Res 2012;40:10417–31.
Vesper O, Amitai S, Belitsky M, Byrgazov K, Kaberdina AC, Engelberg-Kulka H, Moll I. Selective translation of leaderless mRNAs by specialized ribosomes generated by MazF in *Escherichia coli*. Cell 2011;147:147–57.
Wada A, Mikkola R, Kurland CG, Ishihama A. Growth phase-coupled changes of the ribosome profile in natural isolates and laboratory strains of *Escherichia coli*. J Bacteriol 2000;182:2893–9.
Wang J, Dasgupta I, Fox GE. Many nonuniversal archaeal ribosomal proteins are found in conserved gene clusters. Archaea 2009;2:241–51.
Werkman CH. Vitamin effects in the physiology of microorganisms. J Bacteriol 1927;14:335–47.
Wimberly B. Structure of the 30S ribosomal subunit. Nature 2000;407:327–39.
Wollman E. Recherches Sur la bacteriophagie (phenomene de twort-d'Herelle). Ann Inst Pasteur 1925;56:137–64.
Wollman E, Wollman E. Les phases de bacteriophages (facteurs lysogènes). Comptes Rendus Soc Biol Paris 1937;124:931–4.
Woodson S. RNA folding and ribosome assembly. Curr Opin Chem Biol 2008;12:667–73.
Yip WS, Vincent NG, Baserga SJ. Ribonucleoproteins in archaeal pre-rRNA processing and modification. Archaea 2013;614735.
Yoshida H, Maki Y, Furuike S, Sakai A, Ueta M, Wada A. YqjD is an inner membrane protein associated with stationary-phase ribosomes in *Escherichia coli*. J Bacteriol 2012;194:4178–83.
Yusupov M. Crystal structure of the ribosome at 5.5 angstrom resolution. Science 2001;292:883–96.
Zengel JM, Lindahl L. Ribosomal protein L4 and transcription factor NusA have separable roles in mediating terminating of transcription within the leader of the S10 operon of *Escherichia coli*. Genes Dev 1992;6:2655–62.
Zengel JM, Lindahl L. Diverse mechanisms for regulating ribosomal protein synthesis in *Escherichia coli*. Prog Nucleic Acid Res Mol Biol 1994;47:331–70.
Zundel MA, Basturea GN, Deutscher MP. Initiation of ribosome degradation during starvation in *Escherichia coli*. RNA 2009;15:977–83.

CHAPTER 9

Translation—Process and control

Rekha Puria[a], Sumati Rohilla[a], and Savneet Kaur[b]
[a]School of Biotechnology, Gautam Buddha University, Greater Noida, India
[b]Department of Molecular and Cellular Medicine, Institute of Liver and Bilary Sciences, New Delhi, India

The process of translation

The process of converting information carried by the messenger RNA (mRNA) into a sequence of amino acids, to form a polypeptide chain and eventually a protein is called "translation." The information on the mRNA is specified by the codon sequence. Translation is a critical and fundamental process across all forms of life, which allows cells and tissues to maintain homeostasis. It is a very specific, highly regulated and conserved process across prokaryotes, eukaryotes and archaebacteria. Any aberrations in this process cause multiple diseases.

The complete process can largely be divided in three steps: **initiation, elongation, and termination**. Cells have highly organized translation apparatus to carry on with the translation using ribosomal RNAs (rRNA). On the basis of the sedimentation coefficient, the intact ribosomes are 70S in the case of prokaryotes and archaebacteria and 80S in the case of eukaryotes. They are made up of two subunits—small subunit (30S in prokaryotes and 40S in eukaryotes) and the other large subunit (50s in prokaryotes and 60S in eukaryotes). While the rRNA of the small subunit is 16S in the case of prokaryotes and archaebacteria, it is 18S in eukaryotes. Furthermore, the rRNA in the 50S subunit is 23S whereas in human it is 28S and 5.8S. Structurally, the small subunit comprises of head, neck, platform, and body. Basically, the head corresponds to the 3′major domain, the platform is the central domain, and the body corresponds to the 5′domain of 16S rRNA. The neck is a linker between the head and the rest of the small subunit. The mRNA binding site of the ribosomes lies in the small subunit. The ribosome has three binding sites for the transfer RNA (tRNA) known as A, P, and E sites. The A site is for the binding of aminoacyl tRNA (aa-tRNA), the P (peptidyl) site has affinity for peptidyl tRNA, and the E site

Emerging Concepts in Ribosome Structure, Biogenesis, and Function
https://doi.org/10.1016/B978-0-12-816364-1.00006-8

has affinity for deacylated tRNA. The peptidyl transferase center (PTC), GTPase associated center (GAC), and sarcin/ricin-binding loop (SRL) lie on the large subunit (reviewed by Marintchev and Wagner, 2004).

Translation initiation

Translation is the process of the assembly of small and large subunits of the ribosome to make elongation-competent 80S ribosomes, wherein the initiation codon is base-paired with the anticodon of the initiator tRNA (Met-tRNAi^{Met}) in the ribosomal P-site. The whole process essentially uses at least nine eukaryotic initiation factors (eIFs) and some of the recently identified auxiliary factors that support the mechanism of initiation in different ways. Broadly, the complete initiation process can be divided in two steps: (a) the formation of 48S initiation complexes including the base-pairing of the start codon with anticodon of initiator tRNA in the P-site of the 40S ribosomal subunits, and (b) the joining of 48S complexes with 60S subunits (Fig. 9.1).

In general, the formation of 48S complexes involves "scanning" of mRNA to locate the correct start codon. During scanning, the 43S preinitiation complex (comprising of the small ribosome subunit, eIF2-GTP loaded with Met-tRNAi^{Met}, eIF3, eIF1, eIF1A, and eIF5) attaches to the 5′ cap region of mRNAs. The unwinding of the mRNAs at the 5′ terminal secondary structure is done by eIF4A, eIF4B, and eIF4F. The 43S complex scans through the 5′UTR (untranslated region) in the 5′-3′ direction until it identifies the correct initiation codon. After initiation, codon recognition and 48S complex formation, eIF2-GTP gets hydrolyzed to eIF2-GDP through the activity of eIF5 and eIF5B. Subsequently, other eIFs are either released or displaced to allow the joining of a 60S subunit and formation of elongation competent 80S ribosome. Although most mRNAs widely use the scanning mechanism for the identification of the start codon, initiation on a few mRNAs involves different mechanisms, for example, internal ribosome entry sites (IRESs), especially in the case of viruses.

The whole initiation process comprises following major events.

(A) 48S complex formation which includes:
- **(1)** Recycling of ribosomes:
  - Association and antidissociation of ribosomal subunits
- **(2)** Formation of a 43S complex
  - Initiator tRNA recognition
  - eIF2 ternary complex formation

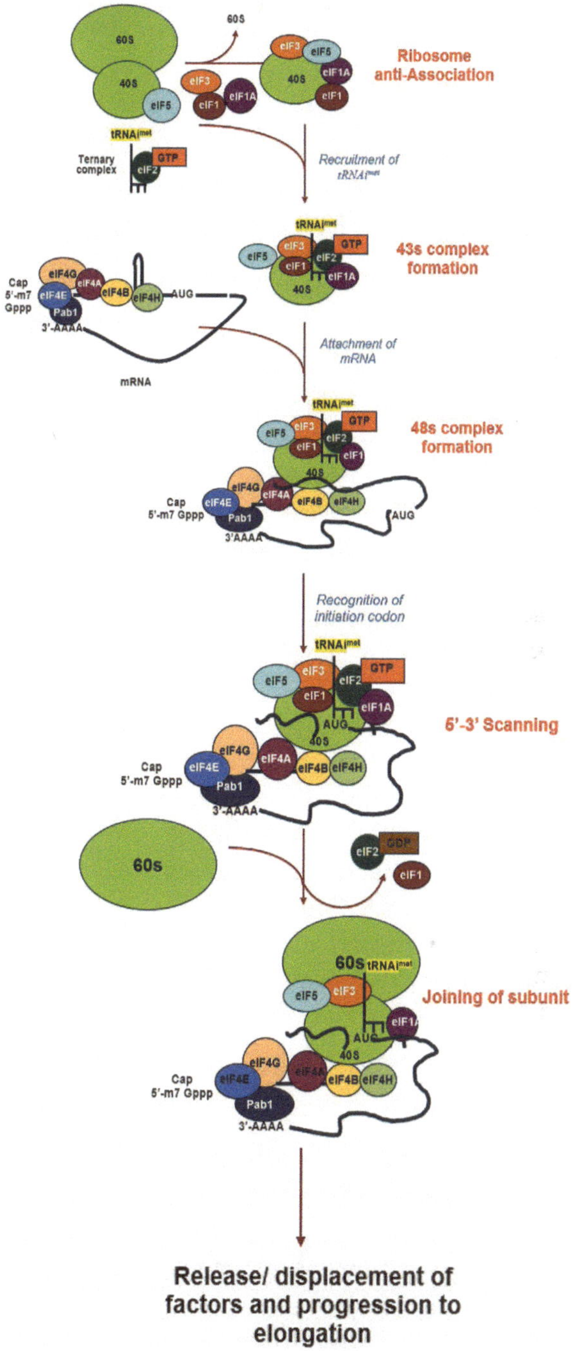

Fig. 9.1 Initiation of translation in eukaryotes. Different steps are (1) Ribosome antiassociation. Includes ribosome recycling of posttermination ribosomes. Antiassociation of ribosomes is ensured by binding of initiation factors eIF1, eIF 1A, eIF3, and eIF5. (2) 43s complex formation. The GTP-eIF2 binds to methionated tRNAi and binds to ribosomal small subunit. (3) 48s complex formation. Involves attachment of 43s complex to eIF4F attached mRNA. (4) 5′-3′ Scanning. Recognition of initiator codon through 5′-3′ scanning of 5′UTR. (5) Joining of subunit. Binding of large subunit to form complete elongation competent ribosome through release or displacement of initiation factors.

(3) Formation of a 48S complex
- Recognition of mRNA
- Canonical/Noncanonical models: Scanning model and IRES
- Initiation codon recognition

(B) Joining of subunit
- Progression to elongation

Recycling of ribosomes

For the initiation of translation, separate ribosome subunits are required. Separation and joining of ribosome subunits during the whole translation is basically a cycling process. Ribosomes used in the beginning for initiation, are derived from ribosomes dissociated after termination known as post-termination ribosome complexes. Dissociation of ribosomes at the end of translation and process of keeping them apart requires several factors associated with translation. Antiassociation ensures the availability of separated ribosome subunits for initiation along with preventing premature assembly of translationally inactive ribosomes.

The postterminal ribosome complexes are basically 80S ribosomal subunits bound to mRNA, P-site deacetylated tRNA, and release factor (eRF1). In bacteria, IF3 is responsible for subunit separation and antiassociation. IF3 competes with the larger subunit for binding to the smaller subunit. Binding of IF3 facilitates the binding of other IFs, IF1, IF2, and IF3 along with initiator tRNA. Similarly, in eukaryotes the antiassociation task is performed by eIF3, whose activity is enhanced by parallel binding of eIF1, GTP-loaded initiator tRNA. eIF2 complex, and small U-rich RNAs (reviewed by Jackson et al., 2010).

Formation of 43S complex

Initiator aa-tRNA recognition

The initiator tRNAs are highly conserved between bacteria, archaebacteria, and eukaryotes and exhibits differences from elongator tRNA used for translation. The differences exist because it has to perform functions significantly distinct from elongator tRNA, for example, (a) it needs to interact with initiation factors but not elongation factors, (b) it needs to recognize and bind directly to the P site, and (c) it needs to aid in correct recognition of the start codon. Various characteristic sequence elements and modifications differentiate initiator tRNA from the elongator methionyl tRNA that adds methionine during peptide chain elongation. The initiator tRNA in bacteria is $tRNA^{fmet}$ while in eukaryotes its $tRNAi^{met}$.

In bacteria, an aminoacylate synthase aminoacylates tRNA with methionine followed by second enzymatic action of addition of formyl group by another enzyme methionyl-tRNA transformylase. While the formylation increases the stability of resulting fmet-$tRNA^{fmet}$ toward diacylation and increases affinity for IF2, in eukaryotes there is no formylation. Instead posttranscriptional modification of tRNAi segregates it from the methionine specific elongator tRNA. First, IF2-GTP binds to the fmet moiety and the acceptor end of tRNA of fmet-$tRNA^{fmet}$, which basically stabilizes binding of fmet-$tRNA^{fmet}$ to the smaller subunit of the ribosome. Similarly, in eukaryotes the function is performed by eIF2 which, in GTP bound state, recognizes the initiator $tRNAi^{Met}$. Other discriminating features of initiator tRNA include the base pair at position 1:72 in the acceptor stem. Bacterial initiator tRNAs contain a C:A mismatch at position 1:72. This allows its methionine to be formylated by methionyl-tRNA transformylase (Lee et al., 1991; Guillon et al., 1992). Interestingly, mutations at this position allow this tRNA to be used for elongation (Seong and RajBhandary, 1987; Drabkin et al., 1998). An A1:U72 base pair is invariant in eukaryotic and archaeal initiator tRNAs (the sole known exception being ψ1:A72 in *Schizosacharomyces pombe*), while mostly the elongator tRNAs of all three domains harbor a G:C base pair in the 1:72 position. There are differences in initiator tRNAs of bacteria and eukaryotes too; for example, the eukaryotic initiator tRNA lacks the T loop's eponymous T, bearing the sequence AU at positions 54–55 instead of the Tψ found in other tRNAs, including the elongator methionyl tRNA. Bacteria possess the Tψ sequence in the T loop of both elongator and initiator tRNAs (Kolitz and Lorsch, 2010).

Formation of the eIF2-GTP ternary complex

eIF3, in association with eIF3j, eIF1, and eIF1A, dissociates post-TCs into free 60S subunits and mRNA- and tRNA-bound 40S subunits. eIF1 promotes release of the tRNA followed by eIF3j-mediated dissociation of mRNA. Subunits remain in dissociated form due to the binding of eIF3, probably eIF1 and eIF1A to the 40S subunit. Thereby in the formation of the 43S initiation complex, the 40S subunit acquires eIF3, eIF1, and eIF1A during the recycling process. The ternary complex of eIF2-GTP-Met-$tRNA^{met}_{i}$ is subsequently recruited to this 40S subunit to make the 43S complex. As mentioned earlier, structurally, the small subunit has three regions, head, body, and platform corresponding to its structural domain 3′major domain, 5′domain, and central domain, respectively. The neck is link between the head and the rest of the 40S subunit. The mRNA binding site of the ribosome is on the small subunit along the neck region. Binding of initiation factors results in the opening of the mRNA entry channel (latch).

Attachment of 43S complexes to mRNA

Fully processed eukaryotic mRNA have an m^7G cap at the 5′end and a poly A tail at the 3′end for stability against degradation by nucleases in the cytoplasm. Interestingly, they also play a crucial role in initiation of translation through controlling the recruitment of the 43S preinitiation complex. Bacteria have Shine-Dalgarno sequences upstream of most start codons. These sequences base pair with the 16S rRNA and localize the 30S subunit to the start codon. Eukaryotes lack the anti-Shine-Dalgarno sequences in their 18S rRNA and thus do not use Shine-Dalgarno sequences to locate initiation codons. In bacteria, the conserved sequence motif Shine Dalgarno sequence (SD) or ribosome binding site (RBS) sequence are present upstream of the start codon. Due to complementarity, it base pairs with the 3′end of 16S rRNA. Regulation of translation in bacteria is usually mediated by mRNA secondary structures and protein binding at or near the SD element or the start codon. In bacteria, mRNA are usually polycistronic in nature. Start site recognition is more complex in eukaryotes compared to bacteria. It is largely because of the 5′cap, 3′poly A tail and 5′UTR and 3′UTR. Secondary RNA structure formed by 5′UTR and binding of regulatory proteins at both UTRs influence translation efficiency. The most accepted recognition sequence for initiation at AUG is the Kozak sequence, GCCA/GCCAUGG, with the most important nucleotides being A or G at position 3 and G at the +4 position. Broadly, the small subunit first binds to the 5′cap-proximal region, followed by scanning until the start codon is reached.

Recognition of initiation codon

Initiation factors play a crucial role in binding to the initiation codon. In bacteria, in addition to AUG, GUG and UUG are also recognized canonical start codons. IF3 can lead to the dissociation of improperly/weakly assembled preinitiation complexes in case of incorrect start site recognition. However, the identification of the start site is done by anticodon interaction with initiator tRNA only. In eukaryotes, translation should begin with recognition by $tRNAi^{met}$ of the appropriate AUG as the first codon. This is ensured by first, existence of this AUG sequence in a highly conserved sequence with purine at −3 and G at +4 positions. These nucleotides play an important role in stabilizing conformations resulting after codon-anticodon base pairing. eIF1 helps the 43S complex to differentiate initiator codon from other codons and removes inappropriately associated ribosomes on other sites. The whole process of initiator codon recognition includes recognition of correct AUG by eIF1, interaction of eIF1A with the 40S subunit, pairing of codon and anticodon sequences, change in conformation of ribosomes followed by displacement of eIF1 from the P site. The displacement of eIF1 relieves repression of GTP hydrolysis and Pi release from eIF2. GTP hydrolysis by eIF5 (GTPase activating protein, GAP) decreases eIF2's affinity for Met-tRNAi^{met} resulting in its dissociation from the 40S subunit. There are two models for recognition of the start site by the 43S initiation complex—(a) scanning model and (b) internal ribosome entry sites (IRES), discussed below.

Scanning model

Several initiation factors are involved in recognition of 5′cap, 3′tail and scanning. The cap binding complex made up of the eIF4F complex facilitates recruitment of small ribosomal subunit 5′proximal region immediately downstream of the 5′cap. The eIF4F complex consists of the cap-binding protein eIF4E, the scaffold protein eIF4G, and the ATP-dependent helicase eIF4A. The scaffold protein eIF4G provides binding sites for eIF4E, eIF4A, poly A-binding protein (PABP), and eIF3. Thus, eIF4G bridges the 5′ and 3′ends of mRNA. The eIF4A helicase activity disrupts secondary structures in the vicinity of the 5′end and along the 5′UTR. The processivity of eIF4A's helicase activity is enhanced by eIF4B and eIF4H. The HEAT (Huntington Elongation factor 3 Protein phosphatase TOR) domain at the N terminus of eIF4G stimulates its helicase activity by aligning DEAD (Asp-Glu-Ala-Asp)-box motifs in both domains. Structurally, the cap region interacts with Trp residues on eIF4E's concave surface and subsequent interactions of the cap stabilize eIF4E's binding to capped mRNA. A segment of eIF4G wraps around eIF4E's N terminus, inducing structural changes that enhance eIF4E's affinity for the cap. Apparently, it appears that eIF4H binds mRNA behind eIF4A and hence eIF4H could stimulate eIF4A's helicase activity. This leads to unwinding and opening of mRNA secondary structures, which promote processive unidirectional eIF4A movement. Thus, recruitment of the 43S complex is ultimately the result of interactions between Cap-eIF4E-eIF4G-eIF3-40S (reviewed by Jackson et al., 2010).

Ribosome scanning of mRNA 5′UTRs The 43S initiation complex is formed at the 40S ribosomal subunit; the ternary cap binding complex is assembled at the 5′end of mRNA. After the attachment of 43S complex, this complex scans the mRNA downstream to reach the start codon. This scanning involves unwinding of secondary structures in the 5′UTR and movement of ribosomes along the mRNA. eIF1A and eIF1 play a crucial role in scanning by the 43S complex, which makes them scanning competent. Like the prokaryotic initiation factor IF1, eIF1A promotes RNA binding to the small ribosomal subunit and hence ensures stable association and highly processive scanning of mRNA. eIF1 helps in recognizing the correct start sites like IF3 and prevents the initiation complex (IC) from remaining stalled at the 5′end or from stopping at incorrect codons. eIF3, which binds mRNAs upstream of the E site, stabilizes binding of eIF1 to the IC and can bind directly to mRNA, providing additional stabilization of the scanning complex on mRNA. The precise mechanism of scanning is still not fully known. Recent studies have shown that scanning occurs in the 5′-3′ direction. In addition, role of other DEAD box family member, DHX29, a novel initiation factor, has been found in efficient scanning of ribosomes through highly structured 5′UTRs by influencing the conformation of mRNA-binding channel at its entrance (Pisareva et al., 2008).

Noncanonical translation initiation in eukaryotes

Internal ribosome entry sites There are a number of alternative mechanisms of initiation that are highly prevalent in viruses and occur in cellular proteins under conditions of stress. The mechanisms are only partially known compared to cap-dependent initiation mechanism. These mechanisms differ in their requirements for initiation factors, 5′cap, etc. Among these the highly studied is the IRES mechanism, which requires a different number of initiation factors without need for 5′cap or free 5′end for recruitment of ribosomes. IRES were first discovered in Poliovirus (PV) and encephalomyocarditis virus (EMCV) of the *Picornaviridae* family (Jang et al., 1988; Pelletier and Sonenberg, 1988). Later these were found in cellular RNAs associated with the stress response, angiogenesis, etc. (Yang and Sarnow, 1997; Walters and Thompson, 2016). IRESs are basically structured/unstructured RNA elements ranging in size from 9 to >1000 nucleotides that facilitate 5′end independent recruitment of ribosomes at internal sites in mRNA. Viral IRESs are classified on the basis of their structure, initiation factor requirements, and mechanism of initiation. Some of the types are explained in Table 9.1. In type 1 IRESs, the 40S ribosomal subunit is recruited to an upstream AUG without mRNA scanning to initiate translation at a

Table 9.1 Different types of internal ribosome entry sites.

S. No	Type	Length	IFs	Examples	Mechanism	Reference
1.	Type I	~450 nt	eIF2, eIF3, eIF4A, eIF4G, ITAF, PCBP2	PV, enterovirus, human rhinovirus type 2	Interaction of IRES with eIF4G enhanced by eIF4A Facilitates recruitment of 43S complex	Pelletier and Sonenberg (1988) and Sweeney et al. (2014)
2.	Type 2	~450 nt	eIF2, eIF3, eIF4A, eIF4G, ITAF PTB	Encephalomyocarditis virus, foot and mouth disease virus	Interaction of IRES with eIF4G enhanced by eIF4A Facilitates recruitment of 43S complex	Sweeney et al. (2014)
3	Type 3	~300 nt	eIF2 eIF3	Hepatitis C virus	eIF3 interacts with IRES and 40S of 43S initiation complex	Fraser and Doudna (2006) and Matsuda and Mauro (2014)
4	Type 4	~200 nt	None	Cricket paralysis virus	Directly binds to 40S subunit without any initiation factors or initiator tRNA	Sasaki and Nakashima (2000), Wilson et al. (2000), and Deniz et al. (2009)

downstream start codon (reviewed by Kwan and Thompson, 2019). Type 2 IRESs have similar initiation factor requirement as Type 1 IRESs, but it lacks scanning and the ribosome-bound AUG serves as the initiator codon. The well-studied HCV-type 3 IRESs initiate translation through stabilized interaction between eIF3 and IRESs and 18S rRNA (Fraser and Doudna, 2006; Matsuda and Mauro, 2014). The type 4 IRESs are found in the Dicistroviridae family, which can recruit ribosomes and initiate translation without any initiation factors or AUG start codon (reviewed by Kwan and Thompson, 2019).

Cellular IRESs Cellular IRESs have been found in mRNAs linked with stress response such as amino acid starvation, hypoxia, DNA damage repair, etc. Interestingly, many of these cellular mRNAs can switch from the cap-dependent to the cap-independent mechanism under stress conditions. Usually, these are less structured and uniformly distributed over the 5′UTR region. IRES trans acting factors (ITAFs) regulate IRESs activity through their RNA chaperone-like activity, e.g., La, poly(RC) binding protein 2 (PCBP2), heterogeneous nuclear ribonucleoprotein (hnRNP), etc. These are either nuclear proteins or proteins that shuttle between the nucleus and the cytoplasm. Studies so far have demonstrated the role of ITAFs in viral tissue and host tropism or regulation of cellular IRESs under different environmental conditions (reviewed by Kwan and Thompson, 2019).

Ribosomal shunting Ribosomal shunting is a cap-dependent mechanism of translation initiation involving bypassing major regions of 5′UTR by ribosomes before reaching the start codon. 5′UTRs lack IRESs and mostly mRNAs with ribosome shunting encode translatable short open reading frames (ORFs). Interestingly, ribosomes remain associated with short ORFs after their synthesis and bypass other ORFs/secondary structures to reinitiate at downstream ORFs. Such a mechanism has been well studied in animal viruses (adenovirus) and plant viruses (cauliflower mosaic virus) and cellular mRNAs such as HSP70 (Sherrill and Lloyd, 2008; Koh et al., 2013, reviewed by Kwan and Thompson, 2019).

Studies on noncanonical translation initiation mechanisms are still in their infancy. There could be interaction between different noncanonical mechanisms of initiation under diverse conditions. However, their significant role under stress conditions in the propagation of viruses, etc. recommends their role in the onset of various diseases including cancer. Thus, an in-depth characterization of these mechanisms would provide novel avenues for the development of therapeutics.

Joining of subunits

Joining of subunits requires interaction between the surfaces of small and large ribosome subunits for which either the initiation factors should be released from the small subunit or can be dissociated for some time to allow the union of the two subunits. Joining of subunits and dissociation of initiation factors eIF1, eIF1A, eIF3, and eIF2 GDP is promoted by highly conserved G protein IF2 and eIF5B. Other initiation factors IF1/eIF1A stabilize the binding of IF2/eIF5B. These factors are released upon joining of ribosomes through the GTPase activity of IF2/eIF5B upon interaction with GAC on the large ribosome subunit.

Reinitiation, initiation at non-AUG codons and leaky scanning

Reinitiation is the ability of attached ribosomes to continue with translation of subsequent ORF without dissociation from mRNA. Reinitiation is commonly observed in bacterial systems as they have a number of ORFs in close vicinity with overlapping start and stop codons. This does not even require the SD sequence. Efficiency of reinitiation decreases with increasing distance from the start codon. Reinitiation is also observed in eukaryotes in cases such as overlapping stop and start codon or reinitiation of scanning at the 3′end by the small subunit. However, this reinitiation is successful only in case of short upstream ORFs as they may still be associated with other IFs required for efficient initiation. Further, the success depends on the ability of the scanning ribosome to acquire the ternary complex of eIF2 before reaching the initiator codon. To a good extent, the concentration of eIF2-GTP. Met-tRNAi determines reinitiation efficiency at subsequent ORFs. In yeast, the translation of general control nondepressible (GCN4) depends on the availability of this complex. GCN4, a transcription factor, regulates the expression of genes associated with amino acid synthesis. It codes for small regulatory upstream ORFs. Unfavorable conditions reduce the level of eIF2GTP Met-tRNAi resulting in nontranslation of upstream upstream open reading frame (uORF 4) and reinitiation at the GCN4 start codon (Sattlegger and Hinnebusch, 2001). This results in the translation of GCN4, which induces the expression of genes required for amino acid synthesis under stress conditions.

Initiation at noncanonical UUG and GUG instead of at AUG is more common in prokaryotes compared to eukaryotes. Primarily, the relative levels of IF3, other IFs and fmet-tRNA^{fmet} determines initiation at noncanonical start codons (Petrelli et al., 2001). In eukaryotes, studies have shown that various mutants of eIF1 and eIF2 can initiate at noncanonical start codons (reviewed by Marintchev and Wagner, 2004).

Leaky scanning is the inability of ribosomes to recognize the correct initiation codon. Instead, the ribosome keeps moving unless it recognizes the next start codon. This may be due to poor Kozak sequence or RNA secondary structures.

Translation elongation

The end of the initiation process leaves an aminoacylated initiator tRNA in the P site of the ribosome and an empty A site, which serves to start the elongation cycle. As illustrated in Fig. 9.2, elongation consists of three steps, decoding of mRNA codon sequence, peptide bond transfer, and translocation.

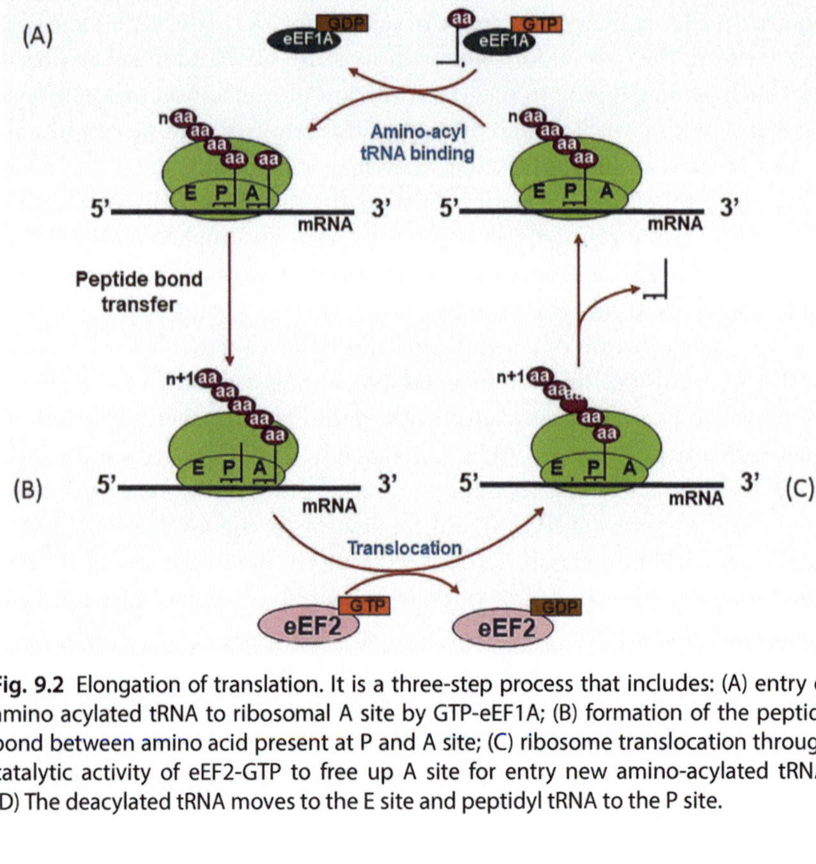

Fig. 9.2 Elongation of translation. It is a three-step process that includes: (A) entry of amino acylated tRNA to ribosomal A site by GTP-eEF1A; (B) formation of the peptide bond between amino acid present at P and A site; (C) ribosome translocation through catalytic activity of eEF2-GTP to free up A site for entry new amino-acylated tRNA. (D) The deacylated tRNA moves to the E site and peptidyl tRNA to the P site.

Decoding of mRNA

The aminoacylated tRNA is brought into the A site as a ternary complex with a GTPase elongation factor, EF-Tu (eEF1A in eukaryotes) and GTP. tRNAs are crucial molecules that are involved in the decoding of codons into amino acids. The anticodon loop in the tRNA recognizes the codons. The D-loop, close to the 5′-end of tRNA, has a dihydrouridine base and the TΨC-loop close to the 3′-end has a sequence of thymine-pseudouridine-cytosine bases. The acceptor arm, at the 3e end of the tRNA, is where the amino acid is attached by an enzyme called aminoacyl-tRNA synthetase. Correct base pairing between the codon on mRNA and the anticodon on tRNA is required for selection of the correct tRNA. This binding results in conformational changes in the ribosome, which stabilizes tRNA binding and triggers GTP hydrolysis by EF-Tu. How codon-anticodon recognition triggers the hydrolysis of GTP by EF-Tu remains an enigma, but presteady-state kinetic studies have proposed that only the correct or cognate tRNA can induce a conformational change in the ribosome. An allosteric three-site model also highlights the induced conformational change. The model states that the affinities of A- and E-site tRNAs are reciprocally coupled. In the presence of E-site tRNA, only the cognate ternary complex has enough affinity for the A site to induce a conformational change in the ribosome on binding, leading to the release of E-site tRNA (Nierhaus, 1990). Ribosomes are key players that control the selectivity of tRNA by direct recognition of the geometry of codon-anticodon base pairing. The ribosome stringently monitors base pairing at the first two positions but is able to tolerate noncanonical base pairs such as a GU wobble base pair at the third position.

This is referred to as genetic code degeneracy at the third position or the "wobble hypothesis" that the same tRNA could recognize codons that differ at the third position, with the third base pair often consisting of certain noncanonical base pairs (Crick, 1966).

EF-Tu in translation elongation

Codon-anticodon recognition triggers the hydrolysis of GTP by EF-Tu. This requires the transmission of a signal from the decoding site on the 30S subunit where codon-anticodon interactions are recognized, to the 50S subunit where the GTPase domain of EF-Tu binds.

Such a signal could be transmitted not only through the ribosome, but also through tRNA itself, since intact tRNA is required for the process (Piepenburg et al., 2000). The precise mechanisms underlying this process are not resolved yet.

The binding of the EF-Tu ternary complex presents the tRNA to the decoding site at a very different angle (Stark et al., 1997). Several structures of EF-Tu have been solved in different binding states by X-ray crystallography, which show that EF-Tu is composed of three domains: nucleotide-binding domain I (GTPase or G domain), which is highly conserved among all GTPases (Bourne et al., 1991; Hilgenfeld, 1995), and domains II and III. The acceptor arm of the tRNA binds mainly with domain III, and some end residues in domains I and II (Dell et al., 1990).

Peptide bond transfer and translocation

GTP hydrolysis leads to the dissociation of EF-Tu from the ribosome and the accommodation of the aminoacyl end of the tRNA into the peptidyl transferase site or A site of the 50S subunit. Accommodation is followed by a rapid peptide bond transfer that involves the deacylation of P-site tRNA and the transfer of the peptide chain to A-site tRNA.

Following peptidyl transfer, the ribosome moves forward by one codon and the protein gets elongated by one amino acid. The pretranslocational ribosome has a deacylated tRNA in the P site and peptidyl tRNA in the A site. To prepare the ribosome for a new round of peptide chain elongation, the tRNAs have to move: the deacylated tRNA needs to be moved from the P site to the E site while the peptidyl tRNA has to move from the A site to the P site. This movement is called as translocation. This movement has to be precise, and at the same time should preserve the reading frame on the mRNA. The ribosome movement consists of an alternating sequence of pause and translocation. This sum of durations of pause and the following translocation is called the ribosome dwell time at any corresponding codon.

Translocation of the tRNAs and mRNA is facilitated by another elongation factor, EF-G, which is also a GTPase. Binding of the GTPase EF-G (eEF2 in eukaryotes) catalyzes the translocation of the mRNA-tRNA complex. The result is a ribosome ready for the next round of elongation, with deacylated tRNA in the E site, peptidyl tRNA in the P site, and an empty A site that is ready to receive the next cognate ternary complex. Translocation can take place even in the absence of mRNA (Belitsina et al., 1981), implying a major role of tRNAs in translocation.

The hybrid state models of translocation

Translocation comprises of two major motions (Frank and Agrawal, 2000; Frank et al., 2007): one is the translocation of the mRNA-$(tRNA)_2$ relative to the large subunit, the other the translocation of the mRNA-tRNAs relative to the small subunit. The first motion is the counterclockwise "ratchet" motion of the small subunit with respect to the large subunit, and the second motion is a reversal of this motion by the small subunit, following GTP hydrolysis. Many theoretical models have been put forward that describe the mechanochemical cycle of the ribosome. It has been long proposed that translocation would lead to "hybrid states" of the tRNA molecules, which would be bound in the A site of the 30S subunit and the P site of the 50S (Bretscher, 1968). An elegant study followed the movement of tRNA during the elongation cycle using base footprinting of the ribosome (Moazed and Noller, 1989).

In this study, when puromycin (a mimic of aminoacyl tRNA in the A site) was added to the ribosomes with an aa-tRNA in the P site, the tRNA footprint on the 50S was lost and replaced by the E site footprint, while the footprint in the 30S subunit was not altered, indicating that the tRNA was in a "P/E" hybrid state. The study also reported that initially there existed an "A/T" hybrid which represented the binding of the ternary complex to the A site of the ribosome. During this state, the aa-tRNA was attached to EF-Tu and when EF-Tu was released, the acceptor arm of the tRNA moved back into the peptidyl transferase site of the 50S. The tRNA now was in the nonhybrid A/A state. After peptidyl transferase, the acceptor arms of tRNAs move to the P and E sites on the 50S subunit, representing the A/P and P/E hybrid states for the tRNA. After translocation by EF-G, the result was an empty A site, and tRNAs in the E and P sites.

"α-ϵ" model for translocation

Another proposed model for translocation is the alternative "α-ϵ" model (Spahn and Nierhaus, 1998). According to this study, the ribosomal contacts of A- and P-site tRNAs do not change during translocation, indicating that there are movable domains within the ribosome that transport the tRNAs through the ribosome. The hybrid states and α-ϵ models can be called mutually exclusive, but recent structural data provide support for some aspects of both models. The discovery of the crystal structure of ribosome now precisely defines the ribosomal contacts of tRNAs in the A, P, and E sites. E-site tRNA in the 50S subunit stays stably bound to the ribosome and requires an active conformational change in the ribosome for its release (Nierhaus, 1990).

Termination of translation

Elongation continues until the ribosome encounters a stop codon on the mRNA and the process enters the termination phase of protein synthesis.

During termination, the stop codon in the mRNA (UAA, UAG, or UGA) enters the A site. The recognition of the stop codons by the ribosomes takes place due to the release factors. In bacteria, two "class I" release factors, RF1 and RF2 (Kisselev and Buckingham, 2000) have been reported. Stop codons UAA and UAG are recognized by RF1 and UGA is recognized by RF2. Unlike prokaryotes, in eukaryotes, the release factor, eRF1, recognizes all three stop codons (Fig. 9.3). A third release factor, called RF3 that belongs to a "class II" release factor binds to the complex of RF1/2 with the ribosome and is a GTPase. The binding of the release factors to a ribosome with the appropriate stop codon in the A site causes the hydrolysis of the ester bond on the P-site peptidyl-tRNA at the peptidyl transferase

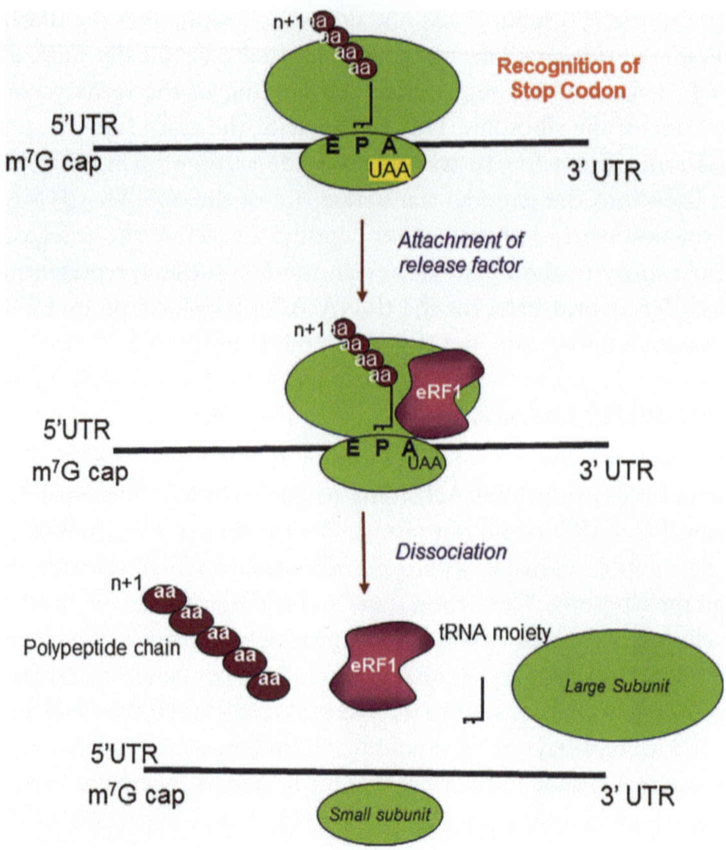

Fig. 9.3 Termination of translation. It starts with the entry of stop codons at ribosomal A site and binding of release factor eRF1. Release factor causes hydrolysis of ester bond on the P site peptidyl-tRNA resulting in release of peptide chain and dissociation of ribosomes.

center. RF3 facilitates a rapid dissociation of RF1 and RF2. The hydrolysis of peptidyl tRNA by RF1/2 is essential for GTP binding to RF3 on the ribosome. This binding causes a conformational change in RF3 and then the dissociation of RF1/2 (Zavialov et al., 2001). GTP hydrolysis is required for subsequent dissociation of RF3. Since there is no amino acid located at the A site, the hydrolysis of the peptide bond holding the polypeptide chain allows the polypeptide chain to be freed from the ribosome; with the release of the polypeptide, the release factor is ejected from the A site, and the empty tRNA is ejected from the P site.

Control of translation

Many components including the mRNA template, tRNAs, ribosomes, trans-acting factors, and regulators affect different aspects of translation elongation and hence protein synthesis.

Aminoacyl tRNA and translation elongation

The rate, efficiency, and accuracy of translation are heavily influenced by the availability of aminoacyl-tRNAs (aa-tRNAs). A microarray-based study has shown that tRNA pools differ among tissues in humans and that there is a significant correlation between relative tRNA abundance and codon usage of highly expressed, tissue-specific genes (Dittmar et al., 2006). Differences between the tRNA landscape leads to changes in the gene expression. For example, proliferating and differentiating cells have been reported to differ in terms of tRNA pool composition (Gingold et al., 2014). Different cell-cycle stages also show different tRNA pools and codon usage (Guimaraes et al., 2013). One reason for differing tRNA pool and codon usage is attributed to histone modifications around the tRNA genes. Additionally, this study also revealed that during proliferation, translation of mRNAs enriched in rare codons is increased.

Modifications in tRNAs

tRNAs have abundant modifications that influence their function and structure (Lorenz et al., 2017; Pan, 2018). The alterations in the anticodon loop are known to affect the decoding efficiency of tRNAs. Modifications in and around the anticodon loop influences the interaction with cognate and near-cognate codons. Ribosome profiling studies in yeast and *C. elegans* has shown that loss of U34 modifications (uridines located at position 34) in the anticodon wobble position leads to stalling of the ribosome and that the loss of such modifications leads to protein homeostasis defects (Nedialkova and Leidel, 2015). DNA methylation of the C38 base, m5C38 mediated by DNA methyl transferase, DNMT2, has been shown to influence the accuracy of translation. This modification is known to confer protection of tRNAs from ribonuclease activity (Tuorto et al., 2015).

Modifications in the wobble position (position 34) and the immediate 30-position of the anticodon (position 37) also influence the decoding function of tRNAs, affecting the accuracy and fidelity of translation (Ranjan and Rodnina, 2016).

Codon usage bias and stability of transcripts

Synonymous codons are used at different frequencies, a property known as "codon usage bias." Ribosome profiling has shed light on several aspects of codon usage and translation. It has been shown that synonymous codons have different decoding rates depending on their usage (Gardin et al., 2014). Codon usage affects elongation rates and it has been reported in yeast that increased ribosome occupancy and its dwell time leads to decreased codon usage (Yu et al., 2015). This however does not hold true for mice. Dwell times and elongation rates could be influenced by codon pairs in different mouse tissues. When a ribosome is translating an mRNA, it interacts with more than one codon at a time. Adjacent codons profoundly affect translation elongation in eukaryotes. A study has identified 17 pairs of inhibitory codons that slowed elongation rates. The majority of these inhibitory codons involve wobble base-pairing leading to poor elongation. Another factor that influences translation elongation rates is amino acid composition of the peptide being synthesized. Negatively charged amino acids are translated faster than positively charged ones and amino acids with smaller side chains are faster than the ones with bigger side chains (Riba et al., 2019).

The codons that are used more often are associated with stable transcripts. These codons are called as optimal codons. They correspond to the cognate tRNA species that are more abundant and undergo faster and efficient translation. Codon optimality influences ribosomal translocation rate and hence translation efficiency (Presnyak et al., 2015). Cells are known to regulate the tRNA landscape and codon usage (Goodarzi et al., 2016; Gingold et al., 2014), and their imbalance can affect translation efficiency and elongation speed (Dana and Tuller, 2012). Codon optimality plays a key role in cellular stress. Codon optimality and tRNA levels are also reported in tumors. For example, a recent study taking advantage of the comprehensive TCGA database showed an altered tRNA landscape in different cancerous tissues, mainly driven by cellular proliferation state (Hernandez-Alias et al., 2021).

Ribosome stalling

Ribosome dwell times can vary based on parameters such as codon optimality, peptide-bond formation efficiency, availability of elongation factors, or nascent chain properties. When the dwell time is so prolonged that the ribosome stops during elongation, it is termed as ribosome stalling. Ribosome stalling can either be physiological regulating correct protein translation, folding, complex assembly, and targeting (Gloge et al., 2014; Panasenko et al., 2019; Shiber et al., 2018) or a signal for the presence of defective mRNAs or altered physiological states. Inhibitory codon pairs and poly(A) tracts within the translated mRNA cause ribosome stalling and reduce protein output. An in vitro reconstituted yeast translation system has been used to demonstrate that inhibitory codon pairs slow elongation rates, which are partially rescued by increased tRNA concentration or by an artificial tRNA not dependent on wobble base pairing. Ribosome profiling data extend these observations by revealing that paused ribosomes with empty A sites are enriched on these sequences, demonstrating a novel mRNA-induced mechanisms of translational stalling in eukaryotic ribosomes [*Molecular mechanism of translational stalling by inhibitory codon combinations and poly(A) tracts*]. A few studies have identified triggers for ribosome stalling. The presence of at least 10 consecutive lysine residues with at least the last two being coded by the AAA codon interacting with the ribosome and the confirmation rearrangement of the decoding center play a role in promoting ribosome stalling in poly(A) motifs such as the poly(A) tail (Chandrasekaran et al., 2019). Another trigger for ribosome stalling is the limitation of amino acids. While the limitation of some amino acids does cause ribosomes to stall during elongation, limitation of other amino acids can lead to a global shutdown of translation initiation due to limited availability of aa-tRNA.

If the stalling is not addressed timely, the trailing translating ribosome collides with the leading/stalled ribosome causing early release of truncated polypeptides and hence demeaning of translation efficiency. To avoid such a situation and hence prevent deleterious effects on cellular physiology and functions, cells have devised several protective mechanisms. One example is the eukaryotic initiation factor 5A (eIF5A; elongation factor-P in prokaryotes), which has been widely described to have an important function during translation elongation. eIF5A, first extracted from rabbit reticulocyte lysate, was identified as a translation initiation factor (reviewed by Park and Wolff, 2018). However, later studies showed it has a substantial role in translation elongation. eIF5A is the sole protein in the eukaryotic proteome to be identified to undergo the posttranslational modification known as hypusination (Dever et al., 2014) at the lysine residue (Lys50 in humans and Lys51 in yeast). Genetic and biochemical assays have shown that this hypusine modification is important for the function of eIF5A protein. eIF5A is described to have a specific role in the elongation of poly-proline motifs (Huter et al., 2017). A poly-proline stretch in a protein could slow down elongation in that particular motif, creating a ribosome stall, where a ribosome with proline in both the P and A sites would have a free E site. In this case, eIF5A is able to recognize the stalled conformation and enter the free E site, stabilizing the prolines and facilitating the peptide-bond formation (Gutierrez et al., 2013).

Ribosome profiling

Ribosome profiling is an emerging technique that has extensively aided in understanding the mechanisms of cellular translation under a variety of environmental conditions. Hence, it is essential to understand the basics of this technique. It is based on the premise that translating ribosomes protect mRNA which remains undigested by ribonucleases. These ribosomes protected fragments (RPF) or ribosome footprints can be sequenced and used for translational analysis of any gene/genome under particular conditions. Basically, they are genome-scale translatome studies which include sequencing of ribosome protected fragments. Ribosome footprints are converted into a library of DNA fragments through reverse transcriptase and analyzed by next-generation sequencing (Ingolia et al., 2012; McGlincy and Ingolia, 2017). Sequencing results decipher the translating transcript and position of each ribosome in its translating at the time of arrest. It is a direct quantitative method to analyze the rate of translation of various transcripts in a cell under any particular condition.

The broad methodology includes blocking of ribosomes on mRNA through different antibiotics like cycloheximide, emetine, chloramphenicol, etc. As translation is a very precise and sensitive process cells should be quickly arrested by freezing, etc. Subsequently, a cytosolic extract is prepared and divided into two fractions—one fraction is digested with ribonuclease while the other fraction is left to be used as control to analyze overall transcript abundance. Different enzymes such as ribonuclease 1 (RNAse1), micrococcal

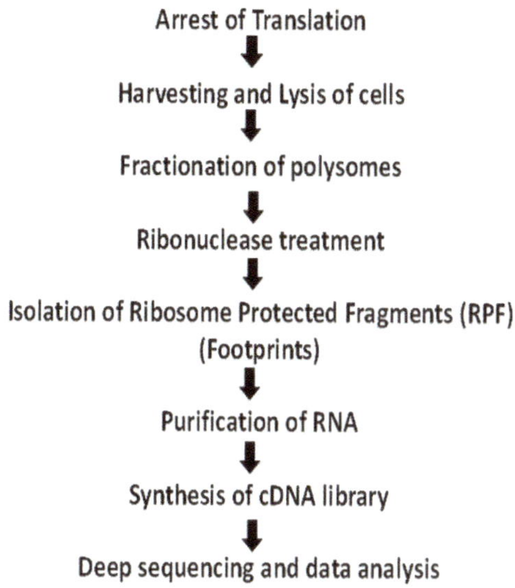

Fig. 9.4 Steps involved in process of ribosome profiling.

nuclease (MNase), and a cocktail of RNAse1 and T1 can be used for the digestion of unprotected sequences. This is followed by the isolation of 80S ribosomes, which contain ribosome protected fragments. Ribosome protected fragments are 20–30 nucleotides long and basically are regions occupied by ribosomes. For sequencing, all fragments are primed at the 3′end and reverse transcribed to generate a DNA library for sequencing (Fig. 9.4). The ratio of RPFs to RNAseq fragments mapped to each transcript is used to estimate relative translation efficiency (TE) (Andreev et al., 2016). The huge amount of data generated through sequencing can be analyzed by using various computational tools such as RUST, RibSeq R, and riboTools.

Application of ribosome profiling

Ribosome profiling has got a number of applications. It has been used for analysis of translation in different cellular compartments, identification of short ORFs, protein variants, and in further understanding of mechanics of translation. Here, we will discuss some of the novel insights provided by ribosome profiling.

Insights into mechanism of translation

Studies using ribose has provided novel insights into the role of various initiation factors and their interactions of translation. Ribosome profiling has shown though interactions of eIF4E with mRNA 5′cap and eIF4G are vital, disruption of this interaction is less inhibitory to global cellular translation (reviewed by Andreev et al., 2016). Further ribosome profiling measurements were performed to identify the role of eIF4A on genomewide protein synthesis. Studies were performed using eIF4A inhibitors such as Pateamine A. It resulted in overall inhibition of translation along with a significant reduction in RPFs on coding ORFs. Interestingly, this effect was unaffected by the length of the 5′leader sequence. Studies with yeast DED1, another DEAD-box translation initiation factor, revealed that reduction of DED1 significantly reduces translation efficiency and it recognizes mRNAs with a long and highly structured 5′end (Sen et al., 2015). The eIF4G recruits eIF4A and stimulates its ATPase activity (Merrick and Pavitt, 2018; Sokabe and Fraser, 2017). It is known that eIF4G is recruited to mRNAs through its interactions with the cap-binding protein eIF4E and poly(A)-binding protein; however, recent studies have shown that in yeast it preferentially binds and promotes the translation of mRNAs with oligo (U) tracts in their UTRs (Zinshteyn et al., 2017). In addition, transcripts that depend on eIF4G also show reduced translation in the absence of the ribosome-associated factor Asc1/RACK1 (receptor for activated protein kinase C1) (Thompson et al., 2016).

Various Riboseq studies have illustrated extensive usage of noncanonical start codons compared to AUG in eukaryotes. However, critical analysis of available data has owed this observation to leaky translation as most of the non-AUG translations are observed upstream of AUG codons (reviewed by Andreev et al., 2016). Interestingly, ribose studies have provided insights into regulatory role of uORFs. Firstly, increases in uROF translation under stress conditions could be result of translation regulation of mRNAs of eIF1, eIF5, and eIF1A. Secondly, differential regulation of eIF2 under stress conditions, e.g., upon integrated stress response, kinases heme-regulated inhibitor (HRI), protein kinase R (PKR), PKR-like endoplasmic reticulum kinase (PERK) or general control non-depressible 2 (GCN2) phosphorylate eIF2. eIF2B forms a highly stable complex with phosphorylated eIF2 resulting in decreased recycling of GTP and inhibition of initiation of translation (Baird and Wek, 2012). Riboseq studies under these conditions supported the fact that hyperphosphoryation of eIF2 inhibits translation at large except in few cases. Interestingly, resistant mRNAs harbored translated uORFs. Usually, these mRNAs exhibit poor translation of ORF and good translation of uORF under normal growth conditions. uORF regulates translation through eIF2 in activation.

Guydosh and Green (2014) unveiled the function of DOM34 (homolog of eRF1) using Riboseq. A comparison ribosome footprint of the yeast WT strain with *dom34* delete strain was performed with the premise that variation in footprints would be observed if dom34 has a role in the dissociation of revealed increased ribosome occupancy at one particular position in HAC1 mRNA. HAC1 mRNA undergoes an unusual event of splicing in cytoplasm resulting in splice variants in cytoplasm. The increased ribosome occupancy at the 3′ end of this suggested stalled ribosomes due to nonfunctional *dom34*. Thus, dom34 is responsible for dissociating ribosomes on truncated transcripts. Another interesting result from this study was the presence of a large number of ribosome footprints in 3′UTR. This suggested that dom34 is also required for release of ribosomes following translation termination.

Insights into translation of noncoding RNAs, especially lncRNA

Recent in depth RNA seq-based studies have demonstrated that very low percentage of a genome codes for genes, which get translated to form proteins. A high percentage of a genome codes for noncoding RNA which plays crucial role in several cellular processes. The inability of these noncoding RNAs (ncRNAs) especially long noncoding RNAs (lncRNAs) despite having a 5′cap and 3′ poly A tail has been owed to short nucleotide length leading to small peptides and absence of start codons. Ribosome profiling-based genome-scale studies have demonstrated the translation potential of several of these ncRNA, leading to a variety of small peptides/proteins in cellular translation. In fact, small peptides coded by these ncRNAs have a crucial role in various diseases including cancer, e.g., a conserved 53 amino acid peptide encoded by lncRNA HOXA cluster antisense RNA 3 (HOXA-AS3) represses colorectal cancer (CRC) growth by controlling alternative splicing of pyruvate kinase M (PKM) (Huang et al., 2017). Intriguingly, recently, Yan, G.R. et al. discovered two lncRNAs that encoded two different functional peptides/proteins in colorectal cancer. The first lncRNA with coding ability is lncRNA LOC90024, which encodes a 130-aa protein that promotes CRC tumorigenesis and progression (Meng et al., 2020). This peptide interacts with serine- and arginine-rich splicing factor 3 (SRSF3) and regulates mRNA splicing. The second lncRNA with coding ability is long intergenic nonprotein coding RNA 266-1 (LINC-00266-1), which encodes a 71-aa peptide called RNA-binding regulatory peptide (RBRP) (Zhu et al., 2020). This peptide is significantly increased in CRC and probably has oncogenic potential.

Expansion of translatome

Ribosome profiling has resulted in the identification of several short translated ORFs enhancing the known translatome of cells (Aspden et al., 2014; Calviello et al., 2016; Menschaert et al., 2013). These large number of functional micropeptides needs to be characterized to decipher their role in cell growth. These micropeptides may interact with proteins to avoid their harmful effects (Ulitsky and Bartel, 2013). Moreover, studies have shown that even synthesized nonfunctional proteins can serve as antigens (Ingolia, 2014). The differential loads of these neoantigens may help in differentiating cancerous cells from noncancerous cells and can be further developed as biomarkers. Further, in-depth studies will help in identifying their role in cancer and development of novel therapeutics interventions that selectively enhance T-cell reactivity against this class of antigens (Schumacher and Schreiber, 2015).

Limitations of ribosome profiling

Undoubtedly ribosome profiling has revolutionized the current understanding of the mechanism of translation in cell growth and various diseases. It is a very precise and quantitative method of estimation of translation. However, the approach has various shortcomings, which can completely skew up the data. Largely, the inaccuracy lies with the procedure used for ribosome profiling. For example, ribonuclease cuts at specific nucleotide sequences and degrades rRNA unequally. Thus, an unbiased ribonuclease such as T1 should be used that preserve the ribosome integrity while digesting polysomes (Gerashchenko and Gladyshev, 2016). Further variations in Riboseq results are due to the use of different types of antibiotics for the stalling of ribosomes. Pretreatment with cycloheximide results in ribosome footprints more concentrated at the start of coding sequences. Furthermore, another antibiotic harringtonine exhibits high codon specificity as it preferentially arrests ribosomes at Lys, Arg, or Tyr codons (reviewed by Andreev et al., 2016). Further variations can result due to different depths of RNAseq used for experiments. With ongoing advances in the understanding of different aspects of techniques, development of more appropriate tools and reliable procedures will be available for studies.

Of note, several significant modifications have been made in original ribosome profiling procedures to gain insights into different aspects of translation. Global translation initiation sequencing (GTI-seq) and further improved quantitative translation initiation sequencing (QTI seq) has been developed for the quantification of ribosomes at the initiation site and detecting the dynamic changes of initiation rate (Gao et al., 2015). Translation complex profile sequencing (TCP-seq) provides information on the intermediates of translation initiation and termination. Ribolace helps in the identification of active ribosomes through use of 3P (an analog of puromycin) which can puromycylate active ribosomes that can be later separated by pull down assays (Clamer et al., 2018).

References

Andreev DE, O'Connor PB, Loughran G, Dmitriev SE, Baranov PV, Shatsky IN. Insights into the mechanisms of eukaryotic translation gained with ribosome profiling. Nucleic Acids Res 2016;45(2):513–26. Epub 2016 Dec 6 27923997.

Aspden JL, Eyre-Walker YC, Phillips RJ, Amin U, Mumtaz MA, Brocard M, Couso JP. Extensive translation of small Open Reading Frames revealed by Poly-Ribo-Seq. elife 2014;3, e03528. 25144939.

Baird TD, Wek RC. Eukaryotic initiation factor 2 phosphorylation and translational control in metabolism. Adv Nutr 2012;3(3):307–21. 22585904.

Belitsina NV, Tnalina GZ, Spirin AS. Template-free ribosomal synthesis of polylysine from lysyl-tRNA. FEBS Lett 1981;131(2):289–92. 7028507.

Bourne HR, Sanders DA, McCormick F. The GTPase superfamily: conserved structure and molecular mechanism. Nature 1991;349(6305):117–27. 1898771.

Bretscher MS. Translocation in protein synthesis: a hybrid structure model. Nature 1968;218(5142):675–6. 5655957.

Calviello L, Mukherjee N, Wyler E, Zauber H, Hirsekorn A, Selbach M, Landthaler M, Obermayer B, Ohler U. Detecting actively translated open reading frames in ribosome profiling data. Nat Methods 2016;13(2):165–70. Epub 2015 Dec 14 26657557.

Chandrasekaran V, Juszkiewicz S, Choi J, Puglisi JD, Brown A, Shao S, Ramakrishnan V, Hegde RS. Mechanism of ribosome stalling during translation of a poly(A) tail. Nat Struct Mol Biol 2019;1132–40. Epub 2019 Nov 25 31768042.

Clamer M, Tebaldi T, Lauria F, Bernabò P, Gómez-Biagi RF, Marchioretto M, Kandala DT, Minati L, Perenthaler E, Gubert D, Pasquardini L, Guella G, Groen EJN, Gillingwater TH, Quattrone A, Viero G. Active ribosome profiling with RiboLace. Cell Rep 2018;25(4):1097–1108.e5. 30355487.

Crick FH. Codon-anticodon pairing: the wobble hypothesis. J Mol Biol 1966;19(2):548–55. 5969078.

Dana A, Tuller T. Determinants of translation elongation speed and ribosomal profiling biases in mouse embryonic stem cells. PLoS Comput Biol 2012;8(11), e1002755. Epub 2012 Nov 1 23133360.

Dell VA, Miller DL, Johnson AE. Effects of nucleotide- and aurodox-induced changes in elongation factor Tu conformation upon its interactions with aminoacyl transfer RNA. A fluorescence study. Biochemistry 1990;29(7):1757–63. 2110000.

Deniz N, Lenarcic EM, Landry DM, Thompson SR. Translation initiation factors are not required for Dicistroviridae IRES function in vivo. RNA 2009;15(5):932–46. Epub 2009 Mar 19 19299549.

Dever TE, Gutierrez E, Shin BS. The hypusine-containing translation factor eIF5A. Crit Rev Biochem Mol Biol 2014;49(5):413–25. Epub 2014 Jul 17 25029904.

Dittmar KA, Goodenbour JM, Pan T. Tissue-specific differences in human transfer RNA expression. PLoS Genet 2006;2(12), e221. Epub 2006 Nov 13 17194224.

Drabkin HJ, Estrella M, Rajbhandary UL. Initiator-elongator discrimination in vertebrate tRNAs for protein synthesis. Mol Cell Biol 1998;18(3):1459–66. 9488462.

Frank J, Agrawal RK. A ratchet-like inter-subunit reorganization of the ribosome during translocation. Nature 2000;406(6793):318–22. 10917535.

Frank J, Gao H, Sengupta J, Gao N, Taylor DJ. The process of mRNA-tRNA translocation. Proc Natl Acad Sci U S A 2007;104(50):19671–8. Epub 2007 Nov 14 18003906.

Fraser CS, Doudna JA. Structural and mechanistic insights into hepatitis C viral translation initiation. Nat Rev Microbiol 2006;5(1):29–38. Epub 2006 Nov 27 17128284.

Gao X, Wan J, Liu B, Ma M, Shen B, Qian SB. Quantitative profiling of initiating ribosomes in vivo. Nat Methods 2015;2:147–53. Epub 2014 Dec 8 25486063.

Gardin J, Yeasmin R, Yurovsky A, Cai Y, Skiena S, Futcher B. Measurement of average decoding rates of the 61 sense codons in vivo. elife 2014;3, e03735. 25347064.

Gerashchenko MV, Gladyshev VN. Ribonuclease selection for ribosome profiling. Nucleic Acids Res 2016;45(2), e6. Epub 2016 Sep 15 27638886.

Gingold H, Tehler D, Christoffersen NR, Nielsen MM, Asmar F, Kooistra SM, Christophersen NS, Christensen LL, Borre M, Sørensen KD, Andersen LD, Andersen CL, Hulleman E, Wurdinger T, Ralfkiær E, Helin K, Grønbæk K, Ørntoft T, Waszak SM, Dahan O, Pedersen JS, Lund AH, Pilpel Y. A dual program for translation regulation in cellular proliferation and differentiation. Cell 2014;158(6):1281–92. 25215487.

Gloge F, Becker AH, Kramer G, Bukau B. Co-translational mechanisms of protein maturation. Curr Opin Struct Biol 2014;24–33. Epub 2013 Dec 12 24721450.

Goodarzi H, Nguyen HCB, Zhang S, Dill BD, Molina H, Tavazoie SF. Modulated expression of specific tRNAs drives gene expression and cancer progression. Cell 2016;165(6):1416–27. 27259150.

Guillon JM, Meinnel T, Mechulam Y, Lazennec C, Blanquet S, Fayat G. Nucleotides of tRNA governing the specificity of *Escherichia coli* methionyl-tRNA(fMet) formyltransferase. J Mol Biol 1992;224(2):359–67. 1373194.

Guimaraes JC, Mittal N, Gnann A, Jedlinski D, Riba A, Buczak K, Schmidt A, Zavolan M. A rare codon-based translational program of cell proliferation. Genome Biol 2013;21(1):44. 32102681.

Gutierrez E, Shin BS, Woolstenhulme CJ, Kim JR, Saini P, Buskirk AR, Dever TE. eIF5A promotes translation of polyproline motifs. Mol Cell 2013;51(1):35–45. Epub 2013 May 30 23727016.

Guydosh NR, Green R. Dom34 rescues ribosomes in 3′ untranslated regions. Cell 2014;156(5):950–62. 24581494.

Hernandez-Alias X, Benisty H, Schaefer MH, Serrano L. Translational efficiency across healthy and tumor tissues is proliferation-related. Mol Syst Biol 2021;16(3), e9275. Erratum in: Mol Syst Biol. 2021 Mar; 17(3):e10097 32149479.

Hilgenfeld R. Regulatory GTPases. Curr Opin Struct Biol 1995;5(6):810–7. 8749370.

Huang JZ, Chen M, Chen D, Gao XC, Zhu S, Huang H, Hu M, Zhu H, Yan GR. A peptide encoded by a putative lncRNA HOXB-AS3 suppresses colon cancer growth. Mol Cell 2017;68(1):171–184.e6. 28985503.

Huter P, Arenz S, Bock LV, Graf M, Frister JO, Heuer A, Peil L, Starosta AL, Wohlgemuth I, Peske F, Nováček J, Berninghausen O, Grubmüller H, Tenson T, Beckmann R, Rodnina MV, Vaiana AC, Wilson DN. Structural basis for polyproline-mediated ribosome stalling and rescue by the translation elongation factor EF-P. Mol Cell 2017;68(3):515–527.e6. 29100052.

Ingolia NT. Ribosome profiling: new views of translation, from single codons to genome scale. Nat Rev Genet 2014;15(3):205–13. Epub 2014 Jan 28 24468696.

Ingolia NT, Brar GA, Rouskin S, McGeachy AM, Weissman JS. The ribosome profiling strategy for monitoring translation in vivo by deep sequencing of ribosome-protected mRNA fragments. Nat Protoc 2012;7(8):1534–50. 22836135.

Jackson RJ, Hellen CU, Pestova TV. The mechanism of eukaryotic translation initiation and principles of its regulation. Nat Rev Mol Cell Biol 2010;11(2):113–27. 20094052.

Jang SK, Kräusslich HG, Nicklin MJ, Duke GM, Palmenberg AC, Wimmer E. A segment of the 5′ nontranslated region of encephalomyocarditis virus RNA directs internal entry of ribosomes during in vitro translation. J Virol 1988;62(8):2636–43. 2839690.

Kisselev LL, Buckingham RH. Translational termination comes of age. Trends Biochem Sci 2000;11:561–6. 11084369.

Koh DC, Edelman GM, Mauro VP. Physical evidence supporting a ribosomal shunting mechanism of translation initiation for BACE1 mRNA. Translation (Austin) 2013;1(1), e24400. 26824018.

Kolitz SE, Lorsch JR. Eukaryotic initiator tRNA: finely tuned and ready for action. FEBS Lett 2010;584(2):396–404. 19925799.

Kwan T, Thompson SR. Noncanonical translation initiation in eukaryotes. Cold Spring Harb Perspect Biol 2019;11(4):a032672. 29959190.

Lee CP, Seong BL, RajBhandary UL. Structural and sequence elements important for recognition of *Escherichia coli* formylmethionine tRNA by methionyl-tRNA transformylase are clustered in the acceptor stem. J Biol Chem 1991;266(27):18012–7. 1917939.

Lorenz C, Lünse CE, Mörl M. tRNA modifications: impact on structure and thermal adaptation. Biomolecules 2017;7(2):35. 28375166.

Marintchev A, Wagner G. Translation initiation: structures, mechanisms and evolution. Q Rev Biophys 2004;37(3–4):197–284. 16194295.

Matsuda D, Mauro VP. Base pairing between hepatitis C virus RNA and 18S rRNA is required for IRES-dependent translation initiation in vivo. Proc Natl Acad Sci U S A 2014;111(43):15385–9. Epub 2014 Oct 13 25313046.

McGlincy NJ, Ingolia NT. Transcriptome-wide measurement of translation by ribosome profiling. Methods 2017;126:112–29. Epub 2017 Jun 1 28579404.

Meng N, Chen M, Chen D, Chen XH, Wang JZ, Zhu S, He YT, Zhang XL, Lu RX, Yan GR. Small protein hidden in lncRNA *LOC90024* promotes "cancerous" RNA splicing and tumorigenesis. Adv Sci (Weinh) 2020;7(10):1903233. 32440474.

Menschaert G, Van Criekinge W, Notelaers T, Koch A, Crappé J, Gevaert K, Van Damme P. Deep proteome coverage based on ribosome profiling aids mass spectrometry-based protein and peptide discovery and provides evidence of alternative translation products and near-cognate translation initiation events. Mol Cell Proteomics 2013;12(7):1780–90. Epub 2013 Feb 21 23429522.

Merrick WC, Pavitt GD. Protein synthesis initiation in eukaryotic cells. Cold Spring Harb Perspect Biol 2018;10(12):a033092. 29735639.

Moazed D, Noller HF. Intermediate states in the movement of transfer RNA in the ribosome. Nature 1989;342(6246):142–8. 2682263.

Nedialkova DD, Leidel SA. Optimization of codon translation rates via tRNA modifications maintains proteome integrity. Cell 2015;161(7):1606–18. Epub 2015 Jun 4 26052047.

Nierhaus KH. The allosteric three-site model for the ribosomal elongation cycle: features and future. Biochemistry 1990;29(21):4997–5008. 2198935.

Pan T. Modifications and functional genomics of human transfer RNA. Cell Res 2018;28(4):395–404. Epub 2018 Feb 20 29463900.

Panasenko OO, Somasekharan SP, Villanyi Z, Zagatti M, Bezrukov F, Rashpa R, Cornut J, Iqbal J, Longis M, Carl SH, Peña C, Panse VG, Collart MA. Co-translational assembly of proteasome subunits in NOT1-containing assemblysomes. Nat Struct Mol Biol 2019;26(2):110–20. Epub 2019 Jan 28 30692646.

Park MH, Wolff EC. Hypusine, a polyamine-derived amino acid critical for eukaryotic translation. J Biol Chem 2018;293(48):18710–8. Epub 2018 Sep 26 30257869.

Pelletier J, Sonenberg N. Internal initiation of translation of eukaryotic mRNA directed by a sequence derived from poliovirus RNA. Nature 1988;334(6180):320–5. 2839775.

Petrelli D, LaTeana A, Garofalo C, Spurio R, Pon CL, Gualerzi CO. Translation initiation factor IF3: two domains, five functions, one mechanism? EMBO J 2001;20(16):4560–9. 11500382.

Piepenburg O, Pape T, Pleiss JA, Wintermeyer W, Uhlenbeck OC, Rodnina MV. Intact aminoacyl-tRNA is required to trigger GTP hydrolysis by elongation factor Tu on the ribosome. Biochemistry 2000;39(7):1734–8. 10677222.

Pisareva VP, Pisarev AV, Komar AA, Hellen CU, Pestova TV. Translation initiation on mammalian mRNAs with structured 5'UTRs requires DExH-box protein DHX29. Cell 2008;135(7):1237–50. 19109895.

Presnyak V, Alhusaini N, Chen YH, Martin S, Morris N, Kline N, Olson S, Weinberg D, Baker KE, Graveley BR, Coller J. Codon optimality is a major determinant of mRNA stability. Cell 2015;160(6):1111–24. 25768907.

Ranjan N, Rodnina MV. tRNA wobble modifications and protein homeostasis. Translation (Austin) 2016;4(1), e1143076. 27335723.

Riba A, Di Nanni N, Mittal N, Arhne E, Schmidt A, Zavolan M. Protein synthesis rates and ribosome occupancies reveal determinants of translation elongation rates. Proc Natl Acad Sci U S A 2019;116(30):15023–32. 31292258.

Sasaki J, Nakashima N. Methionine-independent initiation of translation in the capsid protein of an insect RNA virus. Proc Natl Acad Sci U S A 2000;97(4):1512–5. 10660678.

Sattlegger E, Hinnebusch AG. Separate domains in GCN1 for binding protein kinase GCN2 and ribosomes are required for GCN2 activation in amino acid-starved cells. EMBO J 2001;19(23):6622–33. 11101534.

Schumacher TN, Schreiber RD. Neoantigens in cancer immunotherapy. Science 2015;348(6230):69–74. 25838375.

Sen ND, Zhou F, Ingolia NT, Hinnebusch AG. Genome-wide analysis of translational efficiency reveals distinct but overlapping functions of yeast DEAD-box RNA helicases Ded1 and eIF4A. Genome Res 2015;25(8):1196–205. 26122911.

Seong BL, RajBhandary UL. Mutants of *Escherichia coli* formylmethionine tRNA: a single base change enables initiator tRNA to act as an elongator in vitro. Proc Natl Acad Sci U S A 1987;84(24):8859–63. 3321059.

Sherrill KW, Lloyd RE. Translation of cIAP2 mRNA is mediated exclusively by a stress-modulated ribosome shunt. Mol Cell Biol 2008;6:2011–22. Epub 2008 Jan 14 18195037.

Shiber A, Döring K, Friedrich U, Klann K, Merker D, Zedan M, Tippmann F, Kramer G, Bukau B. Cotranslational assembly of protein complexes in eukaryotes revealed by ribosome profiling. Nature 2018;561(7722):268–72. Epub 2018 Aug 29 30158700.

Sokabe M, Fraser CS. A helicase-independent activity of eIF4A in promoting mRNA recruitment to the human ribosome. Proc Natl Acad Sci U S A 2017;114(24):6304–9. Epub 2017 May 30 28559306.

Spahn CM, Nierhaus KH. Models of the elongation cycle: an evaluation. Biol Chem 1998;379(7):753–72. 9705140.

Stark H, Rodnina MV, Rinke-Appel J, Brimacombe R, Wintermeyer W, van Heel M. Visualization of elongation factor Tu on the *Escherichia coli* ribosome. Nature 1997;389(6649):403–6. 9311785.

Sweeney TR, Abaeva IS, Pestova TV, Hellen CU. The mechanism of translation initiation on type 1 picornavirus IRESs. EMBO J 2014;33(1):76–92. Epub 2013 Dec 15 24357634.

Thompson MK, Rojas-Duran MF, Gangaramani P, Gilbert WV. The ribosomal protein Asc1/RACK1 is required for efficient translation of short mRNAs. elife 2016;5, e11154. 27117520.

Tuorto F, Herbst F, Alerasool N, Bender S, Popp O, Federico G, Reitter S, Liebers R, Stoecklin G, Gröne HJ, Dittmar G, Glimm H, Lyko F. The tRNA methyltransferase Dnmt2 is required for accurate polypeptide synthesis during haematopoiesis. EMBO J 2015;34(18):2350–62. Epub 2015 Aug 13 26271101.

Ulitsky I, Bartel DP. lincRNAs: genomics, evolution, and mechanisms. Cell 2013;154(1):26–46. 23827673.

Walters B, Thompson SR. Cap-independent translational control of carcinogenesis. Front Oncol 2016;6:128. 27252909.
Wilson JE, Powell MJ, Hoover SE, Sarnow P. Naturally occurring dicistronic cricket paralysis virus RNA is regulated by two internal ribosome entry sites. Mol Cell Biol 2000;20(14):4990–9. 10866656.
Yang Q, Sarnow P. Location of the internal ribosome entry site in the 5′ non-coding region of the immunoglobulin heavy-chain binding protein (BiP) mRNA: evidence for specific RNA-protein interactions. Nucleic Acids Res 1997;25(14):2800–7. 9207027.
Yu CH, Dang Y, Zhou Z, Wu C, Zhao F, Sachs MS, Liu Y. Codon usage influences the local rate of translation elongation to regulate co-translational protein folding. Mol Cell 2015;59(5):744–54. Epub 2015 Aug 27 26321254.
Zavialov AV, Buckingham RH, Ehrenberg M. A posttermination ribosomal complex is the guanine nucleotide exchange factor for peptide release factor RF3. Cell 2001;107(1):115–24. 11595190.
Zhu S, Wang JZ, Chen D, He YT, Meng N, Chen M, Lu RX, Chen XH, Zhang XL, Yan GR. An oncopeptide regulates m^6A recognition by the m^6A reader IGF2BP1 and tumorigenesis. Nat Commun 2020;11(1):1685. 32245947.
Zinshteyn B, Rojas-Duran MF, Gilbert WV. Translation initiation factor eIF4G1 preferentially binds yeast transcript leaders containing conserved oligo-uridine motifs. RNA 2017;9:1365–75. Epub 2017 May 25 28546148.

CHAPTER 10

Inhibitors of ribosome biogenesis in prokaryotes and eukaryotes

Sukriti Baweja and P. Debishree Subudhi
Department of Molecular and Cellular Medicine, Institute of Liver and Biliary Sciences, New Delhi, India

Introduction

Ribosomes are megaDalton ribonucleoproteins that have the ability to translate genetic code on mRNAs into proteins. The systemic formation of new ribosomes is important for normal growth and cell division. In prokaryotes, ribosomes are made of 3 ribosomal RNA molecules and 50 other proteins and that are assembled in the same way as a puzzle. Ribosome synthesis in eukaryotic cells involves nearly 250 nonribosomal assembly proteins, and is a high energy-consuming process. Ribosome biogenesis is associated with many other fundamental pathways that control the cell cycle and cellular growth by a complex network that is highly regulated. Therefore, any dysregulation or interference in ribosome biogenesis in any direction affects the turnover rate of cell division and thus results in the uncontrolled growth of cells, leading to disorders associated with developmental processes such as anemia or development of even tumors.

Eukaryotic ribosomes are composed of more than 80 ribosomal factors and 4 rRNAs that are assembled through a highly interwoven process. The mature rRNAs are formed from the transcripts of the precursor which involves many processing steps. Initially, the primary transcripts at specific sites are cleaved and broken into the two precursor ribosomal subunits, the pre-40S and the pre-60S, which are again reorganized into two different pathways. The transportation of the pre-40S occurs in the cytoplasm where the final maturation takes place, whereas the pre-60S particle undergoes rearrangement steps before being exported into the cytoplasm.

The fast growing or activated cells, such as those found in cancer, strongly depend on a higher ribosome biogenesis rate, whereas normal cells can survive even a transitory shutdown of the pathway with the help of their available reservoir. Hence, there is a complete dependency on high

Emerging Concepts in Ribosome Structure, Biogenesis, and Function
https://doi.org/10.1016/B978-0-12-816364-1.00004-4

turnaround time of ribosome biogenesis in cancerous cells which represents a vulnerable point, and any type of perturbation may lead to inhibition. This phenomenon represents another promising therapeutic strategy for the treatment of infections and various malignancies.

Despite the progress in understanding ribosome biogenesis, the complexity and vitality of the maturation process remains unknown. It is extremely important to understand the sequence of events during ribosome biogenesis and its regulation in detail. Previously, many studies have been conducted, predominantly in yeast, to gather fundamental knowledge of ribosome formation which also includes some breakthrough structural insights. However, due to a lack of appropriate tools, these studies only offer static snapshots and therefore were unable to delineate the dynamics of these rapid and interwoven pathways.

The best strategy currently available is the use of small molecular weight chemical compounds as inhibitors to enable researchers to rapidly block the ribosome formation at certain steps and to track the effects within minutes or even seconds. However, there are over 250 maturation factors, for which many potential targets are available, and only a few of the inhibitors of ribosome biogenesis have been identified. Some of the available substances have been reported to interfere with the RNA metabolism and thus can be linked to ribosome biogenesis. Many of the inhibitors have been shown to limit the growth of tumor cells while some are already in clinical use for cancer treatment. This shows that ribosome biogenesis is a promising target for antitumor chemotherapy and thus can provide many new inhibitors for future research. In this chapter, the ribosome biogenesis inhibitors and their biological effects will be discussed, and how ribosome biogenesis is a promising target for drug design for various cancers and other diseases.

Prokaryotic ribosome biogenesis and inhibitors

Ribosome biosynthesis and assembly

The bacterial ribosome is a 2.6-MDa ribonucleoprotein complex that causes protein translation, and forms a 70S particle consisting of 30S and 50S subunits (Stokes et al., 2014). The performance and structure of the ribosome during translation is quite well understood. However the molecular mechanisms that lead to its formation are largely unknown. Ribosome biogenesis proteins transiently bind to ribosomal particles (Bunner et al., 2010) to prevent immature subunits from entering the translation cycle (Strunk et al., 2011, 2012). Ribosome synthesis consumes almost 40% of the

ATP in the cells of rapidly dividing *Escherichia coli*, requiring the modification, transcription, and folding of rRNA transcripts in a sequential manner (Maguire, 2009), In vivo, these events are highly interconnected that take place cooperatively to restrict the assembly of the ribosomes (Holmes and Culver, 2005). Prokaryotes have approximately 60 of such proteins. The cell's translational capability depends on the source of the 30S or 50S regulatory feedback (Gaal et al., 1997;Yamagishi and Nomura, 1988). Genetic probes of ribosome assembly are more prone to various impacts and epistatic phenotypes (Lerner and Inouye, 1991).

Inhibitors of ribosome biogenesis in prokaryotes

The small molecular weight chemical inhibitors are of increasing interest. Such compounds can exert their effects within seconds and be withdrawn or added from cell systems at any time. Small molecules may be administered in doses that achieve different levels of target inhibition, making them useful probes of protein activity (Stokes et al., 2014). Small molecular weight inhibitors can be used as new tools for studying complex pathways, especially those regulated by uncharacterized protein assembly factors. Also, the revelation of novel chemical inhibitors of bacterial ribosome biogenesis could lead to identifying completely new classes of antibiotics. While current available antibiotics provide an accessible platform to understand the molecular mechanism of protein translation in detail, no similar study exist for ribosome biogenesis (Stokes et al., 2014; Comartin and Brown, 2006).

There are some reports on the characterization and identification of small molecular weight chemical compounds as inhibitors for bacterial ribosome biogenesis. They are enriched for molecules that have caused cold-sensitive growth inhibition of the model such as bacterium *E. coli* via a chemical library that includes approved drugs and compounds with well-characterized bioactivity (Stokes et al., 2014). Some investigations have suggested that the genetic defects in ribosome assembly lead to cold-sensitive growth phenotypes (Stokes et al., 2014; Dammel and Noller, 1995). Subsequent research has validated the cold sensitivity of strains from the Keio collection (Baba et al., 2006). Most interesting was the chemical screening that revealed lamotrigine as a drug that has a highly cold-sensitive growth characteristic. This molecule when tested in a dose-dependent manner, led to the accumulation of immature ribosomal subunits without any interference in protein translation. Spontaneous lamotrigine activity suppressors were mapped specifically to translation initiation factor IF2, encoded by INF-B.

These mutations are localized in one of the most evolutionarily divergent domain II of IF2 (not well characterized) that prevent the binding of lamotrigine to IF2 in vitro (Stokes et al., 2014). Lamotrigine is one of the commonly accessible chemical probes of bacterial ribosome biogenesis and is extensively used in prokaryotes.

Ribosome biogenesis can be the target of cold stress

In the research undertaken by Stokes et al. (2014) to determine the relative abundance of genes in each functional class, the number of cold-sensitive genes was divided by the total number of nonessential genes. Interestingly, this approach was helpful in identifying many of the already-known cold-sensitive ribosome biogenesis genes. This also signifies that screening compounds for cold sensitivity would identify only those associated to ribosome synthesis and their biological functions. Thus, the cold-temperature-sensitive growth is recognized as a main characteristic for defects in ribosome biogenesis. Validation of such an enrichment approach was undertaken to screen the nonessential gene deletion strain using the *E. coli* Keio collection (Baba et al., 2006). The clusters of orthologous groups (Tatusov et al., 1997) used for the classification of the cell function of each deleted gene for strains that showed cold sensitivity factor in the top 3.5% (almost 155 clones) which were further examined (Stokes et al., 2014).

Lamotrigine is the most promising cold sensitive chemical compound

The vivid chemical screening including nearly 30,000 compounds by Stokes et al. (2014) was performed to identify molecules that show cold sensitive growth inhibitory characteristics. Interestingly, among these nearly 3500 synthetic molecules were already-approved drugs and synthetic compounds with their known biological functions. All antibiotics were removed due to known mechanisms of biological actions and filtered the active molecules with chemical structure diversity. Subsequently 38 of such active molecules were analyzed at 37°C and 15°C with various doses. One of the most promising drugs that was highlighted was lamotrigine, which showed the largest change with the minimum inhibitory levels when the temperature downshifted. On further investigation with sedimentation analysis, lamotrigine showed a rapid accumulation of nonnative ribosomal particles with apparent sedimentation rates of ~25S (pre-30S) and ~40S (pre-50S) (Stokes et al., 2014; Jomaa et al., 2014; Siibak et al., 2009; Sykes et al., 2010).

Antibiotics that target ribosome biogenesis

(A) *Inhibitors of 30S*

There is a long list of antibiotics, such as aminoglycoside antibiotics, neomycin B, gentamicin, paromomycin, and kanamycin, that are known to bind to the 30S ribosomal subunit. These in turn disrupts the decoding fidelity of mRNA by displacing two adenine residues in helix 44 of the 16S rRNA, which also stabilizes the conformation that normally occurs only when the correct mRNA-tRNA complex is established (Vicens and Westhof, 2003).

A team of researchers at Vernalis Ltd. (Cambridge, United Kingdom) developed a hierarchical screening strategy that helped to identify new ligands that bind to the decoding region of the 30S A-site (Foloppe et al., 2004). In silico digital libraries were screened to select compounds that had less than 7 rotatable bonds with molecular weights of 250–500 Da. Docking of each compound on a 30S A-site was undertaken with software (RiboDock). This software sorted the compounds that had good absorption and distribution, had favorable interaction energies, and a balance between polar and nonpolar contributions. A screening of nearly 2000 compounds was conducted in a binding assay which was able to detect dislocation of a fluorescent paromomycin derivative from an oligonucleotide model of the A-site. To confirm the binding, further analysis was undertaken by nuclear magnetic resonance.

Elsewhere, a team of scientists at Abbott Laboratories (Illinois, United States) also identified novel protein synthesis inhibitors using a high-throughput cell free translation screen (Dandliker et al., 2003). A-72310 and A-692345 were the initial compounds discovered from fluoronaphthyridone, which form an intermediate developed in the program that targeted bacterial type II topoisomerases. However, it was seen that the resultant carboxylic acid which is a fluoroquinolone invariant, is not important for translational inhibitory activity, which was required for permeation into the cells (Clark et al., 2004). In view of these compounds, a class of inhibiting translation and binding to the 30S subunit, resistance to compound class members was bestowed by point mutations in 16S rRNA (Dandliker et al., 2003). Further, protecting the important decoding site residues A1408, A1413, A1492, and A1493 was only altered in footprinting experiments when specific chemical modifications were made in the presence of $tRNA^{Phe}$ (yeast) and poly(U) (Shen et al., 2005).

(B) *Inhibitors of 50S*

Numerous classes of antibiotics were identified which specifically target the 50S ribosomal subunit and these mainly interact with the active hydrophobic site in the peptidyl transferase region (e.g. chloramphenicol, puromycin, linezolid, anisomycin, clindamycin, streptogramin A, and pleuromutilins) (Moore and Steitz, 2005) or with a secondary important hydrophobic cleft near the exit tunnel to a nascent polypeptide (e.g., macrolides, ketolides, streptogramin B, and chloramphenicol) (Berisio et al., 2003). Another antibiotic, sparsomycin, contacts at the substrate of the P-site-bound with its pseudouracil substituent, but its linear part interacts with the active-site hydrophobic sites formed by A2451 and C2452 (Hansen et al., 2003). Blasticidin S, which is a well known peptidyl nucleoside antibiotic with a cytosine linked to a pyranose ring and has an amino acid-like appendage, is a partial mimetic of P-site-bound tRNA (Hansen et al., 2003). Furthermore, to identify how growing peptides are positioned in the tunnel, scientists in a Moscow laboratory generated peptide derivatives (aminooxyacetyl-L-ala-nine-L-alanine methyl ester) of tylosin and desmycosin that were modified at the C6 position of the macrocyclic lactone ring (Sumbatyan et al., 2003). Another important antibiotic, streptogramin, is a cyclic depsipeptide and contains 6–7 amino acids. Interestingly, these antibiotics partially overlap with that of a macrolide binding site which is near the entrance to the peptide exit tunnel in the 50S subunit (Mukhtar and Wright, 2005). In contrast, the cyclic decapeptide antibiotic tyrocidine works by damaging the cellular membranes (Mukhtar et al., 2005). Mukhtar et al. (2005) used a chemoenzymatic approach to generate a compound that is not a labile substrate for the lyase Vgb, making a streptogramin-tyrocidine hybrid in which the ester bond was replaced by an amide. Many of the compounds were made by a chemoenzymatic route or by nonenzymatic methods. Only one of the compounds was found to have weak antibacterial activity against gram-positive bacteria and was confirmed to be resistant to Vgb inactivation. To date, there are no biological or crystallographic data which support that these analogs inhibit cell-free transcription or translation.

Eukaryotic ribosome biogenesis and inhibitors

Ribosome biosynthesis

Ribosome biogenesis in eukaryotes is highly regulated, and any alteration in ribosome formation may lead to severe diseases such as cancer and various other diseases. The ribosome biosynthesis in eukaryotes begins in the nucleolus with pre-rRNA transcription and the initial assembly steps continue in the nucleoplasm and finishes in the cytoplasm. This entire process finishes in a few minutes and is highly dynamic. Some small molecular weight chemical compounds have been identified as promising tools that specifically obstruct either rDNA transcription or distinct steps during the maturation pathway. The benefit of these inhibitors is that they rapidly diffuse into the cell and block their target proteins within a fraction of a second, and hence they are considered excellent tools to investigate ribosome biogenesis in detail.

The ribonucleoprotein complex consists of two subunits (40S and 60S for eukaryotic cytosolic ribosomes) in which the large subunit contains the peptidyl transferase activity and the small subunit harbors the decoding center. Note that each ribosomal subunit comprises of ribosomal RNA (rRNA) core that is covered around by ribosomal proteins (RPs). The ribosome biosynthesis machinery in eukaryotes is massive. This RP assembly with pre-rRNAs and their conversion to mature rRNA takes place by the maturation step that originates in the nucleolus and is finished in the cytoplasm (Baßler and Hurt, 2019; Kofler et al., 2020).

The fact that in all living cells, the normal physiological functioning of ribosomes constitute one of the major activities can be understood by data showing that up to 60% of cellular transcriptions are used for rRNA synthesis (Warner, 1999). The process of ribosome biogenesis requires three RNA polymerases, with polymerase (Pol) I and Pol III primarily dedicated to pre-rRNA synthesis and up to 50% of Pol II activity is needed for the production of mRNAs for r-proteins. Ribosome formation is strongly regulated according to cellular growth and proliferation, which is in turn mediated by various regulatory pathways, including the target of rapamycin complex 1 (mTORC1) (Loewith and Hall, 2011; de la Cruz et al., 2018), MYC, and p53, as detailed in Fig. 10.1. Any hindrance in this interwoven regulatory network may lead to cancer development (Penzo et al., 2019) or neurodegenerative diseases (Mills and Green, 2017).

The incorporation of the individual components during ribosome biosynthesis is well organized and coordinated through multiple proteins that serve as assembly factors (referred to as auxiliary or maturation factors), which includes large number of enzymes, such as methyl-transferases, RNA helicases, ATPases, and GTPases (Sloan et al., 2017). Also, for certain rRNA processing, folding, and modification events (pseudouridinylation or methylation), various snoRNPs are required (Watkins and Bohnsack, 2012). The majority of ribosomal assembly factors and r-proteins are associated with the emerging subunits in the nucleus, which need to be efficiently imported from the cytoplasm that is facilitated by chaperones and importins (Pillet et al., 2017).

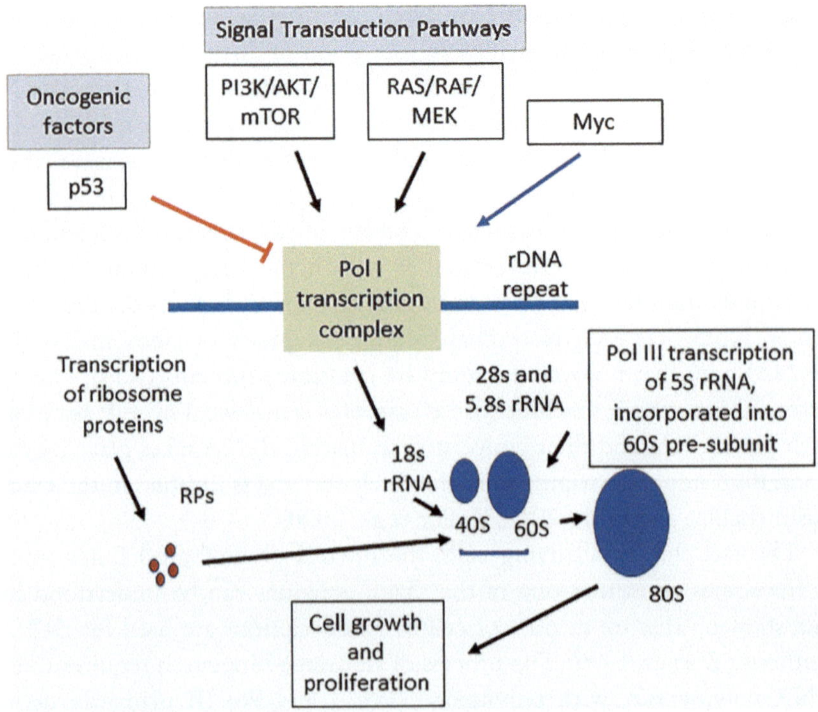

Fig. 10.1 Interwoven regulatory network of ribosome biogenesis.

Ribosome biogenesis inhibitors—Chemical probes

There are many ribosome biogenesis associated factors that are not assigned to distinct maturation steps. The complete elucidation of the steps of the binding process is not very well established. This is mainly because the turn-around time of the individual steps is very short. For example, in the case of a single yeast cell, in a single minute it has to synthesize more than 2000 ribosomes to manage the high demands during exponential growth (de la Cruz et al., 2018). It is a highly vigorous process involving synthesis of the 25S rRNA from the 35S pre-rRNA which takes place within 6 min (Kos and Tollervey, 2010). Therefore, during ribosome assembly, it becomes very difficult to view the dynamic changes with the present available conventional genetics and biochemical methods such as purification of precursor particles from genetically manipulated cells; the onset of genetic disturbances are too slow. The interdependency between the factors is so high that the combination of the fast execution with individual steps makes it difficult to monitor the primary effects.

There are several inhibitors even for the first step of ribosome biogenesis, rDNA transcription, and interestingly, all of which have been detailed in recent years (summarized in Table 10.1, along with their clinical usage for various cancers and disease). The common inhibitors which are recognized include BMH-21, CX-3543, and CX5461 which are known to inhibit polymerase I transcription. The BMH-21 for example, is known to intercalate into GC-rich DNA (Peltonen et al., 2010). CX-3543 (quarfloxin)

Table 10.1 Ribosome biogenesis inhibitors in clinical use for various cancers and diseases.

Drug name	Mechanism of action	Clinical usage
Melphalan	Inhibition of pre-47srRNA	Multiple myeloma, ovarian cancer, melanoma amyloidosis
Cisplatin	Inhibition of RNA Pol I	Testicular cancer, breast cancer, cervical cancer, head and neck cancer, bladder cancer, colorectal cancer
Cyclophosphamide	Inhibition of pre-rRNA and transcription	Nephrotic syndrome
Doxorubicin	Intercalating agent and also inhibits rRNA processing	Breast cancer, bladder cancer, lymphoma, acute lymphocytic leukemia
Mitoxantrone	Inhibition of RNA Pol I	Acute lymphocytic leukemia, prostate cancer, breast cancer
Mitomycin C	Inhibition of rRNA transcription	Gastrointestinal cancer, bladder cancer
Methotrexate	Antimetabolite and inhibition of pre-rRNA	Lymphoma, leukemia, antirheumatic, psoriasis
5-Fluorouracil	Inhibits rRNA processing	Solid tumors
Camptothecin	Topoisomerase inhibitor	Leukemia
Flavopiridol	Kinases inhibitor and inhibits rRNA processing	Acute myeloid leukemia
Roscovitine	Cyclin-dependent kinase inhibitor	Solid tumors (experimental drug)
Bortezomib	Inhibits rRNA transcription	Multiple myeloma
Vinblastine	Mitosis inhibitors and rRNA processing	Hodgkins lymphoma, nonsmall lung cancer, brain cancer, brain cancer, melanoma, testicular cancer

Continued

Table 10.1 Ribosome biogenesis inhibitors in clinical use for various cancers and diseases—cont'd

Drug name	Mechanism of action	Clinical usage
CX5461	Inhibition of RNA Pol I	Hematological cancer, neuroblastoma, solid tumors (clinical trial phase II)
BMH21	Inhibition of RNA Pol I elongation activity	Ovarian cancer (clinical phase II)
BMH9, 22, 23	Inhibition of RNA Pol I and rRNA synthesis	Hematological malignancies (preclinical)
CX3543	Stabilizes p53 and induces apoptosis	Neuroendocrine and carcinoid tumors

inhibits transcription elongation by blocking nucleolin 1 interaction with rDNA (Drygin et al., 2009). CX-5461 prevents the assembly of the transcription factor SL1 to the promotors of rDNA. It has been well described that G-quadruplex DNA structures are stabilized by binding of both CX-5461 and CX-3543 and results in an induction of the DNA damage response (Xu et al., 2017). Interestingly, CX-5461 is known for preferential toxicity to tumor cells in comparison to primary cells, which in turn induces apoptosis and senescence via both independent and p53-dependent mechanisms (Turi et al., 2019). Consequently, for chemotherapeutics, polymerase I inhibitors are more promising chemical compounds that can target cancer cell growth, especially those that are not responding to other treatments. Such inhibitors will pave the way as a major step forward in the clinical treatment of different types of malignancies.

In eukaryotes, out of all the identified small molecular weight compounds that target ribosome maturation, diazaborine is considered to be another promising inhibitor (Pertschy et al., 2004). Diazaborine specifically targets to 60S subunit formation through targeting the cytoplasmic AAA-ATPase Drg1 that contains an N-domain and two ATPase domains (Loibl et al., 2014). Diazaborine masks the hydrolysis process which is required for the rapid release of the assembly factor Rlp24 from pre-60S particles immediately after the nuclear export (Kappel et al., 2012).

However, diazaborine not only prevents the recycling and release of Rlp24, but also of all other assembly factors that carry the pre-60S particle inside the cytoplasm. In turn, these shuttling factors remain confined on the particles that are supposed to be exported and hence, they are unable to re-enter inside the nucleus, which finally leads to blockage of nucleolar maturation steps in the nucleus. The intriguing biological effects of diazaborine on assembly factors from the early nucleolar to late cytoplasmic stages in an organized manner provide a better understanding of the dynamics ribosome biosynthesis.

Among other identified classes of eukaryotic inhibitors of ribosome biogenesis are the ribozinoindoles (Rbins) that inhibit the dynein-like AAA-ATPase Mdn1 in *Schizosaccharomyces pombe* (Kawashima et al., 2016; Galani et al., 2004; Ulbrich et al., 2009). Rea1 is another multifunctional ATPase that is involved in releasing the nucleoplasm factor Ytm1 and Rsa4, most likely in conjunction with Erb1 (Baßler and Hurt, 2019).

Whereas, in accordance with recent studies, there are almost 100 new ribosome biogenesis inhibitors have been reported that blocks the factors at different levels of maturation steps (Awad et al., 2019). Presently, the exact specific targets of these inhibitors are still difficult to track down, but they provide a futuristic tool to delineate the eukaryotic ribosome biogenesis pathway. In the next section we will discuss the advantages of small molecular weight inhibitors which are studied to understand the complex pathway of ribosome synthesis and how they will aid us to determine the dynamics of the assembly process.

Examples of inhibitors

There are few examples of such inhibitors which will be discussed in this section, and we will see how this understanding of molecular mechanism will unravel the dynamics of the ribosome assembly pathway.

Rbin-1

The AAA-ATPase Rea1 from *Saccharomyces cerevisiae* like its ortholog Mdn1 from *Schizosaccharomyces pombe,* is the cause for the release of the WD repeat protein Rsa4 from late nucleoplasm pre-60S particles (Kofler et al., 2020). Such a reaction is required for Rsa4 recycling for successive rounds of ribosome synthesis (Prattes et al., 2019). One of the important structural transitions of pre-60S protein is Rea1 which is present in the nucleoplasm. In turn, Rea1 is mainly involved in the release of the protein Ytm1, which is speculated to be coupled to the extraction of Erb1 (Thoms et al., 2016). Erb1 protrudes into the 60S precursor particle and also acts as the prime protein for various maturation factors (Kofler et al., 2020), which marks Rea1 an important and specific inhibitor.

Rbin-1 has been very useful and also has its derivatives in identifying the mechanistic and structural insights into the catalytic cycle of Mdn1

(Kawashima et al., 2016) and has proved its function in ribosome biogenesis in the case of yeast. Inhibition by both Mdn1 and Rbin-1 results in the accumulation of both 27S and 7S pre-rRNAs that completely blocks the movement of the nucleolar particle in the nucleoplasm (Kawashima et al., 2016). Also, in Rea1 (Bassler et al., 2010), the substrate-binding MIDAS domain of Mdn1 interacts with the MIDO-domain of Rsa4 and, although the binding is weak with that of Ytm1, in order to proceed to a conserved mechanism of release for both WD repeat proteins (Kofler et al., 2020). This blockage stops the biosynthesized particles from entering into the downstream maturation step and result in a decrease of Rsa4 and Ytm1 on nucleoplasm Rix1 particles. However, we need further studies with shorter durations of treatment and kinetic analysis of the drug effects to elucidate this pathway in detail (Kofler et al., 2020).

Diazaborine

To date, one of the best studied inhibitors of ribosome biogenesis is diazaborine. The dynamic effects of inhibition by diazaborine have been investigated from early nucleolar and late cytoplasmic stages using different biological techniques currently available (Loibl et al., 2014). This study depict that treatment with diazaborine blocks recycling of shuttle proteins. However, the primary role of shuttling protein and its depletion on nucleolar maturation requires further investigation for early and late nucleolar particles using Noc2 and Nsa1 as bait proteins. These results provide more insight into the type of particles accumulating upon shuttle protein depletion and how this depletion is transmitted to the earlier stages (Kofler et al., 2020; Loibl et al., 2014).

Ribosome biogenesis inhibitors and cancer

There are many drugs that are used for treating cancer, such as topoisomerase inhibitors, DNA reactive agents, and antimetabolites, that exert their toxic effects by DNA damage or interfering in DNA synthesis (Brighenti et al., 2015). The main idea behind such chemotherapeutic approach is that the cellular function and proliferation and the DNA integrity and duplication are imperative. In proliferating cells, DNA damage is noticed by cell-cycle checkpoint factors that blocks the progression of the cell cycle (Zhou and Elledge, 2000; David-Pfeuty, 2006). The DNA repair mechanism is an essential process for preventing apoptotic cell death. The dividing tumor cells usually function poorly or not at all (Sherr, 1996,

2000) but DNA damage may more often induce apoptosis-associated cell death (Waldman et al., 1997). This is why chemotherapeutic agents are considered to be more effective against cancer cells compared to normal dividing cells. However, these types of drugs frequently inhibit ribosome biogenesis apart from their action on DNA (Burger et al., 2010). Hence, this reduces their specificity for cancer cell elimination. The synthesis of rRNA takes place in both dividing and resting cells (unlike DNA synthesis), the former constitutes a large portion of normal tissues. A seminal study indicates that in certain cases, a specific, nongenotoxic inhibition of rRNA transcription may lead to selective damage to neoplastic cells (Quin et al., 2014; Hein et al., 2013) (Fig. 10.2). The results are associated with the perturbations between ribosome biogenesis and cell proliferation pathway interactions, as well as with the changes in the mechanisms controlling the rate of ribosome biogenesis in cancer cells. This may well explain the specific cytotoxicity of ribosome biogenesis inhibitors against cancer cells (Montanaro et al., 2007). These features could be of significance for the selection of suitable anticancer therapy and for the development of specific rRNA inhibitors.

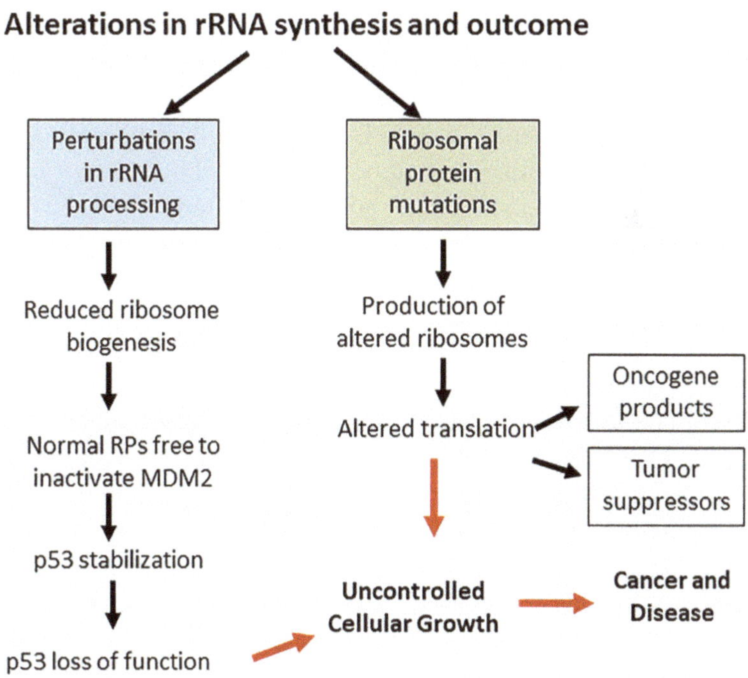

Fig. 10.2 Alterations in ribosome biogenesis and development of cancer.

Ribosome biogenesis has a highly variable turnaround in cancer cells

It is commonly believed that a higher rate of ribosome biogenesis occurs in cancer cells than in normal cells. The high rate specifically in cancer cells could be because neoplastic transformation is frequently characterized by changes tumor suppressor genes and proto-oncogenes (Sherr, 2000) which activates cell division and also triggers the cascade to intensify ribosome biogenesis. In the extracellular signal, specific kinases (MAPK/ERK) activate both Pol I transcription by phosphorylating UBF (Stefanovsky et al., 2001, 2006). In Pol III transcription, TFIIIB is phosphorylated, and ERK further phosphorylates the transcription initiation factor TIF-IA which rearranges itself with the initiation-competent Pol I on the rDNA promoter. Growth factors and mitogenic factors also activate the kinases pathways, which activates MYC, one of the major primary ribosome biogenesis modulators (Zhu et al., 2008).

In turn, the MYC enhances Pol I activity by advancing the recruitment of SL1 to promoters, which stimulates ribosomal protein synthesis by elevating Pol II transcription, and coordinates Pol III transcription by activating TFIIIB (Gomez-Roman et al., 2006). Therefore, the mitogenic growth factor stimulation, which targets the rapamycin (mTOR pathway), also activates both Pol I transcription and Pol III transcription by providing the association of the transcription factors TFIIIB and TFIIIC with 5S rRNA genes (van Riggelen et al., 2010). Furthermore, there is evidence to suggest that tumor suppressor genes products have a detrimental effect on cell proliferation and cell cycle progression, which negatively regulates ribosome biogenesis. As a result, p53 suppresses both Pol I transcription by binding to the selectivity factor SL1—which is required for Pol I recruitment to the rRNA gene promoter (Mayer et al., 2004)—and Pol III transcription. To control the ribosome biogenesis, p53 is activated and this tumor suppressor inhibits the ribosome biogenesis both by inhibiting the UBF recruitment on the transcription complex (Zhai and Comai, 2000), and by suppressing the activity of nucleophosmin (Ayrault et al., 2006). Another important tumor suppressor, pRb, is also known to inhibit ribosome biogenesis. Additionally, PTEN (phosphatase and tensin homolog deleted in chromosome 10) another important tumor suppressor, inhibits the Pol I transcription and activates various signaling events that inhibit cell proliferation and disrupt the SL1 complex (Ayrault et al., 2006). Hence, it demonstrates that cancer cells have a high activity of ribosome biogenesis and it has been observed that many of the genetic alterations occur in cancer cells which upregulate ribosome biogenesis. Many studies have revealed that these changes intensify the variations in human cancers, giving rise to tumors characterized by variability in the turnover rate of ribosome biogenesis, sometimes quite similar to that of normal cells. It is very well known that there is high variability of the nucleolar size in cancer cells (Zhang et al., 2005). Histopathologically, it clarifies the fact that the high variability in the rate of ribosome biogenesis is reflected in differently-sized nucleoli within tumor cells, as the size of the nucleolus is directly proportional to the turnover rate of ribosome biogenesis (Derenzini et al., 2000).

For a long time, the effect of various chemotherapeutic drugs based upon ribosome biogenesis inhibitors has been underestimated. Currently, many targeted drugs used for cancer treatment are associated with either the inhibition of rRNA synthesis or its maturation. In cancer chemotherapy, the introduction of CX-5461, a molecule which selectively inhibits Pol I activity without causing any genotoxicity, may certainly direct more studies aiming to identify and/or produce specific chemical compounds that have these characteristics to target cancer cells. To highlight the recent findings obtained using a small molecular compound, BMH-21, and a 22mer peptide has been shown to stabilize p53 by rDNA transcription inhibition. BMH-21 binds to GC-rich sequences which finally inhibit RNA Pol I activity (Peltonen et al., 2010). BMH-9, BMH-22, and BMH-23 are three additional small molecule compounds that degrade the major catalytic subunit of the Pol I complex through proteasome-dependent degradation (Peltonen et al., 2010). The 22mer peptide interferes with RNA polymerase I and Rrn3 which inhibits rRNA synthesis (Stefanovsky et al., 2001). However, the development of similar compounds appears to be of upmost importance, in accordance with recent findings that cancer cells may acquire metabolic and genetic changes that provides more sensitivity towards the inhibition of rRNA synthesis in normal cells. Also, further validation and dissection of such compounds should be carried out and will be helpful to distinguish between cancers that might significantly benefit from ribosome biogenesis inhibitor therapy, especially for cancers that have a poor chemotherapeutic impact.

Conclusions

Ribosome biogenesis is a dynamic and complex process in a cell with a tight regulatory network. As there is high interlink between ribosome assembly and the cell cycle, the chemical inhibition of this pathway provides the potential to target cancer cells due to a higher demand and supply of freshly-synthesized ribosomes. The slowly dividing cells such as normal cells have lower demands of supply because of the stability of ribosomes, and hence they become less sensitive to inhibition of ribosome biogenesis. Ribosome biogenesis inhibitors appear to be a promising strategy to inhibit the ribosome synthesis in malignant cells. As future anticancer drugs, the ribosome biogenesis inhibitors that are currently in clinical use also have the potential to develop the ribosome biogenesis research field further. The existing extensive research in ribosome biogenesis has provided the basic principles, and a lineation of the chronological order of maturation factor, however the coordination of individual steps still needs further exploration. While the advances in technologies such as cryo-EM analysis have proven the impressive movements and rearrangements during cell maturation, they only provide static snapshots of distinct stages. However, based upon the currently available inhibitors used for research, chemical probing can also be used to investigate the dynamics of more complex ribosome biogenesis pathways even in higher eukaryotes. The vivid information already derived from this small number of specific inhibitors promises to expand our knowledge by applying a much broader tool to study further. The characterization of the new ribosome biogenesis inhibitors is one of the most promising features in translational research and will certainly pave the way to a more detailed understanding of the dynamic nature of this robust and fascinating pathway.

References

Awad D, Prattes M, Kofler L, Rössler I, Loibl M, Pertl M, Zisser G, Wolinski H, Pertschy B, Bergler H. Inhibiting eukaryotic ribosome biogenesis. BMC Biol 2019;17:46.

Ayrault O, Andrique L, Fauvin D, Eymin B, Gazzeri S, Séité P. Human tumor suppressor p14ARF negatively regulates rRNA transcription and inhibits UBF1 transcription factor phosphorylation. Oncogene 2006;25:7577–86.

Baba T, Ara T, Hasegawa M, Takai Y, Okumura Y, Baba M, Datsenko KA, Tomita M, Wanner BL, Mori H. Construction of Escherichia coli K-12 in-frame, single-gene knockout mutants: the Keio collection. Mol Syst Biol 2006;2: 2006.0008.

Baßler J, Hurt E. Eukaryotic ribosome assembly. Annu Rev Biochem 2019;88:281–306.

Bassler J, Kallas M, Pertschy B, Ulbrich C, Thoms M, Hurt E. The AAA-ATPase Rea1 drives removal of biogenesis factors during multiple stages of 60S ribosome assembly. Mol Cell 2010;38:712–21.

Berisio R, Harms J, Schluenzen F, Zarivach R, Hansen HA, Fucini P, Yonath A. Structural insight into the antibiotic action of telithromycin against resistant mutants. J Bacteriol 2003;185:4276–9.

Brighenti E, Treré D, Derenzini M. Targeted cancer therapy with ribosome biogenesis inhibitors: a real possibility? Oncotarget 2015;6(36):38617–27.

Bunner AE, Nord S, Wikström PM, Williamson JR. The effect of ribosome assembly cofactors on in vitro 30S subunit reconstitution. J Mol Biol 2010;398:1–7.

Burger K, Mühl B, Harasim T, Rohrmoser M, Malamoussi A, Orban M, Kellner M, Gruber-Eber A, Kremmer E, Hölzel M, Eick D. Chemotherapeutic drugs inhibit ribosome biogenesis at various levels. J Biol Chem 2010;285:12416–25.

Clark RF, Wang S, Ma Z, Weitzberg M, Motter C, Tufano M, Wagner R, Gu YG, Dandliker PJ, Lerner CG, Chovan LE, Cai Y, Black-Schaefer CL, Lynch L, Kalvin D, Nilius AM, Pratt SD, Soni N, Zhang T, Zhang X, Beutel BA. Novel inhibitors of bacterial protein synthesis: structure-activity relationships for 1,8-naphthyridine derivatives incorporating position 3 and 4 variants. Bioorg Med Chem Lett 2004;14:3299–302.

Comartin DJ, Brown ED. Non-ribosomal factors in ribosome subunit assembly are emerging targets for new antibacterial drugs. Curr Opin Pharmacol 2006;6:453–8.

Dammel CS, Noller HF. Suppression of a cold-sensitive mutation in 16S rRNA by overexpression of a novel ribosome-binding factor, RbfA. Genes Dev 1995;9:626–37.

Dandliker PJ, Pratt SD, Nilius AM, Black-Schaefer C, Ruan X, Towne DL, Clark RF, Englund EE, Wagner R, Weitzberg M, Chovan LE, Hickman RK, Daly MM, Kakavas S, Zhong P, Cao Z, David CA, Xuei X, Lerner CG, Soni NB, Bui M, Shen LL, Cai Y, Merta PJ, Saiki AY, Beutel BA. Novel antibacterial class. Antimicrob Agents Chemother 2003;47:3831–9.

David-Pfeuty T. The flexible evolutionary anchorage-dependent Pardee's restriction point of mammalian cells: how its deregulation may lead to cancer. Biochim Biophys Acta 2006;1765:38–66.

de la Cruz J, Gómez-Herreros F, Rodríguez-Galán O, Begley V, de la Cruz M-CM, Chávez S. Feedback regulation of ribosome assembly. Curr Genet 2018;64:393–404.

Derenzini M, Trerè D, Pession A, Govoni M, Sirri V, Chieco P. Nucleolar size indicates the rapidity of cell proliferation in cancer tissues. J Pathol 2000;191:181–6.

Drygin D, Siddiqui-Jain A, O'Brien S, Schwaebe M, Lin A, Bliesath J, Ho CB, Proffitt C, Trent K, Whitten JP, Lim JK, Von Hoff D, Anderes K, Rice WG. Anticancer activity of CX-3543: a direct inhibitor of rRNA biogenesis. Cancer Res 2009;69:7653–61.

Foloppe N, Chen IJ, Davis B, Hold A, Morley D, Howes R. A structure-based strategy to identify new molecular scaffolds targeting the bacterial ribosomal A-site. Bioorg Med Chem 2004;12:935–47.

Gaal T, Bartlett MS, Ross W, Turnbough Jr CL, Gourse RL. Transcription regulation by initiating NTP concentration: rRNA synthesis in bacteria. Science 1997;278:2092–7.

Galani K, Nissan TA, Petfalski E, Tollervey D, Hurt E. Rea1, a dynein-related nuclear AAA-ATPase, is involved in late rRNA processing and nuclear export of 60 S subunits. J Biol Chem 2004;279:55411–8.

Gomez-Roman N, Felton-Edkins ZA, Kenneth NS, Goodfellow SJ, Athineos D, Zhang J, Ramsbottom BA, Innes F, Kantidakis T, Kerr ER, Brodie J, Grandori C, White RJ. Activation by c-Myc of transcription by RNA polymerases I, II and III. Biochem Soc Symp 2006;73:141–54.

Hansen JL, Moore PB, Steitz TA. Structures of five antibiotics bound at the peptidyl transferase center of the large ribosomal subunit. J Mol Biol 2003;330:1061–75.

Hein N, Hannan KM, George AJ, Sanij E, Hannan RD. The nucleolus: an emerging target for cancer therapy. Trends Mol Med 2013;19:643–54.

Holmes KL, Culver GM. Analysis of conformational changes in 16 S rRNA during the course of 30 S subunit assembly. J Mol Biol 2005;354:340–57.

Jomaa A, Jain N, Davis JH, Williamson JR, Britton RA, Ortega J. Functional domains of the 50S subunit mature late in the assembly process. Nucleic Acids Res 2014;42:3419–35.

Kappel L, Loibl M, Zisser G, Klein I, Fruhmann G, Gruber C, Unterweger S, Rechberger G, Pertschy B, Bergler H. Rlp24 activates the AAA-ATPase Drg1 to initiate cytoplasmic pre-60S maturation. J Cell Biol 2012;199:771–82.

Kawashima SA, Chen Z, Aoi Y, Patgiri A, Kobayashi Y, Nurse P, Kapoor TM. Potent, reversible, and specific chemical inhibitors of eukaryotic ribosome biogenesis. Cell 2016;167:512–24.

Kofler L, Prattes M, Bergler H. From snapshots to flipbook-resolving the dynamics of ribosome biogenesis with chemical probes. Int J Mol Sci 2020;21:2998.

Kos M, Tollervey D. Yeast pre-rRNA processing and modification occur cotranscriptionally. Mol Cell 2010;37:809–20.

Lerner CG, Inouye M. Pleiotropic changes resulting from depletion of Era, an essential GTP-binding protein in Escherichia coli. Mol Microbiol 1991;5:951–7.

Loewith R, Hall MN. Target of rapamycin (TOR) in nutrient signaling and growth control. Genetics 2011;189:1177–201.

Loibl M, Klein I, Prattes M, Schmidt C, Kappel L, Zisser G, Gungl A, Krieger E, Pertschy B, Bergler H. The drug diazaborine blocks ribosome biogenesis by inhibiting the AAA-ATPase Drg1. J Biol Chem 2014;289:3913–22.

Maguire BA. Inhibition of bacterial ribosome assembly: a suitable drug target? Microbiol Mol Biol Rev 2009;73:22–35.

Mayer C, Zhao J, Yuan X, Grummt I. mTOR-dependent activation of the transcription factor TIF-IA links rRNA synthesis to nutrient availability. Genes Dev 2004;18:423–34.

Mills EW, Green R. Ribosomopathies: there's strength in numbers. Science 2017;358, eaan2755.

Montanaro L, Mazzini G, Barbieri S, Vici M, Nardi-Pantoli A, Govoni M, Donati G, Treré D, Derenzini M. Different effects of ribosome biogenesis inhibition on cell proliferation in retinoblastoma protein- and p53-deficient and proficient human osteosarcoma cell lines. Cell Prolif 2007;40:532–49.

Moore PB, Steitz TA. The ribosome revealed. Trends Biochem Sci 2005;30:281–3.

Mukhtar TA, Wright GD. Streptogramins, oxazolidinones, and other inhibitors of bacterial protein synthesis. Chem Rev 2005;105:529–42.

Mukhtar TA, Koteva KP, Wright GD. Chimeric streptogramin-tyrocidine antibiotics that overcome streptogramin resistance. Chem Biol 2005;12:229–35.

Peltonen K, Colis L, Liu H, Jäämaa S, Moore HM, Enbäck J, Laakkonen P, Vaahtokari A, Jones RJ, af Hällström TM, Laiho M. Identification of novel p53 pathway activating small-molecule compounds reveals unexpected similarities with known therapeutic agents. PLoS One 2010;5, e12996.

Penzo M, Montanaro L, Treré D, Derenzini M. The ribosome biogenesis-cancer connection. Cell 2019;8:55.

Pertschy B, Zisser G, Schein H, Köffel R, Rauch G, Grillitsch K, Morgenstern C, Durchschlag M, Högenauer G, Bergler H. Diazaborine treatment of yeast cells inhibits maturation of the 60S ribosomal subunit. Mol Cell Biol 2004;24:6476–87.

Pillet B, Mitterer V, Kressler D, Pertschy B. Hold on to your friends: dedicated chaperones of ribosomal proteins: dedicated chaperones mediate the safe transfer of ribosomal proteins to their site of pre-ribosome incorporation. Bioessays 2017;39:1–12.

Prattes M, Lo YH, Bergler H, Stanley RE. Shaping the nascent ribosome: AAA-ATPases in eukaryotic ribosome biogenesis. Biomolecules 2019;9:715.

Quin JE, Devlin JR, Cameron D, Hannan KM, Pearson RB, Hannan RD. Targeting the nucleolus for cancer intervention. Biochim Biophys Acta 2014;1842:802–16.

Shen LL, Black-Schaefer C, Cai Y, Dandliker PJ, Beutel BA. Mechanism of action of a novel series of naphthyridine-type ribosome inhibitors: enhancement of tRNA footprinting at the decoding site of 16S rRNA. Antimicrob Agents Chemother 2005;49:1890–7.

Sherr CJ. Cancer cell cycles. Science 1996;274:1672–7.

Sherr CJ. The Pezcoller lecture: cancer cell cycles revisited. Cancer Res 2000;60:3689–95.

Siibak T, Peil L, Xiong L, Mankin A, Remme J, Tenson T. Erythromycin- and chloramphenicol-induced ribosomal assembly defects are secondary effects of protein synthesis inhibition. Antimicrob Agents Chemother 2009;53:563–71.

Sloan KE, Warda AS, Sharma S, Entian KD, Lafontaine DLJ, Bohnsack MT. Tuning the ribosome: the influence of rRNA modification on eukaryotic ribosome biogenesis and function. RNA Biol 2017;14:1138–52.

Stefanovsky VY, Pelletier G, Hannan R, Gagnon-Kugler T, Rothblum LI, Moss T. An immediate response of ribosomal transcription to growth factor stimulation in mammals is mediated by ERK phosphorylation of UBF. Mol Cell 2001;8:1063–73.

Stefanovsky V, Langlois F, Gagnon-Kugler T, Rothblum LI, Moss T. Growth factor signaling regulates elongation of RNA polymerase I transcription in mammals via UBF phosphorylation and r-chromatin remodeling. Mol Cell 2006;21:629–39.
Stokes JM, Davis JH, Mangat CS, Williamson JR, Brown ED. Discovery of a small molecule that inhibits bacterial ribosome biogenesis. Elife 2014;3, e03574.
Strunk BS, Loucks CR, Su M, Vashisth H, Cheng S, Schilling J, Brooks 3rd CL, Karbstein K, Skiniotis G. Ribosome assembly factors prevent premature translation initiation by 40S assembly intermediates. Science 2011;333:1449–53.
Strunk BS, Novak MN, Young CL, Karbstein K. A translation-like cycle is a quality control checkpoint for maturing 40S ribosome subunits. Cell 2012;150:111–21.
Sumbatyan NV, Korshunova GA, Bogdanov AA. Peptide derivatives of antibiotics tylosin and desmycosin, protein synthesis inhibitors. Biochemistry (Mosc) 2003;68:1156–8.
Sykes MT, Shajani Z, Sperling E, Beck AH, Williamson JR. Quantitative proteomic analysis of ribosome assembly and turnover in vivo. J Mol Biol 2010;403:331–45.
Tatusov RL, Koonin EV, Lipman DJ. A genomic perspective on protein families. Science 1997;278:631–7.
Thoms M, Ahmed YL, Maddi K, Hurt E, Sinning I. Concerted removal of the Erb1-Ytm1 complex in ribosome biogenesis relies on an elaborate interface. Nucleic Acids Res 2016;44:926–39.
Turi Z, Lacey M, Mistrik M, Moudry P. Impaired ribosome biogenesis: mechanisms and relevance to cancer and aging. Aging (Albany NY) 2019;11:2512–40.
Ulbrich C, Diepholz M, Bassler J, Kressler D, Pertschy B, Galani K, Böttcher B, Hurt E. Mechanochemical removal of ribosome biogenesis factors from nascent 60S ribosomal subunits. Cell 2009;138:911–22.
van Riggelen J, Yetil A, Felsher DW. MYC as a regulator of ribosome biogenesis and protein synthesis. Nat Rev Cancer 2010;10:301–9.
Vicens Q, Westhof E. RNA as a drug target: the case of aminoglycosides. ChemBioChem 2003;4:1018–23.
Waldman T, Zhang Y, Dillehay L, Yu J, Kinzler K, Vogelstein B, Williams J. Cell-cycle arrest versus cell death in cancer therapy. Nat Med 1997;3:1034–6.
Warner JR. The economics of ribosome biosynthesis in yeast. Trends Biochem Sci 1999;24:437–40.
Watkins NJ, Bohnsack MT. The box C/D and H/ACA snoRNPs: key players in the modification, processing and the dynamic folding of ribosomal RNA. Wiley Interdiscip Rev RNA 2012;3:397–414.
Xu H, Di Antonio M, McKinney S, Mathew V, Ho B, O'Neil NJ, Santos ND, Silvester J, Wei V, Garcia J, Kabeer F, Lai D, Soriano P, Banáth J, Chiu DS, Yap D, Le DD YFB, Zhang A, Thu K, Soong J, Lin SC, Tsai AH, Osako T, Algara T, Saunders DN, Wong J, Xian J, Bally MB, Brenton JD, Brown GW, Shah SP, Cescon D, Mak TW, Caldas C, Stirling PC, Hieter P, Balasubramanian S, Aparicio S. CX-5461 is a DNA G-quadruplex stabilizer with selective lethality in BRCA1/2 deficient tumours. Nat Commun 2017;8:14432.
Yamagishi M, Nomura M. Effects of induction of rRNA overproduction on ribosomal protein synthesis and ribosome subunit assembly in Escherichia coli. J Bacteriol 1988;170:5042–50.
Zhai W, Comai L. Repression of RNA polymerase I transcription by the tumor suppressor p53. Mol Cell Biol 2000;20:5930–8.
Zhang C, Comai L, Johnson DL. PTEN represses RNA polymerase I transcription by disrupting the SL1 complex. Mol Cell Biol 2005;25:6899–911.
Zhou BB, Elledge SJ. The DNA damage response: putting checkpoints in perspective. Nature 2000;408:433–9.
Zhu J, Blenis J, Yuan J. Activation of PI3K/Akt and MAPK pathways regulates Myc-mediated transcription by phosphorylating and promoting the degradation of Mad1. Proc Natl Acad Sci U S A 2008;105:6584–9.

CHAPTER 11

Ribosomopathies—A tree of pathologies with many roots and branches!

Sarita and M.G. Sanal
Department of Molecular and Cellular Medicine, Institute of Liver and Biliary Sciences, New Delhi, India

Introduction

Ribosomes are micromachines that are made up of hundreds of parts and are assembled by complex processes. In eukaryotes, these intricate machines translate messenger RNA into proteins have four ribosomal RNAs (18S in the small subunit and 28S, 5.8S, and 5S in the large subunit) and about 80 ribosomal proteins (Peña et al., 2017). More than 200 factors and small RNAs are needed to "assemble" ribosomes in the nucleolus. Some of the structural components also control the processing of precursor ribosomal RNA (pre-rRNA), assembly of the ribosomal subunits, and the translation process itself. It is now well established that not every ribosome has the same protein composition, and it is speculated that certain subpopulations of ribosomes may have unique properties that modify the functions of the proteins they make (Filipovska and Rackham, 2017). Posttranslational modifications of ribosomal proteins as well as posttranscriptional modifications of ribosomal RNA nucleotides could also impact gene expression programs in host cells. Many diseases whose etiology and pathology have remained elusive, are now considered to be ribosome-associated diseases. For example, certain types of Diamond-Blackfan anemia (DBA) are considered to be due to ribosomal defects (Nakhoul et al., 2014). Other major ribosomopathies include North American Indian childhood cirrhosis (a type of progressive familial intrahepatic cholestasis), Treacher-Collins syndrome, Chromosome 5q-syndrome, and isolated congenital asplenia (Chakraborty and Karunasagar, 2015). However, it may be noted that there are problems in defining ribosomal diseases. Perhaps some of these diseases are mere entities "named" by some clinicians depending on phenotypes for the ease of clinical management. In some cases, these disease names can illogical and

Emerging Concepts in Ribosome Structure, Biogenesis, and Function
https://doi.org/10.1016/B978-0-12-816364-1.00008-1

reveals our knowledge gap. There is a lack of consistent correlation between the nature of the mutations and the clinical phenotype. This indicates that yet unidentified factors modulate the phenotypic expression of the primary genetic defect.

Diamond-Blackfan anemia

Diamond-Blackfan anemia (DBA) is an inherited form of childhood anemia characterized by low red blood cell counts. However, other blood cells such as white blood cells and platelet counts are not affected substantially. After cystic fibrosis, DBA is the second most common cause of exocrine pancreatic insufficiency in children. On the other hand, in Fanconi's anemia, all blood cells are affected while in Shwachman-Bodian-Diamond syndrome, mostly neutrophils are affected. The clinical presentation of DBA is quite heterogeneous. So can we call any genetic/epigenetic defect a case of DBA which results in apoptotic death during hematopoiesis or bone marrow failure? The earliest report of what we call DBA was published by Josephs (1936) from Johns Hopkins, United States. However, a more detailed report was published by another team of doctors who provided details of four cases of congenital (erythroid) hypoplastic anemia (Diamond and Blackfan, 1938; Diamond et al., 1961). DBA starts in infancy and requires repeated regular red cell transfusions (Mentzer, 2003). At least, 50% of patients with DBA carry mutations in ribosomal genes. The exome sequencing has identified GATA1 mutations in a proportion of DBA patients with no detectable mutations in their ribosomal genes (Ludwig et al., 2014; Sankaran et al., 2012). GATA1 is a hematopoietic transcription factor that promotes the maturation of red blood cells including the formation of their cytoskeleton and synthesis of hemoglobin. GATA1 defects can cause transient myeloproliferative disorder occurring in Down syndrome and acute megakaryoblastic leukemia. It is important to note that DBA patients also have a higher risk of developing leukemia. GATA1 also plays a critical role in the maturation of blood platelets from megakaryoblasts, promegakaryocytes, and megakaryocytes. The de novo TP53 germline variants can also result in DBA phenotypes (Toki et al., 2018). The erythropoietin (EPO) gene (homozygous missense mutation in exon 5 and a nonribosomal gene) has also been found to be associated with "DBA" in consanguineous Turkish family (Dianzani et al., 1996). Similarly, feline leukemia virus subgroup C receptor 1 (FLVCR1) is a gene coding for a heme exporter. Mice with deficient expression of this gene have some phenotypic resemblance to

DBA patients, including erythroid failure and malformations. In a small proportion of DBA patients, there is a disease linkage to chromosome 1q31, where FLVCR1 gene is mapped (Chiabrando and Tolosano, 2010). The disturbance of tryptophan metabolism was reported in DBA as early as 1953 (Altman and Miller, 1953; Falter and Robinson, 1972). Recent studies have confirmed an abnormal expression of genes involved in amino acid metabolism (Avondo et al., 2009).

Signs and symptoms of DBA: The main feature of DBA is normocytic or macrocytic anemia along with decreased erythroid progenitor cells in the bone marrow which evolves during the neonatal period. The infant presents with pallor and lethargy. The median age at presentation is 8 weeks and the median age of diagnosis is 12 weeks. About 90% of the cases will be presented by the first year of life. The incidence in Europe is around 1/150,000 and there is no sexual or ethnic predisposition. About half of the patients present with low birth weight along with a variety of congenital abnormalities which include cardiac abnormalities, craniofacial malformations, urogenital problems, thumb or upper limb abnormalities, etc. (Ramenghi et al., 1999; Narla and Ebert, 2010).

Triphalangeal thumb is found triphalangeal thumbs in 3% of patients with congenital erythroid hypoplastic anemia (Alter, 1978). Certain facial characteristics often appear in children with DBA. It may consist of two-colored hair (extremely blonde, almost white), a snub nose, wide-set eyes, and a thick upper lip. In addition, the head can be small with almond-shaped eyes and a pointed chin (Costa et al., 2002). DBA patients have a higher risk of developing leukemia/other cancers compared to the general population (Krishnan et al., 1978; Nakhoul et al., 2014; Post et al., 2014). Most of the patients develop anemia in the first year of life, nevertheless, there is a mild phenotype which is often diagnosed later in life.

Genetic heterogeneity of DBA: Approximately 25%–50% of the causes of DBA have been tied to abnormal ribosomal protein genes. The disease is characterized by genetic heterogeneity, affecting different ribosomal gene loci (Table 11.1). DBA was earlier suspected to be an autoimmune disease based on the observation that there is a decrease in the number of erythroid colony-forming units (CFU-E) and was possibly due to a block in erythroid maturation prior to the less mature erythroid burst forming unit (BFU-E) during development (Freedman et al., 1976; Nathan et al., 1978). Congenital mutations in *RPS19* gene encoding ribosomal protein RPS19 were shown for the first time to be associated with DBA—a disorder characterized by hypoplastic, macrocytic anemia (Draptchinskaia et al., 1999;

Table 11.1 Types of Diamond-Blackfan anemia.

Name	Chromosome	Phenotype MIM	Genotype	Protein
DBA1	19q13.2	105650	603474	RPS19
DBA2	8p23-p22	606129		
DBA3	10q22	610629	602412	RPS24
DBA4	15q	612527	180472	RPS17
DBA5	3q29	612528	180468	RPL35A
DBA6	1p22.1	612561	603634	RPL5
DBA7	1p36	612562	604175	RPL11
DBA8	2p25	612563	603658	RPS7
DBA9	6p	613308	603632	RPS10
DBA10	12q	613309	603701	RPS26
DBA11	17p13	614900	603704	RPL26
DBA12	3p24	615550	604174	RPL15
DBA13	14q	615909	603633	RPS29
DBA14	Xp11	300946	300945	TSR2
DBA15	19p13	606164	603685	RPS28
DBA16	17q21	617408	607526	RPL27
DBA17	1q21	617409	603702	RPS27

Miyake et al., 2005). A lower frequency of RPS19 mutations was recently reported in a Japanese cohort of DBA patients (Konno et al., 2010). However, mutations in RPS19 gene account for only in a quarter of the total cases identified as DBA (Gazda et al., 2001, 2006a,b, 2008, 2012; Gustavsson et al., 1997a,b, 1998).

Ribosomal protein-encoding genes, nine of them, are mutated in DBA in 50% of the patients. Approximately in one-fifth of all DBA, there is a family history. In a collaborative study involving six centers, Boria et al. (2010) reported 220 different mutations in nine genes associated with DBA. The DBA Mutation Database now includes data from 355 patients. The majority of these mutations were de novo (125 cases) and the remaining were inherited. RPL5 and RPL11 mutations are more associated with malformations (genotype-phenotype comparison).

The fast-growing technology has made sequencing cheaper and efficient and as a result, the gene discovery because much faster and the number of DBA genes increased at a fast pace adding increasing our understanding of DBA but also the complexity. The overall pattern of DBA genetics may be summarized as follows: uncharacterized > RPS19 > {RPL5, RPL11, RPS26} > {RPL35A, RPS7, RPS10, RPS17, RPS24}, with r-protein gene abnormalities identifiable in about 50%–60% of patients. The failure of the genotype to predict the phenotype makes the disease intriguing. Dramatic examples of the failure of genotype to predict phenotype are observed in family members sharing a known RPS19 mutation. Interestingly, the international consensus recommendations for diagnosis and treatment of DBA suggest consideration of the diagnosis of "nonclassical" DBA in any family member sharing an established DBA r-protein mutation regardless of phenotype! (Farrar and Dahl, 2011) It seems what we know is much less compared to the unknown.

TSR2 gene product binds to RPS26 and is involved in pre-RNA processing. Thus, it plays a role in ribosome biogenesis. Mutations in the TSR2 gene are associated with some rare forms of DBA (Gripp et al., 2014).

Gazda et al. (2012) performed a screen of 79 RP genes in 96 DBA probands. They identified a de novo two-nucleotide deletion in RPL26 in one proband causing a frameshift mutation. RPL26 mutation causes preribosomal RNA (pre-rRNA) maturation defect. Like RPL5, RPL11, and RPS7, RPL26 is also associated with DBA and promotes p53-mediated apoptosis. Landowski et al. (2013) reported a novel large deletion in RPL15, a gene not previously implicated to be causative in DBA. RPL15 is required both for 60S subunit assembly and for efficient cleavage of the internal transcribed spacer 1.

Wang et al. (2015) reported the loss of function mutations in RPL27 and RPS27 resulting in DBA in the Japanese population. They discovered a novel splicing error mutation in RPL27 and frameshift deletion in RPS27 in sporadic patients with DBA which might result in impaired preribosomal RNA processing. In a study of 98 Japanese patients with DBA, they detected probable causative mutations or large deletions in ribosomal protein genes in 56 (55%) of the patients, involving the RPS19 gene in 16 patients, RPL5 in 12, RPS17 in 7, RPL35A in 7, RPL11 in 5, and RPS26 in 4; RPS7, RPS10, RPL27, and RPS27 were each mutated in one patient.

Ball et al. (1996) reported an annual incidence of 5 per million live births from the United Kingdom. In about 12% of the cases, there was a family history. Anemia was present in 13% at the time of birth. The majority of the patients responded to steroids initially (72%) and at the time of the study, 61% were transfusion independent (45% steroid dependent), and 39% required regular transfusions. Anatomic abnormalities were present in 37% and about one-fifth had thumb abnormalities. Interestingly, they observed a seasonal pattern in the birth month distribution of children with sporadic DBA and craniofacial dysmorphism, hinting at a viral etiology.

Willig et al. (1999a) from Life Science Division, Lawrence Berkeley National Laboratory reported 42 probands with DBA resulting from mutations (nonsense, missense, splice sites, and frameshift mutations) in the RPS19 gene. One hundred and seventy-two new families were investigated for mutations in six exons and 5′ untranslated region of the RPS19 gene. Mutations involving the coding sequence of RPS19 or splice sites were found in about one-fifth of the patients whereas mutations in noncoding regions were found in about 5% of patients. However, a lower frequency of RPS19 mutations was observed in Japanese patients (Konno et al., 2010).

Pathogenesis of DBA: DBA was thought to be an autoimmune disease and this hypothesis was supported by its response to corticosteroids. Several immune mechanisms (humoral or cell-mediated immune inhibition by T-cytotoxic or T-helper lymphocytes) were implicated in the pathogenesis of DBA (Da Costa et al., 2018). The early erythroid progenitors are unresponsive to several growth factors (erythropoietin, IL-3, IL-6, GM-CSF), either alone or in the association. Hematopoietic stem cells harvested from DBA patients can differentiate to the granulocyte megakaryocyte upon incubation with IL-3, but not the erythroid compartment (Bagnara et al., 1991).

Protein synthesis in eukaryotes is highly regulated and most of the translational regulation occurs at the level of initiation, the rate-limiting step. Canonical initiation factors (eIFs) and sequence elements within the 5′ and 3′ untranslated regions (UTRs) of the mRNA together with trans-acting proteins and RNAs regulate translation. However, there are other translation start mechanisms in mammalian cells which depend on RNA motifs and specific partner proteins that directly direct the translation such as the 5′ untranslated region (UTR) TOP-motifs, internal ribosome entry segments (IRES), upstream open reading frames, and in the 3′ UTR a number of unique elements in addition to the poly a tail length (Narla and Ebert, 2010).

About a quarter of patients with DBA have mutations in small-ribosome component RPS19. RPS19 autoregulates its expression by binding to its own 5′ UTR TOP region, resulting in an increased translation of the corresponding protein. It is important to note that mTOR pathway differentially regulates the translation of mRNAs which has 5′TOP motif or a TOP-like motif (Bello et al., 2018) (Fig. 11.1). Certain mutations (such as those resulting in DBA) in RPS19 interfere with this autoregulatory process. In addition, RPS19 interacts and promotes the translation of other 40S ribosomal subunit mRNAs (S20, S21, and S24) through a similar binding mechanism,

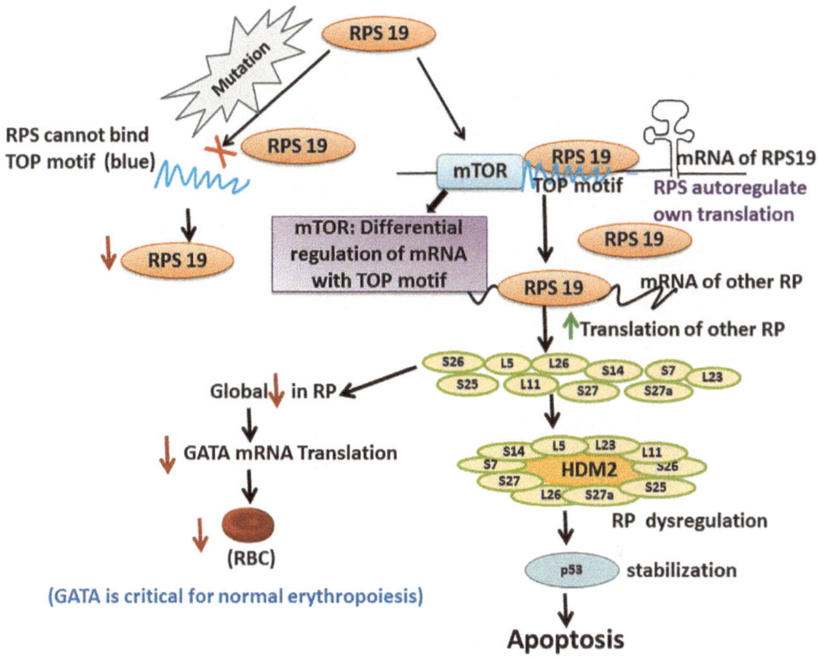

Fig. 11.1 Effect of mutations in small-ribosome component RPS19 of DBA patients. RPS19 autoregulates its expression by binding to its own 5′ UTR TOP region, resulting in an increased translation of the corresponding protein. It important to note that the mTOR pathway differentially regulates translation of mRNAs, which has 5′TOP motif or a TOP-like motif (Bello et al., 2018). Certain mutations (such as those resulting in DBA) in RPS19 interferes with this autoregulatory process. In addition, RPS19 interacts and promotes the translation of other 40S ribosomal subunit mRNAs (S20, S21, and S24) through a similar binding mechanism and therefore a decrease in RPS19 results in global ribosomal abnormalities. HDM2, is a ubiquitin protein ligase that inhibits the transcription of the tumor suppressor p53 and promotes its degradation. This protein decides the stability and activity of the p53 tumor suppressor and thereby decides apoptosis. RP—ribosomal protein. RBC—red blood cell. HDM2—human analog of mouse MDM2 which functions both as an E3 ubiquitin ligase that recognizes the N-terminal transactivation domain of the p53 tumor suppressor and as an inhibitor of p53 transcriptional activation.

and therefore a decrease in RPS19 results in global ribosomal abnormalities (Fig. 11.2). HDM2 (human homolog of MDM2), is a ubiquitin-protein ligase that inhibits the transcription of the tumor suppressor p53 and promotes its degradation. This protein decides the stability and activity of the p53 tumor suppressor. Ribosomal protein L11 inhibits HDM2 function which stabilizes p53. There are signaling pathways through which ribosomal proteins (RPs) transduce nucleolar stress signals via MDM2 to p53

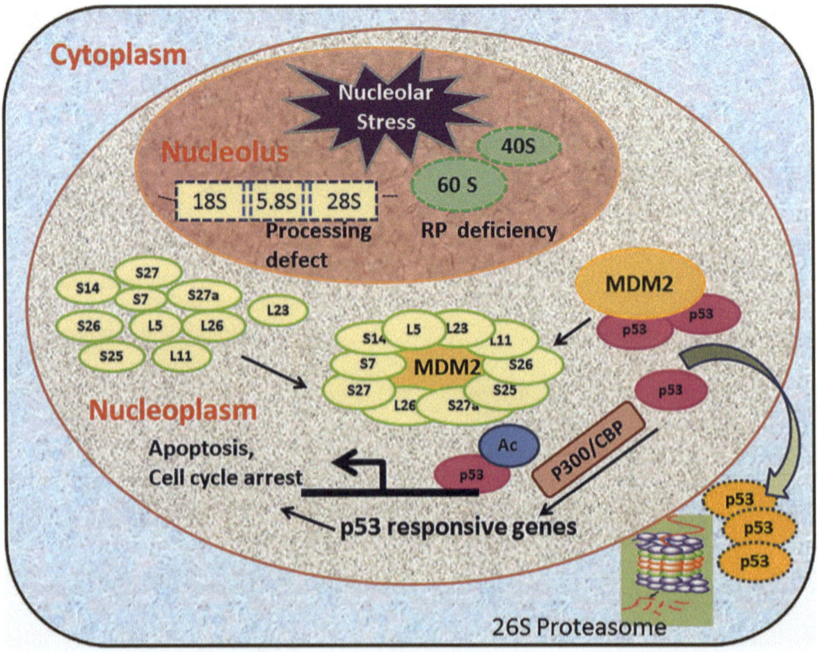

Fig. 11.2 The RPs-MDM2-p53 interaction in response to nucleolar stress. Dysregulated synthesis of ribosomal proteins or defects in ribosomal proteins cause nucleolar stress and ribosomal proteins are released to the nucleoplasm from nucleolus. The free ribosomal proteins will bind to MDM2 and inhibit the MDM2-mediated ubiquitination and degradation of p53, leading to the stabilization and activation of p53. This process integrates ribosome synthesis with cell proliferation and apoptosis.

(Calo et al., 2018). The mechanisms are not clear, but it is believed that the disparity, between ribosome biogenesis and total mRNA levels, leads to the induction of p53 proapoptotic protein (Fig. 11.2). The p53 expressing cells will undergo apoptosis and are lost from the population and over time, p53 deficient cells will accumulate increasing the chance of malignancy, a phenomenon which is associated with DBA. There is evidence that abnormal levels of ribosomes, proapoptotic genes, and transcription factors involved in erythropoiesis are associated with DBA (see below).

GATA1 (gene encoding a critical hematopoietic transcription factor), mutations are a rare cause of DBA. However, this has contributed very much to the understanding of "DBA." A decrease in the GATA1 protein from ribosomal protein haploinsufficiency, the more common cause of DBA, can also lead to decreased GATA1 mRNA translation. The reason could be a higher threshold for initiation of translation of this mRNA in comparison with other mRNAs. The proof is that in primary hematopoietic cells from patients with mutations in RPS19, GATA1 target genes were specifically reduced, indicating that it is the activity rather than the mRNA level, of GATA1 which is decreased in those patients who suffer from DBA which results from mutations affecting ribosomal proteins. We know about haploinsufficiency in DBA, however, it is not clear whether mutated forms of ribosomal/other gene products have a dominant-negative effect (which in fact is possible through multiple mechanisms).

The block in red blood cell formation occurs between the primitive erythroid progenitor cell (BFU-e) stage and the later-stage erythroid progenitor cell (CFU-e) stage. Erythroid blockade results in the aspirate. We do not know why activity the activity of erythrocyte adenine deaminase (eADA), a critical enzyme of the purine salvage pathway, is raised in DBA. Elevated ADA levels are also seen in leukemias, lymphomas, and some immune system disorders (Da Costa et al., 2018).

Gazda et al. (2006a) reported the identification of de novo nonsense and splice-site mutations in another RP, RPS24 in ~2% of RPS19 mutation-negative probands. Using gene expression profiling, they observed the downregulation of multiple ribosomal proteins, genes involved in transcription of erythroid precursors-MYB (Myb-related protein B), upregulation of several proapoptotic genes, including TNFRSF10B and FAS, and dysregulation of many oncogenes in patients with RPS19 defects.

The ribosomal proteins may have totally unrelated/unexpected functions. For example, RPS19 interacts with the basic fibroblast growth factor (b-FGF), an important regulator of cell proliferation, differentiation, and rRNA transcription. B-FGF interacts with free RPS19 in vitro. Unbound extra-ribosomal RPS19 might have a role in cell signaling pathways associated with rRNA transcription. RPS19 is also known to interact with S19BP which a nucleolar protein (Sorensen et al., 2006). The significance of this interaction is not known. All these points to the possibility that ribosomal proteins may have other functions which are cell/tissue-specific which explains why mutations in a widely expressed protein might cause cell/tissue-specific pathology.

Diagnosis of DBA: Anemia before the first birthday with reticulocytopenia, macrocytosis, normal or slightly low neutrophil counts, variable platelet counts, and normal marrow cellularity with a paucity of red cell precursors are classic criteria for the diagnosis of DBA (Burgert Jr et al., 1954). Elevated fetal hemoglobin (HbF) levels and an elevation in erythrocyte adenosine deaminase (eADA) activity are important supporting findings but none are specific to DBA (Whitehouse et al., 1984). It is important to carefully study DBA-affected pedigrees for the presence of members with macrocytosis, elevated HbF levels, and/or increased eADA activity and congenital anomalies (McLennan et al., 1996; Anur et al., 2009).

As noted DBA can be the result of inactivating mutations in ribosomal protein genes, with mutations in 13 of the 80 RP genes causing 50%–60% of cases. The remaining half may have mutations in one of the remaining RP genes or unrelated genes but at much lower frequencies. In families with known mutations which are associated with DBA a candidate gene approach and for sporadic cases or cases where the genetic basis is not established. Whole genome sequencing is perhaps the best option.

Gerrard et al. (2013) applied custom enrichment technology combined with high-throughput sequencing to screen all 80 RP genes. They found inactivating mutations in 88% of DBA patients. Target enrichment combined with high-throughput sequencing is a practical approach for the genetic diagnosis of DBA. Prenatal diagnosis of DBA can be made in women with a family history. If the fetus is affected the pregnancy may be discontinued. Routine blood tests are very important in these patients because of anemia which increases the chances of hematological malignancies.

Decades back the diagnosis of DBA was simple. Often when we understand more, things become difficult and perhaps even more stupid. When we get a deeper understanding, we realize that what we were looking for might not even exist in the form we conceived and perceived! Diagnosis of DBA can be a difficult task because of its heterogeneous clinical presentation and genetic/epigenetic variability. We need to know much more about the DBA (or more correctly the DBA spectrum of diseases) before we can define what exactly we mean by DBA. One practical way is to go by the genetic abnormalities and label it a DBA irrespective of the clinical presentation. With the advent of new sequencing technologies, the cost of WGS has come down tremendously. We have so far underestimated the importance of DBA epigenome and the genetic confounding factors. We can hope that with technological advances, we will continue to identify these factors and further we will be able to understand and define what we mean by DBA. Till then perhaps we may have to depend on some pragmatic and empirical definitions for the practical diagnosis of DBA, which may be a combination of clinical and genetic variables known to be associated with "DBA."

Clinical management of DBA: It is known for a long time that corticosteroids are quite effective in DBA (Kass and Sundal, 1953). This added to the belief that DBA is an autoimmune disease (Gasser, 1951). About 80% of patients respond to steroids initially and nearly half of these treated patients develop cushingoid features, one-fifth develop pathologic fractures, and one-tenth develop cataracts. About half of these patients become refractory to steroids at some point in time.

The toxicity of steroids resulted in an increased dependence on blood and blood products, especially in children. Iron overload and toxicity is the major "side effect" of repeated blood transfusions. However, efficient and less toxic iron chelators facilitated the increasing acceptability of blood transfusion over steroid therapy. In the younger age group, steroids are avoided considering their toxicity. Remissions are observed in DBA patients especially in adolescents and younger adults. Steroid responsiveness is considered a positive predictor of prognosis (Willig et al., 1999b). Pregnancy and the use of birth control pills were found to be associated with relapse. The exact mechanism of corticosteroids in DBA is not clear but it does not involve modulation of ribosomal protein expression. Steroids increase the sensitivity to erythropoietin (Epo) and it results in global changes of gene expression promoting erythroid differentiation (Lipton and Ellis, 2009).

There is not enough evidence for the benefits of cyclosporine or high-dose intravenous methylprednisolone. There is evidence that the erythroid stem cells of DBA patients are refractory, partly or completely, to erythropoietin and therefore Epo will not have much therapeutic benefit.

L-leucine, an amino acid, is a translation enhancer and has some effect in DBA and 5q-syndrome—both have altered ribosome functions. L-leucine is a nutrient regulator of protein synthesis (Bello et al., 2018). Interestingly, L-leucine treatment significantly improved the phosphorylation of RPS6, possibly due to L-leucine-mediated activation of mTORC1. Preclinical studies have shown that protein synthesis can be increased by using high doses of L-leucine (Ovsyannikova et al., 2015).

Another investigational drug is sotatercept (ACE-011). It selectively binds to activin, inhibiting its binding to its receptor. It has rapidly increased hemoglobin in pharmacologic models and in healthy volunteers. It is currently being evaluated in diseases of ineffective erythropoiesis such as β-thalassemia, myelodysplastic syndrome, and DBA (Ear et al., 2013).

Lenalidomide a derivative of thalidomide has antiangiogenic and immunomodulatory effects. It is useful in the treatment of inflammatory diseases, multiple myeloma, and myelodysplastic syndrome. Lenalidomide interacts with ubiquitin E3 ligase cereblon and targets this enzyme to degrade the Ikaros transcription factors KZF1 and IKZF3. Lenalidomide was tried in a patient with DBA, however, it was stopped after 5 weeks due to severe pancytopenia (Joosten et al., 2009).

It was initially intended as a treatment for multiple myeloma, for which thalidomide is an accepted therapeutic treatment. Lenalidomide has also shown efficacy in the class of hematological disorders known as myelodysplastic syndromes.

Cytokines such as stem cell factor (SCF), erythropoietin and interleukin-3, granulocyte-macrophage colony-stimulating factor, and c-kit ligand were experimented for stimulation of bone marrow-derived or blood mononuclear cells from DBA patients (Sieff et al., 1992). Burst-forming unit-erythroid (BFU-E) from all patients responded, however, the responses were not uniform owing to the individual differences in the sensitivity of their BFU-E to SCF (Abkowitz et al., 1991; Carow et al., 1991). In a European multicentric study, IL-3 was given to 40 patients as a daily subcutaneous injection for 12 weeks. However, only three (7.5%) children achieved a significant response, achieving sustained remissions off all therapy (Dunbar et al., 1991). This does not mean IL-3 is an ineffective therapy. Instead, it reveals the genetic and epigenetic heterogenic nature of DBA and the need for personalized medicine based on the genetic roots of the disease.

The only curative therapy in DBA is perhaps cell or gene therapy. A conventional cell therapy approach, such as bone marrow transplantation from a well-matching donor, results in a cure. Umbilical cord blood from matching donors is another option. Umbilical cord blood cell transplantation is associated with more permissive HLA mismatching and decreased incidence of graft versus host disease. It is difficult to find matching bone marrow donors. Conceiving another child with matched HLA (through preimplantation genetic screening) by parents of an affected DBA patient is a good strategy toward a cure. Unrelated cord blood transplant (UCBT) has successfully been used to treat DBA because the probability to find a suitable graft is more for UCBT because of rapid availability and lack of donor risk. The overall survival following UCBT is comparable to the results following related and unrelated donor transplants for DBA as reported by The Center for International Blood & Marrow Transplant Research, or CIBMTR (McFarren et al., 2014; Roy et al., 2005).

Pregnancies in DBA patients need careful follow-up ad timely interventions as they are high-risk pregnancy-maternal complications and fetal death. Medical termination of pregnancy is an option. Alternatively, if signs of anemia are present, utero transfusions and planned early delivery of the fetus may be considered. In utero gene therapy and cell therapy are other advanced (experimental) options.

So far there is no evidence to suggest mesenchymal stem cells from matched siblings/donors who are useful in the treatment of DBA. After all, the definitions of mesenchymal stem cells are not clear or it is highly contextual (Bianco et al., 2013). Gene therapy is making a difference in various genetic diseases. Currently gene delivery is done using retro/lenti/adeno/adeno associated viruses. However, lentivirus and retrovirus can integrate near a promoter sequence of an actively transcribing gene. This is potentially dangerous because there is a chance for activation of some proto-oncogene. Novel efficient DNA delivery systems are the need of the hour. Genome editing tools such as zinc finger nucleases/CRISPR-Cas technology together with high throughput DNA sequencing could bring in revolutionary changes in the management of these diseases (Carroll, 2017).

Other ribosomopathies

Mutations in other genes required for normal ribosome biogenesis have also been implicated in the development of many ribosomopathies, such as 5q deletion syndrome, Bowen-Conradi syndrome, congenital asplenia, Cartilage hair hypoplasia, Treacher Collins syndrome, Schwachman-Diamond syndrome, X-linked dyskeratosis congenital, etc. The details of gene associated with these diseases, their mechanism of action, clinical symptoms, and management strategies are given in Table 11.2.

Suspected ribosomopathies

RBM28—alopecia, neurological defects, and endocrinopathy syndrome (ANE syndrome). Autosomal-recessive ANE syndrome phenotype is characterized by alopecia, progressive neurological defects, and endocrinopathy. By using homozygosity mapping and candidate-gene analysis, it was identified as a loss-of-function mutation in RBM28, which encodes a nucleolar protein. It has an ortholog in yeast (Nop4p) which is known to regulate ribosome biogenesis. Electron microscopy of patient cells proved a marked ribosome depletion and structural abnormalities of the rough endoplasmic reticulum (Nousbeck et al., 2008).

Is cancer a ribosomopathy?

It seems ribosomal defects can originate and accumulate as a cancer makes its progress. An increase in ribosomes could facilitate an increase in protein synthesis which is a requirement for fast-growing cells. It is also possible

Table 11.2 Other ribosomopathies.

Disease	Mutated gene	Chromosome	Symptoms	Mechanism	Treatment
5q deletion syndrome (Ebert et al., 2008)	RPS14, miR-145, miR-146a	Deletion in 5q33.1	Macrocytic anemia Normal or high platelet count hypo-lobulated megakaryocytes	RPS14 deficiency causes a block in the processing of the 18S rRNA, thus reduce the formation of the 40S ribosomal subunit	Lenalidomide
Bowen-Conradi syndrome (Warda et al., 2016)	EMG1	12p13	Severe growth retardation, skeletal dysmorphology, severe prenatal and postnatal growth failure usually leads to death by 1 year of age	EMG1 is a putative methyltransferase that is required for biogenesis of the 40S subunit Mutations in EMG1result in aspartate to glycine substitution at position 86 of the nucleolar RNA methyltransferase EMG1 leading to its aggregation and subsequent degradation	Symptomatic and supportive
Congenital asplenia (Bolze et al., 2013)	RPSA	3p21	Pneumococcal sepsis and meningitis Absence of spleen	RPSA gene required for the processing of the 20S rRNA-precursor to mature 18S rRNA in a late step of the maturation of 40S ribosomal subunits	Antibiotic prophylaxis (transient or lifelong), immunization at the recommended age
Cartilage hair hypoplasia (Thiel and Rauch, 2011)	RMRP (RNase mitochonrial RNA processing) gene	9p13.3	Short-limbed dwarfism; metaphyseal chondrodysplasia; hypoplastic hair; recurrent infections (pneumocystis, CMV)	RMRP gene separates the precursors of the SSU from the LSU pre-rRNA, required for the maturation of the 5.8S rRNA; involved in cell cycle regulation by cleaving the cyclin B2 mRNA and in mitochondrial DNA replication	Management is supportive, with stem cell transplantation being the only curative option

North American Indian childhood cirrhosis (Chagnon et al., 2002)	hUTP4/Cirhin	16q22	Neonatal jaundice Biliary cirrhosis Portal hypertension	hUTP4/Cirhin along with UTP15 and WDR43 are the core components of t-UTP, which is involved in the processing of 18S rRNA in human	Liver transplantation
Treacher Collins syndrome (Dauwerse et al., 2011)	TCOF1		Craniofacial abnormalities (micrognathia, ear deformities, macrostomia, hearing loss, antimongoloid slant)	Treacle is a nucleolar protein involved in transcription of rDNA and methylation of 18SrRNA, mutations lead to truncation of the protein, diminished production of ribosomes, which correlates with decreased proliferation of neural ectoderm and neural crest cells	Symptomatic
Shwachman-Bodian-Diamond syndrome (Boocock et al., 2003)	SBDS	7q11.21	Exocrine pancreatic insufficiency, hematologic abnormalities (esp. neutropenia) neurocognitive impairment GI (esp. liver) abnormalities, bone marrow failure, short stature, leukemia	SBDS is required for the release of eukaryotic initiation factor 6 (eIF6) from the 60S ribosomal subunit	Pancreatic enzyme supplementation, HSCT[a]

Continued

Table 11.2 Other ribosomopathies—cont'd

Disease	Mutated gene	Chromosome	Symptoms	Mechanism	Treatment
X linked dyskeratosis congenita (Heiss et al., 1998)	DKC1/X-DC	Xq28	Mucocutaneous triad (abnormal skin pigmentation, nail dystrophy, and mucosal leukoplakia), pulmonary fibrosis, bone marrow failure	Dyskerin, a nucleolar protein, associates with box H/ACA snoRNPs to catalyze posttranscriptional modification (pseudouridylation) of rRNAs. Dyskerin is also a member of the telomerase complex that is responsible for maintaining telomere length	Oxymetholone HSCT
Prader-Willi syndrome (Cavaillé, 2017)	SNRPN, NDN (necdin) genes	15q11–13	Weak muscles, poor feeding, and slow development	Small nucleolar RNAs (snoRNAs) pair with ribosomal RNA precursors and influence ribosome biogenesis or later their activity. Sometimes abnormal neurological features and behavior	

[a]Hematopoietic stem cell transplantation (HSCT).

that certain mutations might improve the production of the kind of proteins required for malignant cells. However, a defect in a ribosome be the only cause of cancer? It is unlikely. Certain ribosomopathies are associated with cancers (Shenoy et al., 2012). Some of these are featured by "hypoproliferation" (resulting in bone marrow failure and anemia) and at a later phase, they may develop "hyperproliferation" (cancer). This evolution from hypo- to hyperproliferation is paradoxical and is called "Dameshek's riddle" (De Keersmaecker et al., 2015). It is understood that oxidative/cellular stress resulting from ineffective protein synthesis over time can be damaging not only to the cell but also to its environment. In ribosomopathies, the niche surrounding bone marrow stem cells are not normal compared to a healthy bone marrow. This will increase the stress levels in stem cells. Increased stress can cause DNA damage and epigenetic changes. It is known that stem cells when cultured (in vitro) under inappropriate conditions (suboptimal media, matrix/feeder layer) are prone to acquire genetic abnormalities (abnormal karyotypes, mutation, copy number variations, etc.) and epigenetic aberrations. In a study, 625 human nucleolar proteins were analyzed and ~30% of all processing factors identified were already known to be associated with one or more human cancers (Gupta and Warner, 2014).

Conclusions

Ribosomal dysfunction appears to be central to the molecular pathology of the disorder. Ribosomopathies are associated with specific defects in ribosome biogenesis and function which cause distinct clinical phenotypes. Diamond-Blackfan anemia (DBA) is a classical example of ribosomopathy. DBA is a rare congenital bone marrow failure syndrome with impaired erythropoiesis and craniofacial abnormalities. The identification of additional disease alleles and characterization of their functional effects has helped in our understanding of ribosomopathies. Elucidation of molecular mechanisms whereby selective abnormalities in ribosome biogenesis should lead to specific clinical syndromes will hopefully be helpful in developing novel therapeutic strategies for the treatment of ribosomopathies.

References

Abkowitz JL, Sabo KM, Nakamoto B, Blau CA, Martin FH, Zsebo KM, Papayannopoulou T. Diamond-Blackfan anemia: *in vitro* response of erythroid progenitors to the ligand for c-kit. Blood 1991;78:2198–202.

Alter BP. Thumbs and anemia. Pediatrics 1978;62:613–4.

Altman KI, Miller G. A disturbance of tryptophan metabolism in congenital hypoplastic anemia. Nature 1953;172:868.

Anur P, Nemecek ER, Kurre P. The evolving spectrum of 'non-classical' Diamond-Blackfan anaemia—a case of eADA positive pancytopenia in a young adult [Letter]. Brit J Haemat 2009;145:428–30.

Avondo F, Roncaglia P, Crescenzio N, Krmac H, Garelli E, Armiraglio M, Castagnoli C, Campagnoli MF, Ramenghi U, Gustincich S, Santoro C, Dianzani I. Fibroblasts from patients with Diamond-Blackfan anaemia show abnormal expression of genes involved in protein synthesis, amino acid metabolism and cancer. BMC Genomics 2009;10:442.

Bagnara GP, Zauli G, Vitale L, Rosito P, Vecchi V, Paolucci G, Avanzi GC, Ramenghi U, Timeus F, Gabutti V. In vitro growth and regulation of bone marrow enriched CD34+ hematopoietic progenitors in Diamond-Blackfan anemia. Blood 1991;78:2203–10.

Ball SE, McGuckin CP, Jenkins G, Gordon-Smith EC. Diamond-Blackfan anaemia in the U.K.: analysis of 80 cases from a 20-year birth cohort. Brit J Haemat 1996;94:645–53.

Bello E, Kerry J, Singh S, Yip BH, Kušec R, Killick S, Raynaud S, Boultwood J, Pellagatti A. L-leucine increases translation of RPS14 and LARP1 in erythroblasts from del(5q) myelodysplastic syndrome patients. Haematologica 2018;103:e496–500.

Bianco P, Cao X, Frenette PS, Mao JJ, Robey PG, Simmons PJ, Wang CY. The meaning, the sense and the significance: translating the science of mesenchymal stem cells into medicine. Nat Med 2013;19:35–42.

Bolze A, Mahlaoui N, Byun M, Turner B, Trede N, Ellis SR, Abhyankar A, Itan Y, Patin E, Brebner S, Sackstein P, Puel A, Picard C, Abel L, Quintana-Murci L, Faust SN, Williams AP, Baretto R, Duddridge M, Kini U, Pollard AJ, Gaud C, Frange P, Orbach D, Emile JF, Stephan JL, Sorensen R, Plebani A, Hammarstrom L, Conley ME, Selleri L, Casanova JL. Ribosomal protein SA haploinsufficiency in humans with isolated congenital asplenia. Science 2013;340:976–8.

Boocock GR, Morrison JA, Popovic M, Richards N, Ellis L, Durie PR, Rommens JM. Mutations in SBDS are associated with Shwachman-Diamond syndrome. Nat Genet 2003;33:97–101.

Boria I, Garelli E, Gazda HT, Aspesi A, Quarello P, Pavesi E, Ferrante D, Meerpohl JJ, Kartal M, Da Costa L, Proust A, Leblanc T, Simansour M, Dahl N, Fröjmark AS, Pospisilova D, Cmejla R, Beggs AH, Sheen MR, Landowski M, Buros CM, Clinton CM, Dobson LJ, Vlachos A, Atsidaftos E, Lipton JM, Ellis SR, Ramenghi U, Dianzani I. The ribosomal basis of Diamond-Blackfan anemia: mutation and database update. Hum Mutat 2010;31:1269–79.

Burgert Jr EO, Kennedy RLJ, Pease GL. Congenital hypoplastic anemia. Pediatrics 1954;13:218–26.

Calo E, Gu B, Bowen ME, Aryan F, Zalc A, Liang J, Flynn RA, Swigut T, Chang HY, Attardi LD, Wysocka J. Tissue-selective effects of nucleolar stress and rDNA damage in developmental disorders. Nature 2018;554:112–7.

Carow CE, Hangoc G, Cooper SH, Williams DE, Broxmeyer HE. Mast cell growth factor (c-kit ligand) supports the growth of human multipotential progenitor cells with a high replating potential. Blood 1991;78:2216–21.

Carroll D. Genome editing: past, present, and future. Yale J Biol Med 2017;90:653–9.

Cavaillé J. Box C/D small nucleolar RNA genes and the Prader-Willi syndrome: a complex interplay. WIREs RNA 2017;e1417.

Chagnon P, Michaud J, Mitchell G, Mercier J, Marion JF, Drouin E, Rasquin-Weber A, Hudson TJ, Richter A. A missense mutation (R565W) in cirhin (FLJ14728) in North American Indian childhood cirrhosis. Am J Hum Genet 2002;71:1443–9.

Chakraborty A, Karunasagar I. Faulty ribosomes and human diseases: mistakes in "assembly line" going unnoticed? NUJHS 2015;5:87–92.

Chiabrando D, Tolosano E. Diamond Blackfan anemia at the crossroad between ribosome biogenesis and heme metabolism. Adv Hematol 2010;2010:790632.

Costa LD, Fixler J, Berets O, Leblanc T, Willig TN, Mohandas N, Tchernia G. Piebaldism in diamond-blackfan anaemia: a new phenotype? Br J Haematol 2002;119:572.

Da Costa L, Narla A, Mohandas N. An update on the pathogenesis and diagnosis of Diamond-Blackfan anemia. F1000Res 2018;7. pii: F1000 Faculty Rev-1350.

Dauwerse JG, Dixon J, Seland S, Ruivenkamp CA, van Haeringen A, Hoefsloot LH, Peters DJ, Boers AC, Daumer-Haas C, Maiwald R, Zweier C, Kerr B, Cobo AM, Toral JF, Hoogeboom AJ, Lohmann DR, Hehr U, Dixon MJ, Breuning MH, Wieczorek D. Mutations in genes encoding subunits of RNA polymerases I and III cause Treacher Collins syndrome. Nat Genet 2011;43:20–2.

De Keersmaecker K, Sulima SO, Dinman JD. Ribosomopathies and the paradox of cellular hypo- to hyperproliferation. Blood 2015;125:1377–82.

Diamond LK, Blackfan KD. Hypoplastic anemia. Am J Dis Child 1938;56:464–7.

Diamond L, Allen DW, Magill FB. Congenital (erythroid) hypoplastic anemia: a 25 year study. Am J Dis Child 1961;102:403–15.

Dianzani I, Garelli E, Dompè C, Crescenzio N, Locatelli F, Schilirò G, Castaman G, Bagnara GP, Olivieri NF, Gabutti V, Ramenghi U. Mutations in the erythropoietin receptor gene are not a common cause of Diamond-Blackfan anemia. Blood 1996;87:2568–72.

Draptchinskaia N, Gustavsson P, Andersson B, Pettersson M, Willig TN, Dianzani I, Ball S, Tchernia G, Klar J, Matsson H, Tentler D, Mohandas N, Carlsson B, Dahl N. The gene encoding ribosomal protein S19 is mutated in Diamond-Blackfan anaemia. Nat Genet 1999;21:169–75.

Dunbar CE, Smith DA, Kimball J, Garrison L, Nienhuis AW, Young NS. Treatment of Diamond-Blackfan anaemia with haematopoietic growth factors, granulocyte-macrophage colony stimulating factor and interleukin 3: sustained remissions following IL-3. Brit J Haemat 1991;79:316–21.

Ear J, Huang H, Tehrani Z, Sung V, Daniel T, Chopra R, Lin S. RAP-011 efficiently rescues erythropoiesis in zebrafish models of Diamond Blackfan anemia. Blood 2013;122:3702.

Ebert BL, Pretz J, Bosco J, Chang CY, Tamayo P, Galili N, Raza A, Root DE, Attar E, Ellis SR, Golub TR. Identification of RPS14 as a 5q-syndrome gene by RNA interference screen. Nature 2008;451:335–9.

Falter ML, Robinson MG. Autosomal dominant inheritance and amino aciduria in Blackfan-Diamond anemia. J Med Genet 1972;9:64–6.

Farrar JE, Dahl N. Untangling the phenotypic heterogeneity of Diamond Blackfan anemia. Semin Hematol 2011;48:124–35.

Filipovska A, Rackham O. Specialization from synthesis: how ribosome diversity can customize protein function. FEBS Lett 2017;587:1189–97.

Freedman MH, Amato D, Saunders EF. Erythroid colony growth in congenital hypoplastic anemia. J Clin Invest 1976;57:673–7.

Gasser C. Aplastic anemia (chronic erythroblastophthisis) and cortisone. Schweiz Med Wochenschr 1951;81:1241–2.

Gazda H, Lipton JM, Willig TN, Ball S, Niemeyer CM, Tchernia G, Mohandas N, Daly MJ, Ploszynska A, Orfali KA, Vlachos A, Glader BE, Rokicka-Milewska R, Ohara A, Baker D, Pospisilova D, Webber A, Viskochil DH, Nathan DG, Beggs AH, Sieff CA. Evidence for linkage of familial Diamond-Blackfan anemia to chromosome 8p23.3-p22 and for non-19q non-8p disease. Blood 2001;97:2145–50.

Gazda HT, Grabowska A, Merida-Long LB, Latawiec E, Schneider HE, Lipton JM, Vlachos A, Atsidaftos E, Ball SE, Orfali KA, Niewiadomska E, Da Costa L, Tchernia G, Niemeyer C, Meerpohl JJ, Stahl J, Schratt G, Glader B, Backer K, Wong C, Nathan DG, Beggs AH, Sieff CA. Ribosomal protein S24 gene is mutated in Diamond-Blackfan anemia. Am J Hum Genet 2006a;79:1110–8.

Gazda HT, Kho AT, Sanoudou D, Zaucha JM, Kohane IS, Sieff CA, Beggs AH. Defective ribosomal protein gene expression alters transcription, translation, apoptosis, and oncogenic pathways in Diamond-Blackfan anemia. Stem Cells 2006b;24:2034–344.

Gazda HT, Sheen MR, Vlachos A, Choesmel V, O'Donohue MF, Schneider H, Darras N, Hasman C, Sieff CA, Newburger PE, Ball SE, Niewiadomska E, Matysiak M, Zaucha JM, Glader B, Niemeyer C, Meerpohl JJ, Atsidaftos E, Lipton JM, Gleizes PE, Beggs AH. Ribosomal protein L5 and L11 mutations are associated with cleft palate and abnormal thumbs in Diamond-Blackfan anemia patients. Am J Hum Genet 2008;83:769–80.

Gazda HT, Preti M, Sheen MR, O'Donohue MF, Vlachos A, Davies SM, Kattamis A, Doherty L, Landowski M, Buros C, Ghazvinian R, Sieff CA, Newburger PE, Niewiadomska E, Matysiak M, Glader B, Atsidaftos E, Lipton JM, Gleizes PE, Beggs AH. Frameshift mutation in p53 regulator RPL26 is associated with multiple physical abnormalities and a specific pre-ribosomal RNA processing defect in Diamond-Blackfan anemia. Hum Mutat 2012;33:1037–44.

Gerrard G, Valgañón M, Foong HE, Kasperaviciute D, Iskander D, Game L, Müller M, Aitman TJ, Roberts I, de la Fuente J, Foroni L, Karadimitris A. Target enrichment and high-throughput sequencing of 80 ribosomal protein genes to identify mutations associated with Diamond-Blackfan anaemia. Br J Haematol 2013;162:530–6.

Gripp KW, Curry C, Olney AH, Sandoval C, Fisher J, Chong JX, UW Center for Mendelian Genomics, Pilchman L, Sahraoui R, Stabley DL, Sol-Church K. Diamond-Blackfan anemia with mandibulofacial dystostosis is heterogeneous, including the novel DBA genes TSR2 and RPS28. Am J Med Genet A 2014;164A:2240–9.

Gupta V, Warner JR. Ribosome-omics of the human ribosome. RNA 2014;20:1004–13.

Gustavsson P, Skeppner G, Johansson B, Berg T, Gordon L, Kreuger A, Dahl N. Diamond-Blackfan anaemia in a girl with a de novo balanced reciprocal X;19 translocation. J Med Genet 1997a;34:779–82.

Gustavsson P, Willing TN, van Haeringen A, Tchernia G, Dianzani I, Donnér M, Elinder G, Henter JI, Nilsson PG, Gordon L, Skeppner G, van't Veer-Korthof L, Kreuger A, Dahl N. Diamond-Blackfan anaemia: genetic homogeneity for a gene on chromosome 19q13 restricted to 1.8 Mb. Nat Genet 1997b;16:368–71.

Gustavsson P, Garelli E, Draptchinskaia N, Ball S, Willig TN, Tentler D, Dianzani I, Punnett HH, Shafer FE, Cario H, Ramenghi U, Glomstein A, Pfeiffer RA, Goringe A, Olivieri NF, Smibert E, Tchernia G, Elinder G, Dahl N. Identification of microdeletions spanning the Diamond-Blackfan anemia locus on 19q13 and evidence for genetic heterogeneity. Am J Hum Genet 1998;63:1388–95.

Heiss NS, Knight SW, Vulliamy TJ, Klauck SM, Wiemann S, Mason PJ, Poustka A, Dokal I. X-linked dyskeratosis congenita is caused by mutations in a highly conserved gene with putative nucleolar functions. Nat Genet 1998;19:32–8.

Joosten P, Hoogendoorn M, Storm H, Kibbelaar R. Unsuccessful treatment with Lenalidomide in a Diamond-Blackfan anemia patient. Blood 2009;114:4219.

Josephs H. Anemia of infancy and early childhood. Medicine 1936;15:307–451.

Kass A, Sundal A. *Anaemiahypoplastica congenita* (anaemiatypus Josephs-Diamond-Blackfan): report of a case treated with adrenocorticotropin with effect. Acta Paediatr 1953;42:264–74.

Konno Y, Toki T, Tandai S, Xu G, Wang R, Terui K, Ohga S, Hara T, Hama A, Kojima S, Hasegawa D, Kosaka Y, Yanagisawa R, Koike K, Kanai R, Imai T, Hongo T, Park MJ, Sugita K, Ito E. Mutations in the ribosomal protein genes in Japanese patients with Diamond-Blackfan anemia. Haematologica 2010;95:1293–9.

Krishnan EU, Wegner K, Garg SK. Congenital hypoplastic anemia terminating in acute promyelocytic leukemia. Pediatrics 1978;61:898–901.

Landowski M, O'Donohue MF, Buros C, Ghazvinian R, Montel-Lehry N, Vlachos A, Sieff CA, Newburger PE, Niewiadomska E, Matysiak M, Glader B, Atsidaftos E, Lipton JM, Beggs AH, Gleizes PE, Gazda HT. Novel deletion of RPL15 identified by array-comparative genomic hybridization in Diamond-Blackfan anemia. Hum Genet 2013;132:1265–74.

Lipton JM, Ellis SR. Diamond Blackfan anemia: diagnosis, treatment and molecular pathogenesis. Hematol Oncol Clin North Am 2009;23:261–82.

Ludwig LS, Gazda HT, Eng JC, Eichhorn SW, Thiru P, Ghazvinian R, George TI, Gotlib JR, Beggs AH, Sieff CA, Lodish HF, Lander ES, Sankaran VG. Altered translation of GATA1 in Diamond-Blackfan anemia. Nat Med 2014;20:748–53.

McFarren A, Page K, Parikh S, Martin PL, Driscoll TA, Kurtzberg J, Prasad VK. Unrelated umbilical cord blood transplant for Diamond-Blackfan anemia. Biol Blood Marrow Transplant 2014;20(Supplement):S177.

McLennan AC, Chitty LS, Rissik J, Maxwell DJ. Prenatal diagnosis of Blackfan-Diamond syndrome: case report and review of the literature. Prenatal Diag 1996;16:349–53.

Mentzer WC. Louis Diamond and his contribution to haematology. Brit J Haemat 2003;123:389–95.

Miyake K, Flygare J, Kiefer T, Utsugisawa T, Richter J, Ma Z, Wiznerowicz M, Trono D, Karlsson S. Development of cellular models for ribosomal protein S19 (RPS19)-deficient Diamond-Blackfan anemia using inducible expression of siRNA against RPS19. Mol Ther 2005;11:627–37.

Nakhoul H, Ke J, Zhou X, Liao W, Zeng SX, Lu H. Ribosomopathies: mechanisms of disease. Clin Med Insights Blood Disord 2014;7:7–16.

Narla A, Ebert BL. Ribosomopathies: human disorders of ribosomes dysfunction. Blood 2010;115:3196–205.

Nathan DG, Clarke BJ, Hillman DG, Alter BP, Housman DE. Erythroid precursors in congenital hypoplastic (Diamond-Blackfan) anemia. J Clin Invest 1978;61:489–98.

Nousbeck J, Spiegel R, Ishida-Yamamoto A, Indelman M, Shani-Adir A, Adir N, Lipkin E, Bercovici S, Geiger D, van Steensel MA, Steijlen PM, Bergman R, Bindereif A, Choder M, Shalev S, Sprecher E. Alopecia, neurological defects, and endocrinopathy syndrome caused by decreased expression of RBM28, a nucleolar protein associated with ribosome biogenesis. Am J Hum Genet 2008;82:1114–21.

Ovsyannikova GS, Poloznikov AA, Maschan MA, Smetanina NS. Response to L-leucine therapy in patients with Daimond-Blakfan anemia and serum L-leucine concentrations. Blood 2015;126:3619.

Peña C, Hurt E, Panse VG. Eukaryotic ribosome assembly, transport and quality control. Nat Struct Mol Biol 2017;24:689–99.

Post SM, Kantarjian H, Quintás-Cardama A. Biology of adult myelocytic leukemia and myelodysplasia. In: The Molecular Basis of Cancer. 4th ed. Philadelphia, PA: WB Saunders Co.; 2014. p. 421–32.

Ramenghi U, Garelli E, Valtolina S, Campagnoli MF, Timeus F, Crescenzio N, Mair M, Varotto S, D'Avanzo M, Nobili B, Massolo F, Mori PG, Locatelli F, Gustavsson P, Dahl N, Dianzani I. Diamond-Blackfan anaemia in the Italian population. Br J Haematol 1999;104:841–8.

Roy V, Pérez WS, Eapen M, Marsh JC, Pasquini M, Pasquini R, Mustafa MM. Bredeson CN; bone marrow transplantation for Diamond-Blackfan anemia. Biol Blood Marrow Transplant 2005;11:600–8.

Sankaran VG, Ghazvinian R, Do R, Thiru P, Vergilio JA, Beggs AH, Sieff CA, Orkin SH, Nathan DG, Lander ES, Gazda HT. Exome sequencing identifies GATA1 mutations resulting in Diamond-Blackfan anemia. J Clin Invest 2012;122:2439–43.

Shenoy N, Kessel R, Bhagat TD, Bhattacharyya S, Yu Y, McMahon C, Verma A. Alteration in the ribosomal machinery in cancer and hematologic disorders. J Hematol Oncol 2012;5:32.

Sieff CA, Yokoyama CT, Zsebo KM, Trammell J, Andersen JW, Nathan DG, Williams DA. The production of steel factor mRNA in Diamond-Blackfan anaemia long-term cultures and interactions of steel factor with erythropoietin and interleukin-3. Brit J Haemat 1992;82:640–7.

Sorensen V, Nilsen T, Wiedlocha A. Functional diversity of FGF-2 isoforms by intracellular sorting. Bioessays 2006;28:504–14.
Thiel CT, Rauch A. The molecular basis of the cartilage-hair hypoplasia-anauxetic dysplasia spectrum. Best Pract Res Clin Endocrinol Metab 2011;25:131–42.
Toki T, Yoshida K, Wang R, Nakamura S, Maekawa T, Goi K, Katoh MC, Mizuno S, Sugiyama F, Kanezaki R, Uechi T, Nakajima Y, Sato Y, Okuno Y, Sato-Otsubo A, Shiozawa Y, Kataoka K, Shiraishi Y, Sanada M, Chiba K, Tanaka H, Terui K, Sato T, Kamio T, Sakaguchi H, Ohga S, Kuramitsu M, Hamaguchi I, Ohara A, Kanno H, Miyano S, Kojima S, Ishiguro A, Sugita K, Kenmochi N, Takahashi S, Eto K, Ogawa S, Ito E. De novo mutations activating germline TP53 in an inherited bone-marrow-failure syndrome. Am J Hum Genet 2018;103:440–7.
Wang R, Yoshida K, Toki T, Sawada T, Uechi T, Okuno Y, Sato-Otsubo A, Kudo K, Kamimaki I, Kanezaki R, Shiraishi Y, Chiba K, Tanaka H, Terui K, Sato T, Iribe Y, Ohga S, Kuramitsu M, Hamaguchi I, Ohara A, Hara J, Goi K, Matsubara K, Koike K, Ishiguro A, Okamoto Y, Watanabe K, Kanno H, Kojima S, Miyano S, Kenmochi N, Ogawa S, Ito E. Loss of function mutations in RPL27 and RPS27 identified by whole-exome sequencing in Diamond-Blackfan anaemia. Br J Haematol 2015;168:854–64.
Warda AS, Freytag B, Haag S, Sloan KE, Görlich D, Bohnsack MT. Effects of the Bowen-Conradi syndrome mutation in EMG1 on its nuclear import, stability and nucleolar recruitment. Hum Mol Genet 2016;25:5353–64.
Whitehouse DB, Hopkinson DA, Evans DIK. Adenosine deaminase activity in Diamond-Blackfan syndrome. Lancet 1984;324:1398–9.
Willig TN, Draptchinskaia N, Dianzani I, Ball S, Niemeyer C, Ramenghi U, Orfali K, Gustavsson P, Garelli E, Brusco A, Tiemann C, Pérignon JL, Bouchier C, Cicchiello L, Dahl N, Mohandas N, Tchernia G. Mutations in ribosomal protein S19 gene and Diamond Blackfan anemia: wide variations in phenotypic expression. Blood 1999a;94:4294–306.
Willig TN, Niemeyer CM, Leblanc T, Tiemann C, Robert A, Budde J, Lambiliotte A, Kohne E, Souillet G, Eber S, Stephan JL, Girot R, Bordigoni P, Cornu G, Blanche S, Guillard JM, Mohandas N, Tchernia G. Identification of new prognosis factors from the clinical and epidemiologic analysis of a registry of 229 Diamond-Blackfan anemia patients. Pediatr Res 1999b;46:553–61.

CHAPTER 12

Ribosomal profiling—Diversity and applications

Sunita Giri and Vijay Kumar
Department of Molecular and Cellular Medicine, Institute of Liver and Biliary Sciences, New Delhi, India

Introduction

Protein synthesis is the final step in the flow of genetic information during which mRNAs are decoded via translation with the help of cellular nanomachines called ribosomes. Ribosome profiling, also known as "ribosome footprinting" or "Ribo-Seq," has emerged as a powerful technique that provides a "global snapshot" of all the ribosomes actively translating mRNAs in a cell at a given time. It takes advantage of the ability of ribosomes to protect RNAs bound to it by RNAses (Ingolia, 2016). This method has enabled us to monitor the mechanistic details of protein synthesis steps in vivo such as: (a) differential engagement of mRNAs, (b) rate and level of translation, (c) sites of translation initiation, (d) exact positions of ribosomes on mRNAs and their pausing, (e) identification of novel protein-coding genes, (f) dynamics of protein folding, (g) localized protein synthesis in a cell, (h) annotation of the protein-coding capacity of genomes, etc. (Ingolia et al., 2019). Thus ribosome profiling has emerged as a powerful technique that enables genome-wide investigation of gene expression at subcodon resolution.

The method of ribosome profiling was first described in 2009 in the budding yeast model (*Saccharomyces cerevisiae*) to have an in-depth view of translation. It involved deep sequencing of mRNA fragments bound to ribosomes (Ingolia et al., 2009). There are four major steps in ribosome profiling, viz., (1) collection of a physiological sample, (2) inhibition of translation to freeze ribosomes in the act of translation, (3) nuclease digestion to produce ribosome-protected fragments, and (4) isolation of ribosomes and identifying its footprints. It is typically carried out on a split sample, with parallel libraries constructed for measuring mRNA abundance by mRNA-seq. Ribo-Seq complements the information generated by mass spectrometry (MS)-based proteomics and thus offers a better

Emerging Concepts in Ribosome Structure, Biogenesis, and Function
https://doi.org/10.1016/B978-0-12-816364-1.00012-3

understanding of the process of translation. Not surprisingly, ribosome profiling has been applied to refine the existing databases for MS-based proteomics (Menschaert et al., 2013). There are some distinct advantages of using this method, such as:

(a) *Sensitivity and precision of quantification*: It provides a large dynamic range for the detection and quantification of translation in cells. It also allows the measurement of even relatively rare translation events with few sequencing reads.

(b) *Precision of positional information:* It allows precise mapping of the coding region in ORFs and identification of newly transcribed regions in a genome.

(c) *Instantaneous measurements*: It helps in generating rapid information about the dynamic process of translational events as they occur in the cell.

Nowadays, ribosome profiling is a widely used molecular tool for studying gene expression in diverse biological systems from virus and bacteria to higher eukaryotes, including human cells (Ingolia et al., 2011; Li et al., 2012; Stern-Ginossar and Ingolia, 2015). It is believed that the data recovered from cell extracts accurately reflect the in vivo status of translation at the time of the experiment. Although only a few ribosome profiling datasets have been published so far, it has dramatically changed our understanding of translational control as well as in developing quantitative models of translation. Further, it has helped in postulating new hypotheses regarding the origin of protein-coding genes.

Tools and techniques of ribosome profiling

Polysome profiling is considered as a gold standard for the quantitative analysis of translation. As illustrated in Fig. 12.1, the experimental protocols of ribosome profiling include the following steps:

Sample preparation

It involves the processing of biological samples to obtain a postmitochondrial supernatant. The conditions of cell lysis are critical to preserve in vivo ribosome positioning and RNA integrity and the starting biological material should be enough to produce quantifiable ribosome footprints without runoff. Further, drug inhibitors of translation or flash freezing methods like liquid nitrogen and dry ice may be used.

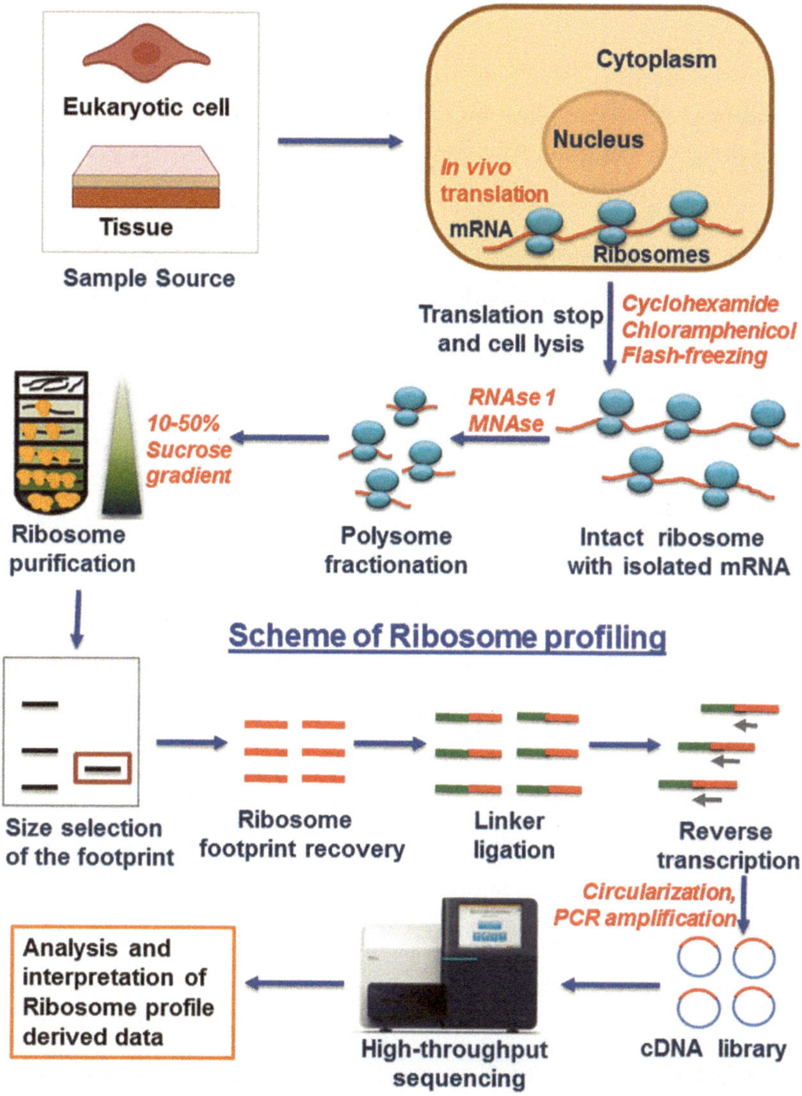

Fig. 12.1 Schematic representation of different steps of ribosome profiling.

RNase protection assay

Also known as "nuclease footprinting," RNase protection is a critical step to protect/spare ribosome-bound regions in mRNAs from unbound regions from digestion by ribonucleases. Reaction time and enzyme concentration are critical to regulate this step. RNase I and micrococcal nuclease (MNase) are commonly used ribonucleases in ribosome profiling experiments.

Isolation of ribosome footprints

This is essentially a ribosome/polyribosome purification step involving different strategies to isolate ribosome-protected fragments. Differential sedimentation of ribosomes through a sucrose cushion by ultracentrifugation is one of the most common strategies to isolate ribosome-enriched fractions. The ribosome fraction is then purified in a denaturing polyacrylamide gel by electrophoresis in order to separate the complexes based on length/size. With the help of appropriate size markers, the positions of complexes are viewed in a dark field transilluminator and the samples are retrieved by cutting pieces of gel corresponding to the 28–30 nt position.

High throughput sequencing

After disrupting the gel slices, precipitation and re-purification of ribosome footprints, the samples are used for library preparation and sequencing. Library preparation involves steps that are common in many high-throughput sequencing experiments, like end repair, 3′ adaptor ligation, reverse transcription and cDNA purification, circularization of cDNA, and PCR amplification. After checking the length and concentration of the ribosome footprints library, the samples are sequenced. Ribosome profiling, based on deep sequencing of ribosome-protected mRNA fragments, has proven to be powerful in defining ribosome positions on the entire transcriptome.

Bioinformatics analysis

Finally, advanced bioinformatics tools are applied to ensure quality control of the raw data, mapping, count normalization, and gene expression levels (Ingolia et al., 2012).

Details of the ribosome profiling protocol can be found elsewhere (Esposito et al., 2010).

Other critical parameters

(a) *Use of translation inhibitors*

Proper selection of the translation initiation site on mRNAs is crucial for the production of desired protein products. Therefore precise determination of translation initiation codons across the entire transcriptome has been a fundamental and long sought goal toward understanding translational regulation (Lee et al., 2012). Ribosomal scanning is a well-accepted model for start codon selection. It is a general assumption that the first AUG codon in mRNAs serves as the start site for translation (Fig. 12.2A)

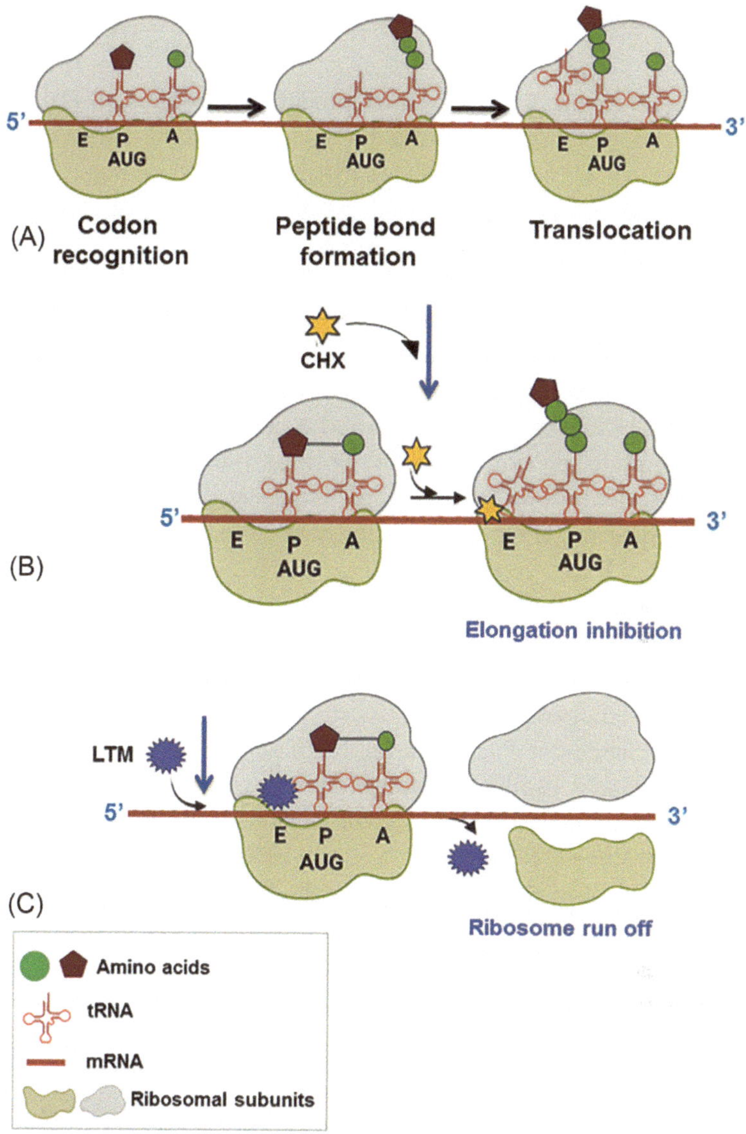

Fig. 12.2 Inhibition of translation at initiation and elongation stages. (A) The small subunit of the ribosome has three binding sites: an aminoacyl site (A), a peptidyl site (P), and an exit site (E). The initiator tRNA molecule carrying the amino acid methionine binds to the AUG start codon (codon recognition) of the mRNA transcript at the ribosome's P site where it will become the first amino acid incorporated into the growing polypeptide chain (peptide bond formation). After the peptide bond is formed, the ribosome translocates, thus causing the tRNA to occupy the E site. The tRNA is then released to the cytoplasm to pick up another amino acid. As the initiator tRNA moves to the peptidyl (P)-site, it renders the E-site transiently accessible; (B) Cycloheximide (CHX) binds to the exit (E) site of the large ribosomal subunit, close to the position where the 3′ hydroxyl group of the deacylated transfer RNA (tRNA) normally binds. CHX, thus, prevents the release of deacylated tRNA from the E-site. Thereby it blocks the translocation step of the elongation phase; (C) Lactimidomycin (LTM), on the other hand, binds to the free E-site and acts preferentially on the initiating ribosome thereby inhibiting the end of initiation phase but not the elongating ribosome. Thereby, LTM halts translation initiation while allowing elongating ribosomes to run off.

(reviewed in Kozak, 2002). Ribosome profiling in combination with specific translation inhibitors (Table 12.1) has proven to be a powerful tool toward understanding the underlying mechanisms of translation initiation and unambiguous identification of alternative start sites, especially those associated with overlapping ORFs (Ingolia et al., 2011).

Cycloheximide (CHX) is widely used in ribosome profiling of eukaryotic cells because it specifically targets translation elongation and stabilizes ribosomes on mRNAs (Lee et al., 2012). There is a significant accumulation of ribosomes at the initiation sites of actively translated mRNAs in the presence of CHX. Both biochemical and structural studies have revealed that CHX binds to the exit (E) site of the large ribosomal subunit and subsequently blocks ribosomal translocation. However, binding of CHX to the E-site alone does not affect translocation while binding of both CHX and deacylated tRNA stalls translocation (Fig. 12.2B).

Lactimidomycin (LTM) is another potent translation inhibitor isolated from Streptomyces spp. (Ju et al., 2009). Being larger in size than CHX, LTM does not bind to the E-site and thus does not affect the peptide bond formation but efficiently stalls the 80S ribosome at the AUG start codons (Fig. 12.2C) (Schneider-Poetsch et al., 2010).

Harringtonine is another translation initiation inhibitor which binds to the peptidyl transferase center in disassembled 60S large ribosomal subunits. It preferentially arrests ribosomes at Lys, Arg, or Tyr codons and poisons the entire cellular pool of ribosomes within a few minutes. It has been specifically used to mark the translation initiation sites in mRNAs, while a combination of harringtonine first and cycloheximide at different time points has been used for measuring the speed of translation elongation (reviewed in Stern-Ginossar and Ingolia, 2015).

Puromycin drives premature termination, thereby removing most elongating ribosomes and leaving footprints predominantly at sites of initiation (Fritsch et al., 2012). Further, puromycin can be used in combination with LTM in order to stabilize initiating ribosomes while depleting elongating ribosomes in cells (Gao et al., 2014). Translation inhibitors like chloramphenicol and 5′-guanylyl imidodiphosphate have also been used to reveal specific aspects of translation.

In addition to antibiotics and chemical inhibitors, the process of translation can also be stopped by flash freezing using liquid nitrogen or dry ice.

(b) *Precautions for handling RNA samples*

Special precautions must be taken when working with RNA. Since RNA has the normal 2′ hydroxy group, it makes the phosphodiester

Table 12.1 List of translation inhibitors.

Translation inhibitors	Targets	Mechanism of action
Harringtonine	Binds to 60S subunit and form 80S ribosomal complex with the initiator tRNA	Blocks aminoacyl-tRNA binding in the A-site and peptide formation, depletes elongating ribosomes from mRNAs, thereby halting ribosomes at initiation codons
Lactimidomycin	Binds to the 80S ribosome already assembled at the initiation codon and permits the formation of the first peptide bond	Blocks translocation by occupying the empty Exit (E)-site of initiating ribosomes thereby inhibit initiation
Puromycin	Binds to the ribosomal A site and participates in peptide bond formation, producing peptidyl-puromycin. It does not engage in translocation and quickly dissociates from the ribosome	Premature termination of (polypeptide synthesis) elongating ribosomes
Cyclohexamide	Binds to the exit (E)-site of the large ribosomal subunit, close to the position where the 3′ hydroxyl group of the deacylated transfer RNA (tRNA) normally binds thus prevents the release of deacylated tRNA from the E-site	Inhibit elongation speed, i.e., ribosomal eEF2-mediated translocation
Anisomycin (Flagecidin)	Interferes with protein synthesis by inhibiting peptidyl transferase or the 80S ribosome	Inhibits eukaryotic protein synthesis, i.e., inhibit elongation
Chloramphenicol	Binds to 23S rRNA of the 50S ribosomal subunit	Blocks the peptidyl transfer step of elongation in both bacteria and mitochondria
Emetine	Irreversibly binds the ribosomal 40S subunit, inhibiting aminoacyl-tRNA transfer but does not prevent transpeptidation or translocation of peptidyl-tRNA from acceptor to donor sites	Inhibits elongation speed, i.e., translocation
Tigecycline	Binds to the 30S ribosomal subunit of bacteria and thereby block the interaction of aminoacyl-tRNA with the A site of the ribosome	Exhibits bacteriostatic activity and prevent peptide elongation

bond unstable and susceptible to nucleophilic attack and self-hydrolysis. Therefore proper care must be taken to avoid contamination with RNases that are found everywhere, including on the laboratory worker's hands and in airborne microorganisms. All reagents and equipment must be specially treated to inactivate RNases prior to use. RNase-free glassware or plasticware should be used for preparing as well as storing all buffers. Glassware can be baked, whereas plasticwares and water for preparing reagents need to be treated with 0.1% (*v*/*v*) diethyl pyrocarbonate (DEPC) to eliminate environmental RNases (Hu and Coller, 2013). However, use of protease cocktails containing EDTA (a Mg^{++} chelator and inhibitor of metalloproteases) should be avoided because Mg^{++} is required for the association of small and large ribosomal subunits (Pringle et al., 2019).

Quality control of mRNA

Cells may make mistakes during gene expression at the mRNA level. To ensure error-free translation, cells have evolved several quality control mechanisms in the nucleus and the cytoplasm to minimize the damage as a consequence of such mistakes. Along with multiple quality controls that occur during mRNA biogenesis, each transcript undergoes a rigorous quality control (QC) check to ensure the integrity of protein-encoding gene expression. The improperly processed mRNAs or mRNAs without termination codons are immediately degraded in order to prevent the expression and accumulation of aberrant proteins (Behm-Ansmant and Izaurralde, 2006; Isken and Maquat, 2007). The QC responses during translation are signaled by slow movement or stalling of elongating ribosomes that could be reflected in the form of translational repression, mRNA decay, ribosome rescue, and/or degradation of newly synthesized proteins (Fig. 12.3). The canonical mRNA decay is primarily signaled by slow-moving ribosomes, whereas other errors in mRNAs signal more comprehensive QC via nonsense-mediated decay, nonstop decay, and no-go decay pathways (D'Orazio and Green, 2021; Isken and Maquat, 2007). Examples of some key QC mechanisms associated with translation are given below:

Canonical decay of mRNA and connections to translation elongation

The canonical mRNA decay occurs as a cotranslational process and the majority of mRNAs are initially degraded by this process (Hu et al., 2009;

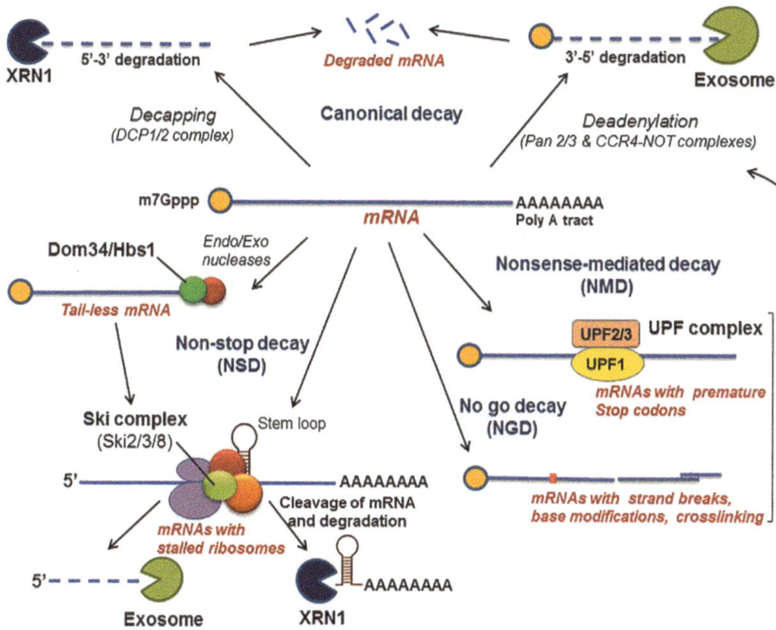

Fig. 12.3 Quality control of mRNA in the cell. The canonical mRNA decay is primarily signaled by slow-moving ribosomes, whereas other errors in mRNAs signal a more comprehensive quality control (QC) via nonsense-mediated decay, nonstop decay, and no-go decay pathways. Multiple macromolecular complexes and nucleases participate in the RNA degradation process. The canonical mRNA decay occurs as a cotranslational process and the majority of mRNAs are initially degraded by this process. It begins with a shortening and removal of the poly-A tail (deadenylation) followed by decapping and degradation by exonucleases. Other errors in mRNAs signal a more comprehensive QC via nonsense-mediated decay, nonstop decay, and no-go decay pathways.

Pelechano et al., 2015). It begins with a shortening and removal of the poly-A tail (deadenylation) with the help of Pan2-3 and CCR4-NOT complexes that carry exonucleases at their core. Pan2 in the Pan2-3 complex functions as the exonuclease, whereas Caf1 and Ccr4 are nucleases of the CCR4-NOT complex (Chen and Shyu, 2011; Wahle and Winkler, 2013; Webster et al., 2018; Yi et al., 2018). The Pan2-3 complex trims longer (>150 nt) poly-A tails in mammalian cells. Following poly-A shortening, CCR4-NOT stimulates the ATPase activity of Dhh1 (a DEAD Box protein) to engage decapping regulators for removal of 5′ Cap. It also signals the major 5′ to 3′ exonuclease XRN1 to rapidly decay the mRNA (Mugler et al., 2016).

Nonsense-mediated decay of mRNA

Nonsense-mediated decay (NMD) is a surveillance pathway for mRNAs that participates in diverse cellular functions. Its main role is to maximize error-free gene expression by eliminating transcripts that carry some mutations or premature stop codons. Proteins expressed from such aberrant mRNAs could have deleterious gain-of-function or dominant-negative activity (Chang et al., 2007; He et al., 2008; Hori and Watanabe, 2008). The NMD-mediated degradation of target mRNAs in yeast and human cells is initiated by removal of the cap structure by the DCP1–DCP2 decapping complex. Following decapping, the mRNA is exposed to degradation by 5′–3′ exonuclease XRN1 (Lejeune et al., 2003; Medina et al., 2014). Alternatively, the decay of mRNAs with a premature termination codon can be accelerated by deadenylation and degradation with the help of 3′–5′ exonuclease activity of exosomes (Couttet and Grange, 2004; Lejeune et al., 2003; Mitchell and Tollervey, 2003). Three trans-acting factors called "upframeshift (UPF)" proteins constitute the trimeric core complex of the NMD pathway, namely UPF1, UPF2, and UPF3 (Gehring et al., 2005). The formation of this conserved core complex is supported by genetic, biochemical, and structural evidences. UPF1 is a RNA helicase with a key effector function in the NMD pathway UPF2, which carries three MIF4G (middle portion of eIF4G) domains and directly binds to UPF3 (an RNA-binding protein with an RNP domain). Assembly of the trimeric complex triggers the phosphorylation of UPF1 by the serine-threonine kinase SMG1 (Yamashita et al., 2001). The surveillance complex then recruits RNA degradation enzymes for a rapid decay of the target mRNA (Conti and Izaurralde, 2005; Lejeune and Maquat, 2005). The absence of any of these factors can stabilize the aberrant mRNAs and inhibit the NMD pathway.

No-go decay of mRNAs with disruptions within the ORF

Like DNA, damage in RNA can also take place in the form of strand breakage, cross-linking, or chemical modification of nucleotides, all of which could result in problematic mRNAs with disruptions within the ORF (Yan and Zaher, 2019; Tuck et al., 2020). The disruptions in mRNAs could be encoded or environmental but the damage inflicted by the environment is more common. Encoded disruptions in the mRNA could include strong natural pauses on specific peptide motifs, collections of rare codons, or some secondary structural elements (Doma and Parker, 2006). The regulatory purpose of such disruptions is unclear, but it seems to impede ribosomal elongation during translation and the incomplete gene products expressed by these mRNAs are often toxic to cells. mRNAs containing a stall in the middle of an ORF are targeted for decay. This process is broadly referred to as "No Go Decay" or "NGD" (Harigaya and Parker, 2010). The predominant pathway for NGD involves an endonucleolytic mRNA cleavage. The decay intermediates are subsequently processed by exonuclease XRN1 and the exosome.

Nonstop decay of mRNAs

Nonstop decay (NSD) is a cellular mechanism of mRNA surveillance to detect mRNA molecules without a stop codon and prevent their translation (Klauer and van Hoof, 2012; Venkataraman et al., 2014). Further, NSD seems to involve mRNA decay factors that are different from NMD. Degradation of mRNAs via the NSD pathway is strongly dependent on a multiprotein Ski complex consisting of three main proteins: the RNA helicase Ski2, Ski3, and Ski8. The Ski complex recognizes the stalled ribosomes in the poly-A tail region and effectively pulls mRNAs from a translating ribosome until the mRNA dissociates from the ribosome entirely. As the Ski complex pulls the mRNA from the ribosome, the exosome binding factor Ski7 (bound to the complex) can thread it directly into the exosome for degradation. Therefore the levels of mRNAs containing internal stalls in their ORFs are unaffected on deletion of the Ski complex but are restored on deletion of Xrn1.

Degradation of truncated mRNAs

The Ski complex described in the previous section also participates in the degradation of cytoplasmic mRNAs that lack a poly-A tail. These mRNA species are generated by endonucleolytic (e.g., Ire1 or Cue2) or exonucleolytic processing of normal transcripts. When a ribosome encounters the 3′ end of a tailless mRNA, Dom34:Hbs1 is thought to recognize the empty A-site and catalyze ribosome subunit dissociation or ribosome rescue (Shoemaker et al., 2010; Tsuboi et al., 2012). The truncated mRNA with a free 3′ end (and without a poly-A tail and a ribosome) is used as a substrate for the Ski complex and progressive mRNA degradation.

Understanding the mechanism and regulation of translation

The process of translation involves three consecutive steps, viz., initiation, elongation, and termination. According to the canonical model of translation, it begins soon after the formation of a 43S preinitiation complex (PIC) involving methionyl-tRNAi, the small ribosomal subunit (40S in eukaryotes and 30S in prokaryotes), and eukaryotic initiation factor 2 (eIF2), among other factors, is recruited to the m7G-capped 5′ ends of mRNAs (reviewed in Gobet and Naef, 2017). The PIC scans the 5′-untranslated regions (UTR) of mRNAs until the AUG start codon is recognized. Consequently, the large subunit (60S in eukaryotes and 50S in prokaryotes) is recruited and the resulting 80S complex is ready for decoding the genetic information and elongation of the peptide chain (reviewed in Hinnebusch and Lorsch, 2012). The prevalence of alternative start sites was previously not appreciated. However, evidence for shorter and potentially less stable protein products from non-AUG initiation has emerged in parallel. Finally, translation terminates when the ribosome recognizes one of the stop codons (UGA, UAG, or UAA). Ribosome profiling has allowed the capture of different steps of translation with amazing accuracy.

Ribosome profiling of initiating ribosomes

Translation is thought to be mainly regulated at the initiation step via global or mRNA-specific mechanisms. Therefore identifying translation initiation sites and initiating factors on transcripts are crucial for annotating translated reading frames. Ribosome profiling has been adapted to find translation initiation sites (TIS) by trapping and footprinting initiating ribosomes with the specialized translation inhibitors. The first experimental mapping of TIS was performed in murine mesenchymal stem cells using translation inhibitor harringtonine (Ingolia et al., 2011). This inhibitor binds to the 60S subunit and blocks the binding of aminoacyl-tRNA at the A-site and peptide formation. In this study, it was observed that the majority (65%) of these transcripts had more than one detectable TIS with 16% containing four or more sites. In fact, the unexpected ribosome occupancy seen in many parts of the transcripts could be due to non-AUG initiation in a substantial part of the population (Ingolia et al., 2012). In another experiment, human cytomegalovirus-infected cells were treated with harringtonine or LTM. It was observed that > 98% of the initiation sites detected using harringtonine were also detected using LTM (Stern-Ginossar et al., 2012). Although the mechanism of action of the two drugs is different, both arrested ribosomes mostly at the same locations. In many mRNAs, translation initiates on very short ORFs, which are unlikely to produce functional peptides and thus could be a gene expression noise.

In addition to the translation initiation sites, a number of initiation factors also participate to regulate translation during the initiation phase. The mechanistic target of the rapamycin complex 1 (mTORC1) pathway serves as the master regulator of translation, mostly through phosphorylation of eIF4E binding proteins (4E-BPs) and their subsequent release from eIF4E, allowing the formation of the eIF4F complex (Thoreen et al., 2012). In addition, mTORC1 regulates other factors such as ribosomal protein S6 kinase as well as ribosome biogenesis and tRNA synthesis (Hsieh et al., 2012).

Non-AUG translation—In many prokaryotes, codons such as GUG, CUG, and UUG are frequently used as initiator codons. However, a recent study suggests that over 74% of the upstream TISs in mammalian transcripts are non-AUG codons, often associated with short uORFs (Lee et al., 2012). Translation initiation from non-AUG codons may produce proteins with N-terminal extensions and sometimes unpredictable functions. For example, the long isoform of the tumor suppressor PTEN may be secreted or have altered intracellular distribution and activity inside the cell (Gil et al., 2016). In yeast, initiation at non-AUG codons is observed as a response to starvation and coincides with the presence of uORFs in the 5′-UTRs by ribosomal footprint (Eastman et al., 2018).

Alternate protein isoforms—Ribosome profiling is also suitable for detecting protein isoforms with initiation at alternative sites both upstream and downstream of the annotated protein-coding ORFs. Protein isoform variants can also be generated by pre-mRNA splicing. Analysis of 4994 genes expressed in mouse embryonic stem cells identified 570 genes with potential N-terminal extensions and 870 with N-terminal truncations (Ingolia et al., 2011). Distinct protein isoform variants can display different and even opposing functions, similar to many dominant-negative alleles. For example, internal translation initiation in the transcription factor C/EBP produces truncated mutants without a trans-activation domain but with DNA-binding domains. These mutants compete with full-length C/EBP for DNA binding and inhibit transactivation function (Ossipow et al., 1993). The frequent truncations observed in ribosome profiling data suggest that inhibitory isoforms may be quite widespread (Michel and Baranov, 2013) and highlight its usefulness in improving the existing annotations.

Ribosome profiling of elongating ribosomes

Growing evidences suggest that the translation elongation rate plays a major role in the cotranslational folding of nascent peptide chains and thus fine-tuning of gene expression. During the elongation phase, ribosomes undergo substantial conformational changes to accommodate incoming aminoacyl-tRNAs and slide along the mRNA template (Wu et al., 2019). As a reason, mRNAs with slow-moving ribosomes tend to yield better folded and more stable proteins. Moreover, the ribosome elongation rate can also influence the mRNA half-life, thereby regulating gene expression. Ribosome profiling has helped in throwing new insights into the elongation process of translation.

Analysis of elongation: Once the translation has been initiated, the ribosome starts synthesizing protein through the elongation process. Thus ribosome profiling of elongating ribosomes is important for studying differential gene expression, measuring global and local translation elongation rates, and identifying novel genes and gene products. In these experiments, cells are pulse-treated with translation elongation inhibitors to immobilize and faithfully capture elongating ribosomes in their in vivo translational positions prior to cell lysis. Cycloheximide has been widely used in the elongating ribosome profiling studies. However, simple liquid nitrogen freezing as well as other antibiotics such as "emetine" in eukaryotes and "chloramphenicol" in bacteria have also been used. The elongation rate for mammalian cells using ribosome profiling has been estimated at five amino acids per second (Wu et al., 2016).

Analysis of differential gene expression: Knowing changes in gene expression in a cell in response to changing conditions is essential for understanding the genetic determinants of phenotypical behavior. Though microarray and RNA-seq techniques have been extensively used for measuring differential gene expression, the correlation between mRNA and protein levels has been insufficient and inaccurate for predicting protein expression in the cell. This could happen when a ribosome is stalled on an mRNA or translation is limited to upstream open reading frames (uORFs) hindering translation of the main protein product ORF. Further, two transcripts expressed at the same level but of different length would produce a different number of short reads aligning to them, as the number of reads is proportional to the length of the transcript. Since ribosomes broadly translate mRNAs at a similar elongation rate, conversion of the absolute number of footprints into ribosome density is used for estimating translation rates. Besides, the triplet periodicity property of ribosome profiling and the generation of subcodon profiles are also used (Michel and Baranov, 2013). Thus ribosome profiling studies combined with the application of some computational tools on the Ribo-seq and RNA-seq data have been successfully used for getting better insights into the differential translation of transcripts.

Role of tRNA landscapes—The tRNA landscape of cells (including the availability of aminoacyl-tRNAs and specific modifications on tRNAs) is considered to be a major determinant of variation in elongation rate. Not surprisingly, an altered tRNA landscape is frequently observed in malignant cells. In fact, the analysis of the gene-specific elongation velocity index in different cell types has shown that the elongation rates are closely linked to their respective tRNA composition (Lian et al., 2016). Consequently, mRNAs with slow elongation indices generally produce correctly folded and more stable proteins. Thus a reduction in elongation velocity may be necessary for the maintenance of cancer cells. Besides an upregulated tRNAs seen in highly metastatic breast cancer cells could promote the stability and translation of transcripts that are enriched for the corresponding codons (Goodarzi et al., 2016). It means that there could be reshaping of protein expression through modulation of ribosome occupancy and transcript stability. Further, in renal cell carcinoma, there is high ribosome occupancy on proline codons due perhaps to the limiting availability of aminoacylated tRNA for proline (Goodarzi et al., 2016).

Thus ribosome profiling not only can measure tRNA levels, low availability of amino acids in cells, and compensatory mechanisms in tumors, but can also reveal the impact of ribosome elongation on the fate of transcripts and protein expression (Goodarzi et al., 2016; Ingolia et al., 2011; Lian et al., 2016; Loayza-Puch et al., 2016; Rooijers et al., 2013; Wu et al., 2016).

Ribosome profiling of translation termination

Termination of mRNA translation occurs when a stop codon (UAG, UAA, or UGA) enters the A site of the ribosome, leading to successful completion of protein synthesis. In eukaryotes, translation termination is mediated by two release factors (RFs), viz., eRF1 and eRF3, which form a ternary eRF1/eRF3-guanosine triphosphate (GTP) complex. Here eRF1 recognizes the stop codon, whereas eRF3 promotes the hydrolysis of the ester bond in peptidyl-tRNA in the P site and mediates the release of the nascent peptide and disassembly of ribosomes (Hellen, 2018). Failure of translation termination results in ribosome read-through and misincorporation of a near-cognate tRNA and elongation of a peptide chain on the mRNA. Thus ribosome profiling has been useful in elucidating the determinants of translation termination as opposed to the continued elongation process of stop codon read-through in vivo. In addition to canonical RFs, it has helped in uncovering the function of Dom34 in yeast—a distant homolog of release factor eRF1 (Guydosh and Green, 2014). Dom34 was shown to dissociate the stalled ribosomes in vitro along with Hbs1 (a homolog of eRF3) and Rli1. While Dom34 facilitated termination/recycling on certain intact mRNAs, Rli1 appeared to be the principal recycling factor since the Rli1-depleted yeast cells show accumulation of 80S ribosomes at stop codons (Guydosh and Green, 2014).

The ribosome is also known to change its reading frame during translation through a process called "programmed ribosomal frameshifting" that allows controlled slippage of the translating ribosome to alternative reading frames. The translational frameshifting is frequently observed during translation in many viruses but infrequent in higher eukaryotes (Mikl et al., 2020). For example, ribosome profiling in the ciliate *Euplotes* exhibits a widespread frameshifting at stop codons. Here the termination of translation is dependent on the proximity of a stop codon to the 3′ mRNA end, whereas internal stop codons induce either +1 or +2 frameshifts depending on the preceding codon (Lobanov et al., 2017).

Diversity and applications of ribosome profiling

First developed in yeast, ribosome profiling is now established as a global translatome analysis technology that is navigating important biological discoveries in various living systems. This is exemplified by some key observations given below:

Effect of stress on translation

Protein synthesis is an expensive anabolic process and therefore is sensitive to the available nutrition, in particular, amino acids. Therefore response to cellular stress is considered to be a hallmark of global translation repression attributed mainly to the inhibition of translation initiation. To study the effect of nutrient deprivation on protein synthesis, yeast cells were deprived of amino acids for 20 min and changes at the translational level were measured. Elevated expression of nearly one-third of the 3769 genes and a dramatic sevenfold increase in the translation of GCN4 mRNA were observed by ribosome profiling, which otherwise could not be observed earlier by routine polysome profiling. This example illustrated a distinct advantage of ribosome profiling over polysome profiling, which allowed distinguishing mRNAs with efficiently translated coding regions from those where only the 5′UTRs were translated (Ingolia et al., 2019). In another study, the translational response to oxidative stress was investigated in yeast cells treated with hydrogen peroxide, and many genes whose expression was altered at the transcriptional and translational levels were identified using ribosome profiling (Gerashchenko et al., 2012). The results showed a marked increase of ribosome occupancy in the upstream open reading frames (uORFs) as well as changes in ribosome densities along coding sequences which could be blocked in the presence of translation inhibitors such as cycloheximide.

In mammalian cells, stress often triggers an unexpected global ribosome pausing as shown by stacking of ribosomes during translation. In case of heat shock, ribosomes seem to accumulate in the first 200 nucleotides of transcripts apparently caused by formation of misfolded proteins (Liu et al., 2013a). Glucose and oxygen deprivation also lead to increased ribosome density in the 5′ ends of mRNAs often showing differential translation efficiency of some genes such as genes related to oxidative phosphorylation (Andreev et al., 2015a). Interestingly, the upstream open reading frames of some mRNAs maintain a constant translation level that provides resistance to stress response (Andreev et al., 2015b). Further, in murine tumor cells, the methylation of tRNAs and their subsequent cleavage seems to be a regulatory mechanism to control of translation under stress conditions (Blanco et al., 2014).

Cell cycle and translation regulation

Proliferating cells require a broad range of biochemical mechanisms to regulate the dynamic process of the cell cycle. Use of specific inhibitors allows the arrest of proliferating cells in different phases of the cell cycle. For example, a single thymidine block could arrest cells in the S phase, whereas a double thymidine block is known to arrest cells in the late G1 phase. On the other hand, nocodazole is known to interfere with microtubule polymerization and spindle formation and arrest cells before metaphase. Ribosome profiling of HeLa cells grown in the presence of specific cell cycle inhibitors such as thymidine or nocodazole indeed revealed a wide range of translational control (Stumpf et al., 2013). In another study, a cyclin-dependent kinase 1 (CDK1) inhibitor RO-3306 was used to arrest RPE-1 cells in the G2 phase. When the inhibitor was removed, the cells resumed balanced growth in a relatively short time after release and nearly half of them progressed to the next phase (Aramayo and Polymenis, 2017). Interestingly, the ribosome occupancy also exhibits fluctuations during its different phases of the cell cycle. For example, ribosome profiling in cells treated with silvestrol (a specific eIF4A inhibitor) revealed a novel 5′-UTR "sequence signal" that worked independently of the mTOR-eIF4E-eIF4G pathway (Rubio et al., 2014). On the other hand, eIF4A regulated mostly cell cycle-related genes with long and structured 5′-UTRs or the G-quadruplex (Wolfe et al., 2014).

The length of the poly (A) tail in mRNAs has been considered to be a key factor influencing their stability and translation regulation. However, the ribosome profiling combined 3′-end sequencing during the cell cycle revealed a limited impact of the poly (A) tail length on translation regulation. On the contrary, the translation efficiency of genes was linked to shorter tails (of less than 20 nucleotides) in the transcripts (Park et al., 2016).

Ribosome profiling during viral infections

Viruses as obligate intracellular parasites need the host cell machineries of their host cells for their genome replication, particle assembly, and release of new viral progeny. Many viruses have evolved enzymatic machinery required for: (a) replication or transcription of their genome; (b) shutting down host machinery to favor viral translation; (c) triggering the decay of host mRNAs and transferring their 5′-7-methylguanosine cap on viral transcripts. Also, some viruses have evolved "internal ribosome entry sites" (IRES) in their genome/transcripts for directly recruiting ribosomes for translation and mechanisms for initiation at non-AUG codons, programmed ribosomal frame-shifting, or stop codon read-through. Ribosome profiling may offer new parameters for better annotation and understanding of viruses and their complexity. In fact, as illustrated below, ribosome profiling has been extensively used to annotate gene expression and understand disease biology during viral infections (Stern-Ginossar and Ingolia, 2015).

Human cytomegalovirus (HCMV)—HCMV is a beta-herpes virus that infects most humans and may cause severe disease in newborns and immune compromised adults. The 240 kb DNA genome of HCMV is estimated to express between 165 and 252 proteins. Ribosome profiling and systematic transcript analysis have been used to prepare a comprehensive map of the polypeptides encoded by the viral genome and understand their temporal expression during infection (Stern-Ginossar et al., 2012). The findings were full of surprises as this led to the identification of: (i) internal ORFs lying within existing ORFs either in frame (resulting in N-terminally truncated translation products) or out of frame (resulting in previously unknown polypeptides); (ii) upstream ORFs lying upstream of canonical ORFs; (iii) ORFs within transcripts antisense to canonical ORFs; and (iv) previously unidentified short ORFs encoded by distinct transcripts. Besides, many of these ORFs started at non-AUG codons (such as CUG) and several long RNAs lacked canonical ORFs.

Kaposi's Sarcoma-associated Herpesvirus (KSHV)—This virus belongs to the gamma herpesvirus family and is the etiologic agent of Kaposi's sarcoma. KSHV genome exists as a linear dsDNA (165–170-kb in length) in the virus capable of encoding nearly 100 viral proteins as well as several undocumented short ORFs and putative protein isoforms. However, RNA sequencing combined with ribosome profiling during the lytic cycle allowed accurate mapping of the viral genome. The current annotation of the genome shows over 90 genes that encode for viral proteins and both long and small noncoding RNAs (Arias et al., 2014).

Vaccinia virus (*VACV*)—VACV is a prototype poxvirus that was used as the vaccine to eradicate smallpox. It has a linear, double-stranded DNA genome approximately 190 kbp in length with more than 200 annotated ORFs. Using ribosome profiling, most of the VACV's known ORFs were identified and additional putative ORFs were annotated; most occurred within previously known ORFs, and a few were found in untranslated leader sequences, antisense strands, and intergenic regions (Yang et al., 2015).

Bacteriophage λ—This bacterial virus is one of the most extensively studied entities and has been a primary model for understanding basic modes of genetic regulation. Ribosome profiling has been used to examine the progress of *λ* gene expression during phage development (Liu et al., 2013b). Several ORFs were discovered and a few of those appeared to be expressed at levels similar to those of well-known prophage-specific genes. Regulation of late gene expression in *λ* and related phages occurs through transcription antitermination.

A known frameshift between the *λ* genes G and T results in a fusion protein between these reading frames when approximately 3.5% of the ribosomes slip back by one nucleotide at the so-called slippery sequence. This slippage results in the accumulation of ribosomes over this sequence perhaps to ensure availability of mRNA for the translation of other parts of the message and also provide stability. Thus functional ORFs might be retained in regions of the genome where transcription occurs, even if no functional polypeptide is produced.

The data obtained from ribosome profiling suggest that it is possible to quantify the rate of viral protein production throughout infection and thus provide better insights into biological functions and disease progression. This is especially useful in the case of large DNA viruses, in which viral proteins are expressed with well-defined kinetics and temporal fashion. However, in the case of RNA viruses, a polyprotein expressed from a single mRNA is often used to generate multiple proteins in equimolar amounts by proteolytic cleavage. In this case, ribosome profiling can help in estimating fine differences in the production rates of different viral proteins whether required for enzymatic or structural functions.

Ribosome profiling during parasitic infections

This powerful technique of ribosome profiling has been extensively used to understand the developmental regulation of protein production and new gene discovery in several protozoan parasites, including trypanosomes *Trypanosoma brucei* and *Trypanosoma cruzi,* and malarial parasite *Plasmodium falciparum.* Ribosome profiling has been successfully used to identify new protein-coding sequences (CDSs) and mark the boundaries of previously annotated CDSs in *T. brucei*. Further, it indicated that translation efficiencies vary widely between genes and for the same gene across stages. For example, in *T. brucei,* the Tb927.7.180 gene shows ~13-fold more mRNA expression in the bloodstream form (the actively dividing long slender form in the mammalian host) than in the procyclic cultured form. However, the ribosome profiling exhibited 70-fold more protein production indicating a change in the expression of gene changes during development (Parsons and Myler, 2016). Analysis of the functional categories of genes suggested a high expression of structural proteins of the cytoplasmic ribosome. In another study of *T. cruzi* epimastigotes and metacyclic trypomastigotes, it was observed that nearly 2200 CDSs were translated in epimastigotes, but not in metacyclic trypomastigotes, whereas 526 CDSs were exclusively translated in metacyclic forms allowing a direct comparison between dividing (epimastigotes) and nondividing (metacyclic trypomastigotes) insect stages (Smircich et al., 2015). Interestingly, the latter group was enriched for families of surface proteins expressed in metacyclic forms such as trans-sialidases, gp63 protease, and MASP proteins.

Translational regulation is also considered important for different stages of the life cycle of *Plasmodium* spp., especially the asexual erythrocytic stages. Ribosome profiling from *P. falciparum* revealed no apparent change in the global translation efficiencies in any particular stage of the life cycle. Further, a majority of the upregulated genes were predominantly related to merozoite egress and invasion function, with peak mRNA in schizonts and peak translational efficiency in rings (Caro et al., 2014). The timing of translation is fine-tuned by the binding of RNA binding protein *Pf*Alba1 to a subpopulation of asexual stage mRNAs (Vembar et al., 2015). It would be interesting to apply the ribosome profiling method for sporozoite (from the mosquito salivary gland) or gametocyte stages of Plasmodium, where there is striking evidence of translational repression of specific mRNAs.

Ribosome profiling in *Escherichia coli*

The ribosome profiling method has also been developed and validated to study the translation status of mRNAs and its variations in *E. coli* as it allows parallel quantification of the specific translational response to changing gene expression. The studies were conducted for the translational regulation of the *cysZ* gene in the "arabinose operon" that encodes a transporter protein for sulfate uptake and cysteine synthesis (reviewed in Schleif, 2000) and the regulation of the *lacZ* gene under the "lac operon" that codes for a β-D-galactosidase enzyme is involved in lactose and other β-galactoside catabolism (reviewed in Osbourn and Field, 2009). The arabinose operon is under the transcriptional control of the P_{BAD} promoter and is inducible under high arabinose and low glucose conditions. The lac operon is under the regulatory control of the lac repressor that can be induced in the presence of high lactose. When these genes were induced, the increase in *lacZ* mRNAs was considerably higher than the noninducible genes, whereas the increase was only marginal for *cysZ* mRNAs compared to the noninducible genes. Nevertheless, a similar overall translation of both *cysZ* and *lacZ* mRNAs was observed due to a higher ribosome loading suggesting cotranscriptional regulation of translation (Zhang et al., 2014). Thus cotranscriptional regulation of *cysZ* and *lacZ* translation contributes to physiological adaptation when cells regulate the mRNA level of genes to adapt to environmental changes. These observations confirmed that the ribosome profiling method is useful in measuring the translational response of a gene following changes in its mRNA level in the cell.

Summary and conclusions

The advent of Ribosome profiling has uncovered key concepts and mechanisms underlying the regulation of protein production in different organisms and systems. It has also explained the broad range of mechanisms impacting translation initiation and elongation, as well as the interplay between these different steps and their implications in disease. Besides, the experiments in mammalian cells and tissues have revealed genome-wide and dynamic regulation of translation.

Ribosome profiling has also made it possible to monitor translation with unprecedented precision and deep insights. It provides fine details of protein synthesis, including information about the protein per se like its starts and stops, rate of synthesis, and also about its regulation. Such detailed and quality information is unlikely to be generated by just measuring the mRNA levels. Besides, the latter method is laborious and normally yields qualitative information about protein synthesis. Thus ribosome profiling has enabled us to identify and measure micro-variations in the translation process, total versus individual transcripts, and association of specific translation factors. It has also allowed us to measure localized translation within cells with the help of biotin-tagged ribosomes and identification of locally translated transcripts and their subcodon usage. Thus ribosome profiling has bridged the existing technological gap between our abilities to simultaneously quantify the transcriptome and the proteome through its genome-wide applications.

References

Andreev DE, O'Connor PB, Zhdanov AV, Dmitriev RI, Shatsky IN, Papkovsky DB, Baranov PV. Oxygen and glucose deprivation induces widespread alterations in mRNA translation within 20 minutes. Genome Biol 2015a;16:90.

Andreev DE, O'Connor PB, Fahey C, Kenny EM, Terenin IM, Dmitriev SE, Cormican P, Morris DW, Shatsky IN, Baranov PV. Translation of 5′ leaders is pervasive in genes resistant to eIF2 repression. Elife 2015b;4, e03971.

Aramayo R, Polymenis M. Ribosome profiling the cell cycle lessons and challenges. Curr Genet 2017;63:959–64.

Arias C, Weisburd B, Stern-Ginossar N, Mercier A, Madrid AS, Bellare P, Holdorf M, Weissman JS, Ganem D. KSHV 2.0: a comprehensive annotation of the Kaposi's sarcoma-associated herpesvirus genome using next-generation sequencing reveals novel genomic and functional features. PLoS Pathog 2014;10, e1003847.

Behm-Ansmant I, Izaurralde E. Quality control of gene expression a stepwise assembly pathway for the surveillance complex that triggers nonsense-mediated mRNA decay. Genes Dev 2006;20:391–8.

Blanco S, Dietmann S, Flores JV, Hussain S, Kutter C, Humphreys P, Lukk M, Lombard P, Treps L, Popis M, Kellner S, Hölter SM, Garrett L, Wurst W, Becker L, Klopstock T, Fuchs H, Gailus-Durner V, Hrabě de Angelis M, Káradóttir RT, Helm M, Ule J, Gleeson JG, Odom DT, Frye M. Aberrant methylation of tRNAs links cellular stress to neurodevelopmental disorders. EMBO J 2014;33:2020–39.

Caro F, Ahyong V, Betegon M, DeRisi JL. Genome-wide regulatory dynamics of translation in the Plasmodium falciparum asexual blood stages. Elife 2014;3, e04106.

Chang YF, Imam JS, Wilkinson MF. The nonsense-mediated decay RNA surveillance pathway. Annu Rev Biochem 2007;76:51–74.

Chen CY, Shyu AB. Mechanisms of deadenylation-dependent decay. Wiley Interdiscip Rev RNA 2011;2:167–83.

Conti E, Izaurralde E. Nonsense-mediated mRNA decay: molecular insights and mechanistic variations across species. Curr Opin Cell Biol 2005;17:316–25.

Couttet P, Grange T. Premature termination codons enhance mRNA decapping in human cells. Nucleic Acids Res 2004;32:488–94.

Doma MK, Parker R. Endonucleolytic cleavage of eukaryotic mRNAs with stalls in translation elongation. Nature 2006;440:561–4.

D'Orazio KN, Green R. Ribosome states signal RNA quality control. Mol Cell 2021;81:1372–83.

Eastman G, Smircich P, Sotelo-Silveira JR. Following ribosome footprints to understand translation at a genome wide level. Comput Struct Biotechnol J 2018;16:167–76.

Esposito AM, Mateyak M, He D, Lewis M, Sasikumar AN, Hutton J, Copeland PR, Kinzy TG. Eukaryotic polyribosome profile analysis. J Vis Exp 2010;(40):1948.

Fritsch C, Herrmann A, Nothnagel M, Szafranski K, Huse K, Schumann F, Schreiber S, Platzer M, Krawczak M, Hampe J, Brosch M. Genome-wide search for novel human uORFs and N-terminal protein extensions using ribosomal footprinting. Genome Res 2012;22:2208–18.

Gao X, Wan J, Liu B, Ma M, Shen B, Qian SB. Quantitative profiling of initiating ribosomes in vivo. Nat Methods 2014;12:147–53.

Gehring NH, Kunz JB, Neu-Yilik G, Breit S, Viegas MH, Hentze MW, Kulozik AE. Exon-junction complex components specify distinct routes of nonsense-mediated mRNA decay with differential cofactor requirements. Mol Cell 2005;20:65–75.

Gerashchenko MV, Lobanov AV, Gladyshev VN. Genome-wide ribosome profiling reveals complex translational regulation in response to oxidative stress. Proc Natl Acad Sci USA 2012;109:17394–9.

Gil A, López JI, Pulido R. Assessing PTEN Subcellular Localization. Methods Mol Biol 2016;1388:169–86.

Gobet C, Naef F. Ribosome profiling and dynamic regulation of translation in mammals. Curr Opin Genet Dev 2017;43:120–7.

Goodarzi H, Nguyen HCB, Zhang S, Dill BD, Molina H, Tavazoie SF. Modulated expression of specific tRNAs drives gene expression and cancer progression. Cell 2016;165:1416–27.

Guydosh NR, Green R. Dom34 rescues ribosomes in 3′ untranslated regions. Cell 2014;156:950–62.

Harigaya Y, Parker R. No-go decay: a quality control mechanism for RNA in translation. Wiley Interdiscip Rev RNA 2010;1:132–41.

He F, Amrani N, Johansson MJ, Jacobson A. Qualitative and quantitative assessment of the activity of the yeast nonsense-mediated mRNA decay pathway. Methods Enzymol 2008;449:127–47.

Hellen CUT. Translation termination and ribosome recycling in eukaryotes. Cold Spring Harb Perspect Biol 2018;10, a032656.

Hinnebusch AG, Lorsch JR. The mechanism of eukaryotic translation initiation: new insights and challenges. Cold Spring Harb Perspect Biol 2012;4, a011544.

Hori K, Watanabe Y. Chapter 8. In vivo analysis of plant nonsense-mediated mRNA decay. Methods Enzymol 2008;449:165–76.

Hsieh AC, Liu Y, Edlind MP, Ingolia NT, Janes MR, Sher A, Shi EY, Stumpf CR, Christensen C, Bonham MJ, Wang S, Ren P, Martin M, Jessen K, Feldman ME, Weissman JS, Shokat KM, Rommel C, Ruggero D. The translational landscape of mTOR signalling steers cancer initiation and metastasis. Nature 2012;485:55–61.

Hu W, Coller J. Polysome analysis for determining mRNA and ribosome association in Saccharomyces cerevisiae. Methods Enzymol 2013;530:193–206.

Hu W, Sweet TJ, Chamnongpol S, Baker KE, Coller J. Co-translational mRNA decay in Saccharomyces cerevisiae. Nature 2009;461:225–9.

Ingolia NT. Ribosome footprint profiling of translation throughout the genome. Cell 2016;165:22–33.

Ingolia NT, Ghaemmaghami S, Newman JR, Weissman JS. Genome-wide analysis in vivo of translation with nucleotide resolution using ribosome profiling. Science 2009;324:218–23.

Ingolia NT, Lareau LF, Weissman JS. Ribosome profiling of mouse embryonic stem cells reveals the complexity and dynamics of mammalian proteomes. Cell 2011;147:789–802.

Ingolia NT, Brar GA, Rouskin S, McGeachy AM, Weissman JS. The ribosome profiling strategy for monitoring translation in vivo by deep sequencing of ribosome-protected mRNA fragments. Nat Protoc 2012;7:1534–50.

Ingolia NT, Hussmann JA, Weissman JS. Ribosome profiling: global views of translation. Cold Spring Harb Perspect Biol 2019;11, a032698.

Isken O, Maquat LE. Quality control of eukaryotic mRNA: safeguarding cells from abnormal mRNA function. Genes Dev 2007;21:1833–56.

Ju J, Rajski SR, Lim SK, Seo JW, Peters NR, Hoffmann FM, Shen B. Lactimidomycin, iso-migrastatin and related glutarimide-containing 12-membered macrolides are extremely potent inhibitors of cell migration. J Am Chem Soc 2009;131:1370–1.

Klauer AA, van Hoof A. Degradation of mRNAs that lack a stop codon: a decade of nonstop progress. Wiley Interdiscip Rev RNA 2012;3:649–60.
Kozak M. Pushing the limits of the scanning mechanism for initiation of translation. Gene 2002;299:1–34.
Lee S, Liu B, Lee S, Huang SX, Shen B, Qian SB. Global mapping of translation initiation sites in mammalian cells at single-nucleotide resolution. Proc Natl Acad Sci USA 2012;109:E2424–32.
Lejeune F, Maquat LE. Mechanistic links between nonsense-mediated mRNA decay and pre-mRNA splicing in mammalian cells. Curr Opin Cell Biol 2005;17:309–15.
Lejeune F, Li X, Maquat LE. Nonsense-mediated mRNA decay in mammalian cells involves decapping, deadenylating, and exonucleolytic activities. Mol Cell 2003;12:675–87.
Li GW, Oh E, Weissman JS. The anti-Shine-Dalgarno sequence drives translational pausing and codon choice in bacteria. Nature 2012;484:538–41.
Lian X, Guo J, Gu W, Cui Y, Zhong J, Jin J, He QY, Wang T, Zhang G. Genome-wide and experimental resolution of relative translation elongation speed at individual gene level in human cells. PLoS Genet 2016;12, e1005901.
Liu B, Han Y, Qian SB. Cotranslational response to proteotoxic stress by elongation pausing of ribosomes. Mol Cell 2013a;49:453–63.
Liu X, Jiang H, Gu Z, Roberts JW. High-resolution view of bacteriophage lambda gene expression by ribosome profiling. Proc Natl Acad Sci USA 2013b;110:11928–33.
Loayza-Puch F, Rooijers K, Buil LC, Zijlstra J, Oude Vrielink JF, Lopes R, Ugalde AP, van Breugel P, Hofland I, Wesseling J, van Tellingen O, Bex A, Agami R. Tumour-specific proline vulnerability uncovered by differential ribosome codon reading. Nature 2016;530:490–4.
Lobanov AV, Heaphy SM, Turanov AA, Gerashchenko MV, Pucciarelli S, Devaraj RR, Xie F, Petyuk VA, Smith RD, Klobutcher LA, Atkins JF, Miceli C, Hatfield DL, Baranov PV, Gladyshev VN. Position-dependent termination and widespread obligatory frameshifting in Euplotes translation. Nat Struct Mol Biol 2017;24:61–8.
Medina DA, Jordán-Pla A, Millán-Zambrano G, Chávez S, Choder M, Pérez-Ortín JE. Cytoplasmic 5′-3′ exonuclease Xrn1p is also a genome-wide transcription factor in yeast. Front Genet 2014;5:1.
Menschaert G, Van Criekinge W, Notelaers T, Koch A, Crappé J, Gevaert K, Van Damme P. Deep proteome coverage based on ribosome profiling aids mass spectrometry-based protein and peptide discovery and provides evidence of alternative translation products and near-cognate translation initiation events. Mol Cell Proteomics 2013;12:1780–90.
Michel AM, Baranov PV. Ribosome profiling: a Hi-Def monitor for protein synthesis at the genome-wide scale. Wiley Interdiscip Rev RNA 2013;4:473–90.
Mikl M, Pilpel Y, Segal E. High-throughput interrogation of programmed ribosomal frameshifting in human cells. Nat Commun 2020;11:3061.
Mitchell P, Tollervey D. An NMD pathway in yeast involving accelerated deadenylation and exosome-mediated 3′- ->5′ degradation. Mol Cell 2003;11:1405–13.
Mugler CF, Hondele M, Heinrich S, Sachdev R, Vallotton P, Koek AY, Chan LY, Weis K. ATPase activity of the DEAD-box protein Dhh1 controls processing body formation. Elife 2016;5, e18746.
Osbourn AE, Field B. Operons. Cell Mol Life Sci 2009;66:3755–75.
Ossipow V, Descombes P, Schibler U. CCAAT/enhancer-binding protein mRNA is translated into multiple proteins with different transcription activation potentials. Proc Natl Acad Sci USA 1993;90:8219–23.
Park JE, Yi H, Kim Y, Chang H, Kim VN. Regulation of poly (A) tail and translation during the somatic cell cycle. Mol Cell 2016;62:462–71.
Parsons M, Myler PJ. Illuminating parasite protein production by ribosome profiling. Trends Parasitol 2016;32:446–57.

Pelechano V, Wei W, Steinmetz LM. Widespread co-translational RNA decay reveals ribosome dynamics. Cell 2015;161:1400–12.
Pringle ES, McCormick C, Cheng Z. Polysome profiling analysis of mRNA and associated proteins engaged in translation. Curr Protoc Mol Biol 2019;125, e79.
Rooijers K, Loayza-Puch F, Nijtmans LG, Agami R. Ribosome profiling reveals features of normal and disease-associated mitochondrial translation. Nat Commun 2013;4:2886.
Rubio CA, Weisburd B, Holderfield M, Arias C, Fang E, DeRisi JL, et al. Transcriptome-wide characterization of the eIF4A signature highlights plasticity in translation regulation. Genome Biol 2014;15:476.
Schleif R. Regulation of the L-arabinose operon of Escherichia coli. Trends Genet 2000;16:559–65.
Schneider-Poetsch T, Ju J, Eyler DE, Dang Y, Bhat S, Merrick WC, Green R, Shen B, Liu JO. Inhibition of eukaryotic translation elongation by cycloheximide and lactimidomycin. Nat Chem Biol 2010;6:209–17.
Shoemaker CJ, Eyler DE, Green R. Dom34:Hbs1 promotes subunit dissociation and peptidyl-tRNA drop-off to initiate no-go decay. Science 2010;330:369–72.
Smircich P, Eastman G, Bispo S, Duhagon MA, Guerra-Slompo EP, Garat B, Goldenberg S, Munroe DJ, Dallagiovanna B, Holetz F, Sotelo-Silveira JR. Ribosome profiling reveals translation control as a key mechanism generating differential gene expression in Trypanosoma cruzi. BMC Genomics 2015;16:443.
Stern-Ginossar N, Ingolia NT. Ribosome profiling as a tool to decipher viral complexity. Annu Rev Virol 2015;2:335–49.
Stern-Ginossar N, Weisburd B, Michalski A, Le VT, Hein MY, Huang SX, Ma M, Shen B, Qian SB, Hengel H, Mann M, Ingolia NT, Weissman JS. Decoding human cytomegalovirus. Science 2012;338:1088–93.
Stumpf CR, Moreno MV, Olshen AB, Taylor BS, Ruggero D. The translational landscape of the mammalian cell cycle. Mol Cell 2013;52:574–82.
Thoreen CC, Chantranupong L, Keys HR, Wang T, Gray NS, Sabatini DM. A unifying model for mTORC1-mediated regulation of mRNA translation. Nature 2012;485:109–13.
Tsuboi T, Kuroha K, Kudo K, Makino S, Inoue E, Kashima I, Inada T. Dom34:hbs1 plays a general role in quality-control systems by dissociation of a stalled ribosome at the 3′ end of aberrant mRNA. Mol Cell 2012;46:518–29.
Tuck AC, Rankova A, Arpat AB, Liechti LA, Hess D, Iesmantavicius V, Castelo-Szekely V, Gatfield D, Bühler M. Mammalian RNA decay pathways are highly specialized and widely linked to translation. Mol Cell 2020;77:1222–36.
Vembar SS, Macpherson CR, Sismeiro O, Coppée JY, Scherf A. The PfAlba1 RNA-binding protein is an important regulator of translational timing in *Plasmodium falciparum* blood stages. Genome Biol 2015;16:212.
Venkataraman K, Guja KE, Garcia-Diaz M, Karzai AW. Non-stop mRNA decay: a special attribute of trans-translation mediated ribosome rescue. Front Microbiol 2014;5:93.
Wahle E, Winkler GS. RNA decay machines: deadenylation by the Ccr4-not and Pan2-Pan3 complexes. Biochim Biophys Acta 2013;1829:561–70.
Webster MW, Chen YH, Stowell JAW, Alhusaini N, Sweet T, Graveley BR, Coller J, Passmore LA. mRNA deadenylation is coupled to translation rates by the differential activities of Ccr4-not nucleases. Mol Cell 2018;70:1089–100.
Wolfe AL, Singh K, Zhong Y, Drewe P, Rajasekhar VK, Sanghvi VR, Mavrakis KJ, Jiang M, Roderick JE, Van der Meulen J, Schatz JH, Rodrigo CM, Zhao C, Rondou P, de Stanchina E, Teruya-Feldstein J, Kelliher MA, Speleman F, Porco Jr JA, Pelletier J, Rätsch G, Wendel HG. RNA G-quadruplexes cause eIF4A-dependent oncogene translation in cancer. Nature 2014;513:65–70.
Wu B, Eliscovich C, Yoon YJ, Singer RH. Translation dynamics of single mRNAs in live cells and neurons. Science 2016;352:1430–5.

Wu CC, Zinshteyn B, Wehner KA, Green R. High-resolution ribosome profiling defines discrete ribosome elongation states and translational regulation during cellular stress. Mol Cell 2019;73:959–70.

Yamashita A, Ohnishi T, Kashima I, Taya Y, Ohno S. Human SMG-1, a novel phosphatidylinositol 3-kinase-related protein kinase, associates with components of the mRNA surveillance complex and is involved in the regulation of nonsense-mediated mRNA decay. Genes Dev 2001;15:2215–28.

Yan LL, Zaher HS. How do cells cope with RNA damage and its consequences? J Biol Chem 2019;294:15158–71.

Yang Z, Cao S, Martens CA, Porcella SF, Xie Z, Ma M, Shen B, Moss B. Deciphering poxvirus gene expression by RNA sequencing and ribosome profiling. J Virol 2015;89:6874–86.

Yi H, Park J, Ha M, Lim J, Chang H, Kim VN. PABP cooperates with the CCR4-NOT complex to promote mRNA deadenylation and block precocious decay. Mol Cell 2018;70:1081–8.

Zhang L, Jiang W, Nan J, Almqvist J, Huang Y. The *Escherichia coli* CysZ is a pH dependent sulfate transporter that can be inhibited by sulfite. Biochim Biophys Acta 2014;1838:1809–16.

Index

Note: Page numbers followed by *f* indicate figures and *t* indicate tables.

Q

R

S

Kolibri 360 GmbH
Ferdinand-Jühlke-Straße 7
99095 Erfurt, Deutschland
produktsicherheit@kolibri360.de